EARLY HISTORY OF COSMIC RAY STUDIES

ASTROPHYSICS AND SPACE SCIENCE LIBRARY

A SERIES OF BOOKS ON THE RECENT DEVELOPMENTS
OF SPACE SCIENCE AND OF GENERAL GEOPHYSICS AND ASTROPHYSICS
PUBLISHED IN CONNECTION WITH THE JOURNAL
SPACE SCIENCE REVIEWS

VOLUME 118
PROCEEDINGS

EARLY HISTORY
OF
COSMIC RAY STUDIES

Personal Reminiscences with Old Photographs

Edited by

YATARO SEKIDO
Nagoya University, Nagoya, Japan

and

HARRY ELLIOT
Imperial College, London, U.K.

D. REIDEL PUBLISHING COMPANY

A MEMBER OF THE KLUWER ACADEMIC PUBLISHERS GROUP

DORDRECHT / BOSTON / LANCASTER

Library of Congress Cataloging in Publication Data

Main entry under title:

Early history of cosmic ray studies.

(Astrophysics and space science library ; v. 118)
Includes index.
1. Cosmic rays—Research—History—Addresses, essays, lectures.
I. Sekido, Yataró, 1912-   . II. Elliot, H. III. Series.
QC484.9.E19   1985   539.7'223   85–11763

ISBN-13: 978-94-010-8899-2   e-ISBN-13: 978-94-009-5434-2
DOI: 10.1007/978-94-009-5434-2

Published by D. Reidel Publishing Company,
P.O. Box 17, 3300 AA Dordrecht, Holland.

Sold and distributed in the U.S.A. and Canada
by Kluwer Academic Publishers,
190 Old Derby Street, Hingham, MA 02043, U.S.A.

In all other countries, sold and distributed
by Kluwer Academic Publishers Group,
P.O. Box 322, 3300 AH Dordrecht, Holland.

CONTENTS

Part 5 Showers: Local and Extensive
Terrestrial Generation of Penetrating Radiations

Part 6 Primaries and Secondary Products
Composition of Each Generation

Part 7 Further in Particle Physics

Part 8 Further in Space Physics

PREFACE

On the occasion of the International Conference on Cosmic Rays held in Kyoto in August 1979 five aged members of the cosmic ray fraternity, H. Elliot, V.L. Ginzburg, B. Peters, Y. Sekido, and J.A. Simpson met together as a dinner party devoted to the enjoyment of Japanese cuisine and reminiscences of our younger days. This pleasant occasion called to mind the many friends of our own age as well as some eminent seniors not present at the conference whose recollections would have further enriched and enlivened our evening. By the time the dinner came to an end we had agreed that the compilation of a more extensive collection of personal reminiscences would be an interesting and worthwhile undertaking. Accordingly, the next day we held an editorial meeting to draw up a list of potential authors and two of us, the present editors, started work on the project.

In putting the book together our intention has been to try to capture and record through these personal accounts something of the atmosphere, the excitement and the frustrations of research in cosmic rays as experienced at first hand by some of the practitioners in the field. It has never been our intention that it should comprise a systematic history of the subject. Neither, unfortunately, can it be a fully representative collection since practical limits to the size of the volume alone would preclude that. Also some important potential contributors have not felt able to produce articles in the time available whilst others, regrettably, are no longer with us.

We wish to thank all those who have entered so enthusiastically into the project and provided articles at rather short notice and we hope and believe that their contributions will be of interest not only to all those who have worked on cosmic ray physics in the past but also to the wider community wishing to see through the eyes of those most closely concerned something of the way in which research is done and progress in scientific understanding achieved. An introduction has been included for the benefit of the wider reading public. So many articles were contributed that we felt that they should be classified for convenience, sake into eight parts like the sessions of an imaginary symposium. The divisions are by no means clearcut because no such classification was envisaged when the editors first approached prospective authors. All the contributions were written in the period from late 1979 to early 1981.

By good fortune the International Symposium on the History of Elementary Particles was held at Fermilab, Batavia, Illinois, USA, in May 1980, and the Proceedings of that meeting* will be published almost simultaniously with the present book of cosmic ray reminiscences. We wish to express our cordial thanks to those concerned for thier cooperation in agreeing to publish the two articles by Professor C. D. Anderson and Professor D. Skobeltzyn in both the Proceedings and the present volume.

It is with the greatest regret that we have to report that five of the authors have died

---

* Proceedings of the Symposium on History of Elementary Particles, edited by L. Brown and L. Hoddeson, which will be published by Cambridge University Press.

since their contributions to this volume were written. Professor Agasi N. Charakhchan died on 8 March 1981, Professor Georg Pfotzer on 24 July 1981, Professor Hideki Yukawa on 8 September 1981, Professor Walter Heinrich Heitler on 15 November 1981 and Professor Sergei N. Vernov on 10 October 1982. We take this opportunity of offering our condolences to their relatives and colleagues.

We wish to record our thanks to the organisers of the Kyoto Conference, 1979 which gave us the opportunity for planning this book and to all those, especially Professor K. Murakami, who helped in its compilation. A part of the publishing cost was defrayed from the funds by the Japanese Ministry of Education, Science, and Culture.

The Editors
London, November 10, 1982   *H. Elliot*
Nagoya, November 29, 1982   *Y. Sekido*

PARTICIPATING AUTHORS

Hannes Alfvén
  Nobel Prize Recipient, Royal Institute of Technology, Stockholm, Sweden

Carl D. Anderson
  Nobel Prize Recipient, California Institute of Technology, Pasadena, USA

Pierre Auger
  12 Rue Emile Faguet, Paris, France

Erich R. Bagge
  University of Kiel, Kiel, West-Germany

Hugh Carmichael
  9 Beach Avenue, Deep River, Canada

Agasi N. Charakhchan
  Lebedev Physical Institute Acad. Sci., Moscow, USSR

Alexander, E. Chudakov
  Institute for Nuclear Research Acad. Sci., Moscow, USSR

Geroge W. Clark
  Massachusetts Institute of Technology, Cambridge, USA

Nicolai A. Dobrotin
  Lebedev Physical Institute Acad. Sci., Moscow, USSR

Harry Elliot
  Imperial College, London, UK

Eugene L. Feinberg
  Lebedev Physical Institute Acad. Sci., Moscow, USSR

Scott E. Forbush
  Carnegie Institution of Washington, Washington (DC), USA

Vitaly L. Ginzburg
  Lebedev Physical Institute Acad. Sci., Moscow, USSR

Naum L. Grigorov
  Moscow State University, Moscow, USSR

Walter H. Heitler
  University of Zürich, Zürich, Switzerland

Gerge B. Khristiansen
  Moscow State University, Moscow, USSR

Serge A. Korff
  New York University, New York, USA

Devendra Lal
  Physical Research Laboratory, Ahmedabad, India

Louis Leprince-Ringuet
  Ecole Polytechnique, Paris, France

Marian Mięsowicz
  Institute of Nuclear Physics, Kraków, Poland

H. Victor Neher
  California Institute of Technology, Pasadena, USA

Edward P. Ney
  University of Minnesota, Minneapolis, USA

Sergei I. Nikolsky
  Lebedev Physical Institute Acad. Sci., Moscow, USSR

Minoru Oda
  Institute of Space and Astronautical Science, Tokyo, Japan

Yash Pal
  Space applications Center, Ahmedabad, India

Bernard Peters
  Danish Space Research Institute,

Lundtoftevej, Lyngby, Denmark

Georg Pfotzer
Max-Planck-Institut für Aeronomie,
Katlenburg-Lindau, West-Germany

Jerzy Pniewski
University of Warsaw, Warsaw,
Poland

George D. Rochester
University of Durham, Durham, UK

Bruno Rossi
Massachusetts Institute of Technology, Cambridge, USA

Yataro Sekido
Nagoya University, Nagoya, Japan

John A. Simpson
University of Chicago, Chicago,
USA

Dimitry V. Skobeltzyn
Lebedev Physical Institute Acad.
Sci., Moscow, USSR

Sergei A. Slavatinsky
Lebedev Physical Institute Acad.
Sci., Moscow, USSR

Antal J. Somogyi
Central Research Institute of Physics Acad. Sci., Budapest, Hungary

Rudolf Steinmaurer
University of Innsbruck, Innsbruck,
Austria

Mituo Taketani
1-16-26 Shakujiidai, Nerima,
Tokyo, Japan

Masa Takeuchi
Yokohama National University,
Yokohama, Japan

Sergei N. Vernov
Moscow State University, Moscow,
USSR

John G. Wilson
University of Leeds, Leeds, UK

Hideki Yukawa
Nobel Prize Recipient, Kyoto University, Kyoto, Japan

Georgi T. Zatsepin
Lebedev Physical Institute Acad.
Sci., Moscow, USSR

INTRODUCTION

This joint introduction is for the benefit of those readers with no first hand acquaintance with cosmic ray physics. To avoid any preconceptions which might be given by an historical introduction of the usual kind the introductions written independently by Mr. E and by Mr. S were combined in a style of dialogue exposing the discrepancies between them on their views as well as their emphases.

S - Cosmic ray is an interesting display of the nature with an unique feature, the longest and the finest line in the world. It may be long enough to catch a special property of the biggest space, the Universe, and fine enough to dissect the smallest matter, the elementary particle. Such a feature was, however, not apparent in the early days when cosmic rays appeared on the scene as exotic strangers from space.

E - It is now almost seventy years ago that V. F. Hess made the historic balloon flight on August 8, 1912 which opened up the new vistas in physics and astronomy which form the subject matter of this book. Following that flight, in which he used electroscopes to measure the ionization as a function of altitude, Hess concluded that the increase in ionization with height which he had observed could best be explained by a penetrating radiation incident on the atmosphere from above. It was to be some years, however, before this simple but bold conclusion came to be generally accepted as correct.

S - That wonderful balloon flight by Hess turned out to be one of the last notable scientific achievements of the Austro-Hungarian Empire. Only two years were to elapse before the Archduke Franz Ferdinand, heir to the Habsburg throne, would be killed at Sarajevo. During the First World War the study of the newly discovered radiation was continued by Hess himself and by the German physicist W. Kolhörster. Since 1921 the existence of the radiation was examined and confirmed by the work of R. A. Milikan in the United States. Because of the very great penetrating power of the radiation Milikan concluded that it must be gamma rays of exceedingly high energy since this type of radiation was known to interact only weakly with matter. Milikan introduced in 1926 the name cosmic rays which was eventually adopted by the scientific community.

Doubt as to whether the radiation was high energy photons was raised by the discovery in 1927 by the Dutch physicist J. Clay that the intensity of the radiation was dependent on latitude and therefore in some measure controlled by the earth's magnetic field. This doubt was reinforced in 1929 by the work of the German physicists W. Bothe and W. Kolhörster. By an ingenious experiment of coincidence counting, they were able to show that the cosmic rays at *sea level* were predominantly charged particles rather than photons. Stories until this time are in **Part 1** of this book except D. V. Skobeltzyn's work, which is in **Part 2** together with some reminiscences on their own memories around 1930 as shown in the table.

Part		1910's	1920's	1930's	1940's	1950's
1		————————————				
2	(remote days)		———————			
3	(positron, mesotron)			———————		
4	(sea, mountain, etc.)			——————————————— · · · · · · · · · · · · · · · ·
5	(showers)			—————————————————————
6	(primaries, secondaries)			—————————————————————
7	(particle physics)			· · · · · —————————————
8	(space physics)			· · · · · · · · · —————————

E - Although Bothe and Kolhörsters' results showed that the cosmic rays at ground level were charged particles and Clay's work indicated that the primary radiation incident on the atmosphere was probably in the form of charged particles, it was not until the early 1930's that this could be finally established. This was achieved by a combination of accurate measurements of the cosmic ray intensity as a function of latitude and direction of arrival together with M. S. Vallarta and others' theoretical developments in the motion of charged particles in a dipole magnetic field. In this way it was finally established that the bulk of the primary radiation at the top of the atmosphere consisted of positively charged particles rather than high energy photons as had been supposed.

S - Including A. H. Compton's work on the latitude effect, various experiments of intensity measurement were done since 1930's using the Earth as a tool and are mentioned in **Part 4**. The positively charged primary particles described above were supposed to be protons by W. F. G. Swann. Experimental results by manned and sounding balloons in the stratosphere favored this view already before the Second World War.

E - During the 1930's interest in cosmic rays grew rapidly and there was a great increase in work on the nature and energy spectrum of the primary particles and of the interactions in the atmosphere. Although clearly of great astronomical importance because of the extraordinary high energy of individual particles (from 10^9 eV upwards) and because the total energy flux carried by the cosmic rays is comparable with that of starlight the place and mode of origin of the radiation remained a mystery. An astronomical oddity, the cosmic rays were an ever present reminder of the existence of extreme physical processes in the Universe which remained unidentified and seemingly isolated from the general body of astrophysical knowledge.

S - As a solution to the mystery of the origin of the radiation Milikan proposed that the energy came from the annihilation of atoms somewhere in the depths of space, a view which he continued to hold for a long time. This explanation was based on the notion that cosmic rays were high energy photons which we now know to be incorrect. One of the secret dwellings of the mystery of cosmic ray origin was likely to lie in the gap between the classical theories of electromagnetism and hydrodynamics. A Swedish cosmic ray physicist H. Alfvén filled this gap to open the way for the systematic study of cosmic plasma.

E - The lack of detailed knowledge of the composition of the primary radiation and of the way in which the particles are accelerated did not, however, hinder the study of their interactions with matter and for the following twenty years up to the early 1950's the study of high energy interactions and fundamental particles was the almost exclusive prerogative of the cosmic ray physicist. Following the discovery of the positive electron by C. D. Anderson in 1932 there came in rapid succession the discovery of the muon, pion and a veritable host of short lived unstable particles.

S - The exciting events in 1930's on the discovery of positron and mesotron are mentioned in **Part 3**. Also in the early 1930's a new cosmic ray phenomenon called 'Cosmic Ray Showers' was discovered by Anderson, P. M. S. Blackett and G. Occhialini, B. Rossi and others. The word "shower" means simultaneous incidence of two or more cosmic ray particles. As mentioned in **Part 5** shower implies the production of secondary cosmic rays on the Earth and provided a method of investigating on the interaction of cosmic rays with matter. One kind of shower, called a cascade shower, was explained in 1930's as electromagnetic interaction of electrons and photons, though another kind was also found in this period.

Since the mid-1930's there has been a rapid increase in cosmic ray research and many different aspects of the work during this period are described in various articles. In order to prevent confusion on the names of new particles and the various components of the cosmic radiation it is necessary to describe, in more detail one aspect of this development. In 1935 P. Auger and others showed that the sea level cosmic ray particles consisted of two components with different penetrating power, namely the soft and hard components. In 1937 the particles of the soft component were identified, with the help of theory developed by H. J. Bhabha, W. Heitler and others, as electrons and positrons while Anderson discovered a new particle, the mesotron, which corresponded to the hard component. The mesotron, subsequently renamed meson, was at first thought to be the new particle (U-particle) which had been theoretically predicted by H. Yukawa in 1935. In fact they were not the same and this became quite clear in 1947 as mentioned in **Part 7** when C. F. Powell obtained a cosmic ray track in a nuclear emulsion showing one type of meson decaying into another. The parent meson in this process was called the π-meson or pion and corresponded to Yukawa's particle whilst the decay product called the μ-meson or muon proved to be Anderson's mesotron. Consequently we now know that muons are the particles of the hard component at ground level which are decay products of pions whilst their parent pions are found in the stratosphere and upper troposphere generated in the collisions between the primary cosmic rays and oxygen and nitrogen unclei. The bulk of primary particles was protons as supposed by Swann. Studies on various components in primaries and various secondary products are mentioned in **Part 6**.

E - In this context it was not until the late 1950's that man-made accelerators regained their ascendency over the cosmic accelerator in the field of fundamental particle physics and by a curious coincidence, just as the particle physicists were transferring their attention from cosmic rays to laboratory accelerators, the interest in the astrophysical and solar-terrestrial aspects of cosmic rays began to grow. As mentioned in **Part 8** balloons capable of carrying heavy loads to great altitude soon to be followed by earth satellites and space probes made possible the direct study of the primary cosmic rays. At the same time the discovery that charged particles were sometimes accelerated up to cosmic ray energies in association with solar flares and the subsequent realisation that radio emission from the Galaxy and many other exotic celestial objects such as supernova remnants is synchrotron radiation generated by relativistic election gyrating in magnetic fields has led to the integration of cosmic rays into the broad new field of high energy astrophysics. Furthermore, the discovery of planetary magnetospheres has made the direct in situ investigation of acceleration processes in naturally occurring plasmas and magnetic fields a practical possibility. Still more recently the emergence of the new science of gamma ray astronomy holds out the prospect of investigating distant energetic proton populations through the production in proton-proton collisions of π^{0} mesons which decay into pairs of gamma rays. These gamma rays are already providing our first glimpses of the high energy protons accelerated in remote astronomical objects just as we have been able to learn about the distant energetic electron populations through their radiation at radio and X-ray wavelengths.

S - Nevertheless, in spite of these recent developments in cosmic ray astronomy and laboratory accelerators many cosmic ray physicists continue to study high energy interactions and the production of fundamental particles because cosmic rays still provide the highest energy particles available for such work.

E - Over the past seven decades the development of cosmic ray physics has been of particular importance and fascination because of the variety and fundamental nature of the topics embraced which range in scale from the sub nuclear world of fundamental particles to vast dimensions of cosmic space. It is a subject in which there has always been the excitement of new discoveries and the expectation of the unexpected just round the corner. The changing pattern of activity and the achievements of those working in the cosmic ray field is well documented in the Proceedings of the sixteen biennial cosmic ray conferences which have been held in many countries of the world since 1947 under the auspices of the Cosmic Ray Commission of the IUPAP. As we have explained in the Preface it was at the sixteenth such conference held in Kyoto in 1979 that the idea of compiling this book of personal reminiscences emerged.

S - Over the past years no fewer than five scientists have received Nobel prizes for their work on cosmic rays namely Hess (1936), Anderson (1936), Blackett (1948), Powell (1950) and Bothe (1954). Two of the authors of articles in this book, apart from Anderson, are Nobel Laureates in physics namely Yukawa (1949) and Alfvén (1970).

I believe that the reminiscences collected together in this volume will be of lasting interest and value and I can only hope that our somewhat journalistic introduction written for the benefit of the wider reading public will not be thought to detract from the articles which follow.

Y. Sekido

H. Elliot

PART 1   THE BEGINNINGS

It was in his young days
*   at the age of twenty nine*
When he met with her, a
*   charming stranger from space*

That was in the calm sky
*   of his native land*
Which met with war
*   twice after that*

Leaving his home country
*   with the dear instrument*
He still looks at the image
*   of her native place*

Composed by Yataro Sekido
Revised by Hannah Peters

NATIVE PLACE OF COSMIC RAYS

PICTORIAL PROLOGUE

on August 16ᵗʰ 1952

Victor F Hess

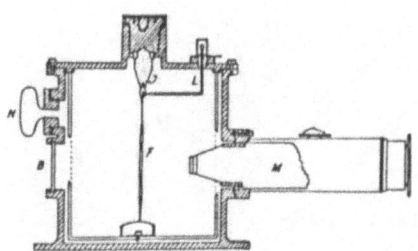

V. F. Hess, discoverer of cosmic rays, looking through the telescope of his ion chamber used for his discovery of cosmic rays in 1912. Design of chamber: from *Physikal. Zs.* **14**, 1135 (1913).
<Ed> Displaced from p. 14, Fig. 9, contributed by Y. Sekido.

3

Y. Sekido and H. Elliot (eds.), Early History of Cosmic Ray Studies, 3–8.
© *1985 by D. Reidel Publishing Company.*

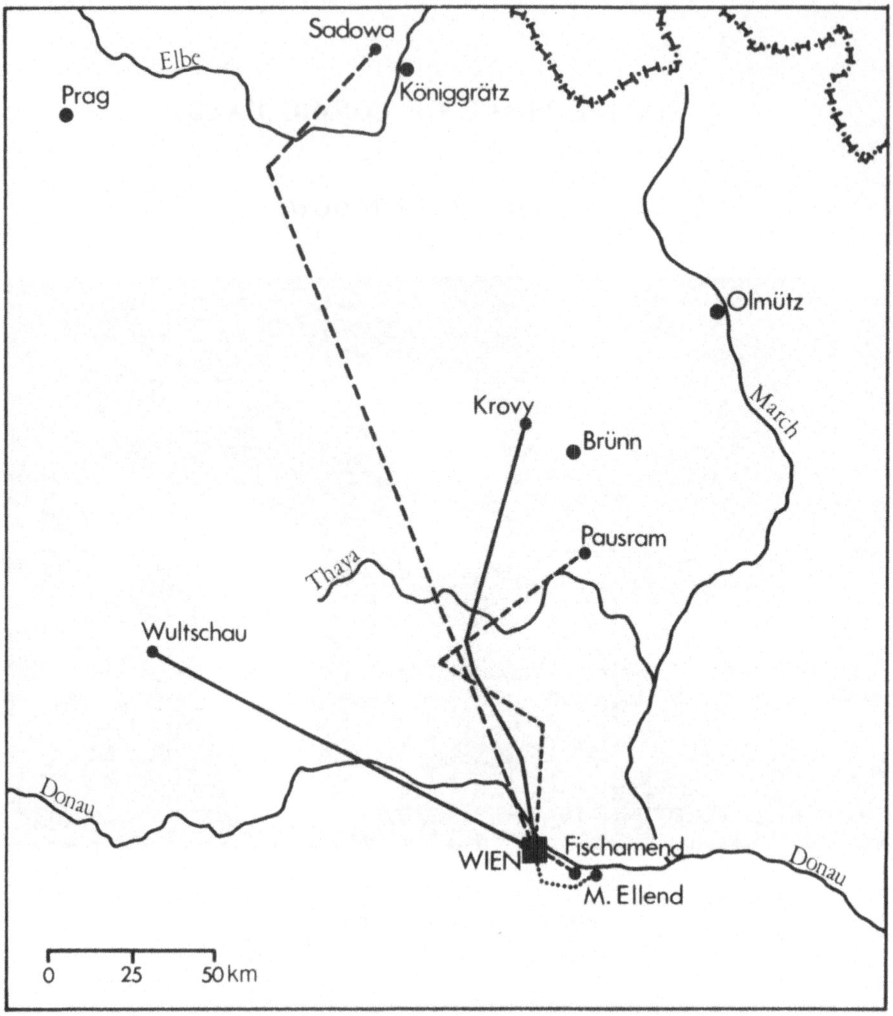

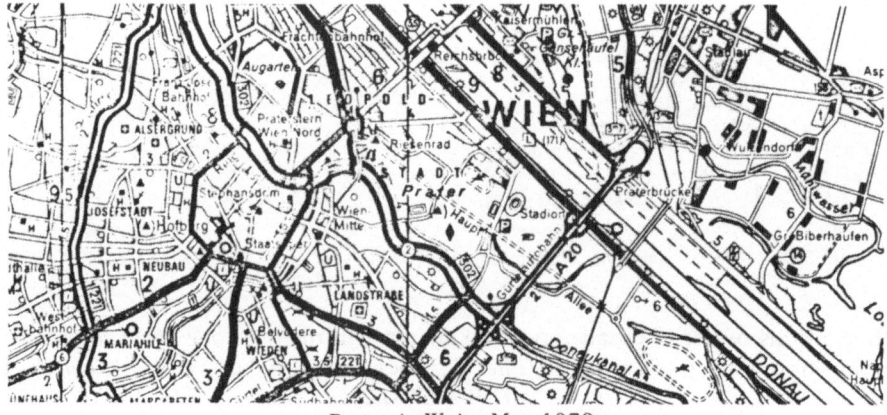

Ballonfluge von Prof. Hess von Wien aus.

<Ed> Route-map of Hess' test flights, contributed by R. Steinmaurer. See p. 17.

Prater in Wein, Map 1979.

Aeronautisches Gelände im Wiener Prater, von dem aus V. F. Hess in den Jahren 1911/12 seine ersten Freiballon-Forschungsfahrten unternommen hatte. (Courtesy of Heeresge-schichtliche Museum, Vienna)

<Ed> Contributed by R. Steinmaurer. See p. 17.

5

M der Fahrt: 7   **am:** 19. Juni 1912
Verein: Österreich
Ballon: Baumgarten   **Grösse:** 1120 cbm   **Füllung:** Leucht.

Führer: Dr Hess
Mitfahrende: (Solanyau) solo
Dauer: 1 St. 45 Min.   **Länge:** 25 km (in der Luftlinie gemessen)
Dchschn.Geschw.: 15 km i.d.St.

Zeit	Höhe m	Ballast-vorrat	Fahrt-richtung	Geschwindigkeit km i.d.St.	Temperatur trock. / feucht	Ortsbestimmung	Wetter und Wolken	Bemerkungen
17 00	1	11 Sd 2 kg	—	—	— / —	*[handwritten]*	*[symbol]*	Start.
17 24	460	10½	SE	10	/	*[handwritten]*	*[symbol]*	
17 38	770	10	"	—	/	*[handwritten]*		
17 54	1010	10	"	10 – 15	/	*[handwritten]*	"	
18 07	1040	10	"	12 – 15	/	*[handwritten]*	"	
18 10	920	9¼	"	"	/	*[handwritten]*	"	
18 24	1260	9	"	"	/	*[handwritten]*	"	
18 33	690	8	"	"	/	*[handwritten]*	"	
18 55	550	8	"	—	/	*[handwritten]*		*[handwritten]* gelandet

Unterschrift des Führers: *Dr Hess*

Aus Hess' Ballonbordbuch.

<Ed> Written by Hess on 19 July 1912 in flight from Wien to Fischamend. Contributed by R. Steinmaurer. See p. 17.

Hess on gondola in 1912 probably in test flight. The date and place is not clear at present.
⟨Ed⟩ Contributed by R. Steinmaurer. See p. 17.

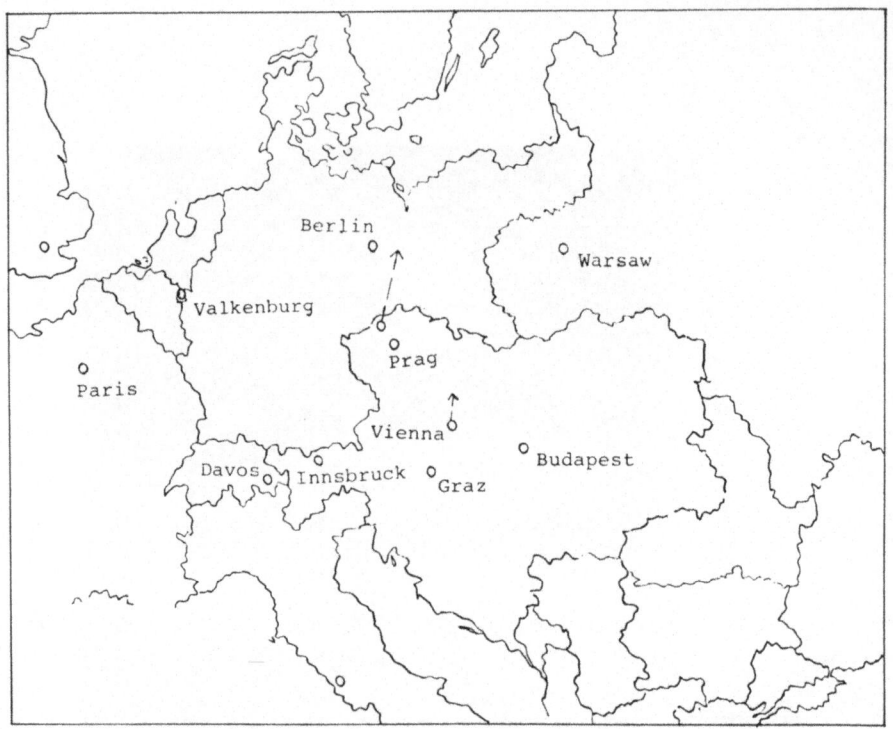

Central Europe in 1912—13.

Hess' birth place: near Graz. Balloon launch place (1912): Wien, Aussig near Prag. Station network (1913): Graz (Benndorf), Davos (Dorno), Wien (Hess), Innsbruck (v. Schweidler), Valkenburg (Wulf).

HESS' OLD PREDECESSORS

Yataro SEKIDO[*]

In 1940 Y. Nishina *et al.* (1941), including myself, wrote a text book on cosmic rays in Japanese. The first chapter dealt with its history since 1900 prepared by Nishina himself. The first sentence was "Cosmic rays were discovered by chance like many other great discoveries". I could understand that chance is a necessary condition for discovery but was not satisfied. Since I wrote my own book (Sekido, 1944), I have tried to extend the genealogy of cosmic ray studies from 1900 back to very old times. I was impressed in finding many instances of what I call a double discovery by a single person, which cannot be explained by mere chance but implies the importance of the discoverer's spirit. And, I felt a strong similarity in their spirits, i.e. a pure and deep interest in strangeness without any motive from need. Thus, the whole story seems to be one of the typical profiles of pure physics.

Cosmic rays were discovered in 1912 from the studies in 1900 on the dark current of electricity (Nishina *et al.,* 1941; Miehlnickel, 1938). The dark current was discovered in 1785 by the discoverer of the law of electric force, C.A.Coulomb (1785). It was found when he was examining some leakage trouble experienced in the course of his experiment on the law of electric force with a new sensitive torsion balance invented by himself. His double discovery is a natural result of his attitude towards the experiment to study the strange nature of electricity according to a guiding principle of pure physics. Electricity was discovered by the prospector of this principle of physics, Thales of Miletus, as R.A.Millikan (1935) assigned a double honor to him when describing the two works of Thales.

I think that Thales' double honor is also a double discovery, for his prospect is the result of his discovery of a certain point of view, in which an aspect of nature has been revealed with rational and beautiful outline. When he saw the profile of nature from this particular point, the phenomena described by him such as radiant heat, light, magnetism and electricity (Millikan, 1935) probably appeared to compose the beautifully simple outline inside which every complication was included as details. This standpoint provides the criterion of strangeness, and any strange phenomenon on this outline should be earnestly studied for the fine adjustment of his standpoint. This attitude was probably what made him turn a mere information on amber, probably informed by a necklace polisher, to the recognition of its important strangeness. Probably he felt that amber was so strange as if it had a soul. According to Diogenes Laertius, Aristotle and Hippias say that he (Thales) gave a share of soul even to inanimate (lit. soulless) objects, using Magnesian stone and amber as indications. Thales' word "soul" is not a mythic element of nature but seems to me to point at an intrinsic motive power due to the wonderful organization of nature. Hippias lived only a hundred years later than Thales and systematised wide materials of informations. Probably he was a desk worker while Thales was a wonderfully active man of practice. G.S.Kirk and J.E.Raven (1957, 1975) seem to suggest that Hippias had the knowledge

* Nagoya University (emeritus), Nagoya, Japan.

Y. Sekido and H. Elliot (eds.), Early History of Cosmic Ray Studies, 9–15.
© *1985 by D. Reidel Publishing Company.*

that "amber becomes magnetic when rubbed". This implies a classification of souls or a recognition that the type of motive power of amber and that of Magnesian stone were of the same kind. I can hardly imagine that Thales did not try any experiment with his amber necklace before attaining to this recognition in spite of the instability of amber's power. That experiment was probably the beginning of physics.

Technologies already existed. For example, there was a need to predict solar eclipses. Thales did so in 585BC, but there was no physics nor rationalist philosophy before his discovery of the above described point of view. Moralist philosophy appeared more than a hundred years earlier in Greece and India, i.e. the poetry of Hesiod and Upanishads, respectively. Almost simultaneously with Thales, a materialist appeared in India. It was a transient reaction against metempsychosis and did not induce physics in ancient India. However, I had a chance to learn that the rather traditional Bhagavad Gita (ca 300 AD) was a favourite with H.J.Bhabha (Fig.1). H.Yukawa told me that W.Heisenberg studied the moralist philosophy of Lao-tsu, which appeared in China little later than India and Greece. In Greece, Thales' rationalist philosophy was succeeded by many thinkers, while the first stranger in physics was left alone for a long time after Thales, the oldest predecessor of Hess. How does amber lose its strange power after rubbing? This question was asked by Coulomb in 1785, though electrical conduction of air around a hot solid (Richardson, 1921) was known since 1740.

All undertakings indeed are
clouded by defects as fire by
smoke–
Bhagavad Gita   **H.J. Bhabha**

Fig. 1

One of the two laws of Coulomb was the law of electric current in air $i = $ const E, which preceeded Ohm's law. In 1850, Matteucci found a strange characteristic of this dark current $i(E) \rightarrow i_s$ if E was sufficiently large (Matteucci, 1849, 1850). The saturation current i_s provided a measure of ionizing agent after 1896, when J.J.Thomson and E.Rutherford found gaseous ions to be responsible for electric conduction through a gas, though Thomson had started his research on this problem since about 1884 stimulated by Faraday's concept of an ion (Thomson). On the other hand, W. Linss observed in 1887 the leakage of electricity from a charged body exposed to open atmosphere (Linss, 1887), and an Austrian physicist Franz Exner (1887) planned an international observation network of atmospheric electricity by devising a portable electroscope

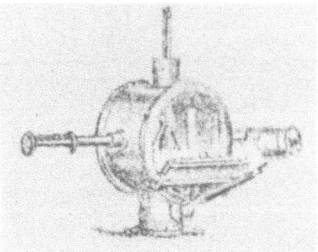

Fig. 2

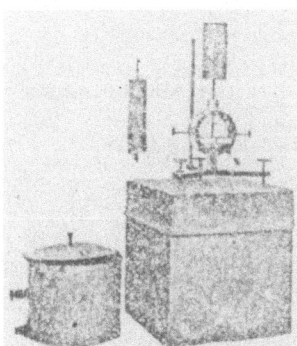
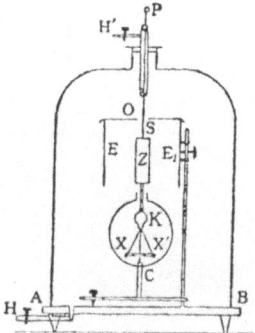

Fig. 3

(Fig.2). J.Elster and H.Geitel, who had been studying unipolar electrification around a hot wire (Richardson, 1921) in 1882–89 at a middle school of Wolfenbüttel in Germany, were excited by him since about 1888, and concluded in 1898 that the atmospheric leakage was also due to the motion of ions after their confirmation of the saturation current (Elster and Geitel, 1899a,b). During their observation, they noticed that wind carried densely ionized air, then Geitel enclosed the instrument in a big bell jar in 1900 (Fig.3) and found that the dark current still remained. C.T.R.Wilson obtained the same result at the same time. His observation was also a dark current in a closed vessel, but he arrived at this experiment from a different root, where lived another "soul", the growth of an inanimate object (Fig.4).

After a friction machine was brought to Japan by a Dutch merchant, a Japanese inventor Gennai Hiraga made a friction machine in 1776 from his technological interest. Donsai Hashimoto, probably the first Japanese physicist, once translated Egbert Buys' book (1771) and then wrote in 1810–13 a book titled "Erekiteru, Kyurigen" (Hashimoto, 1813, 1942) which means "Electricity, the Source of Physics" to introduce the guiding principle of pure physics by describing many experiments done by himself with a friction machine. One of them, probably his original experiment, showed condensation of moisture on the inside wall of a glass bottle by connecting a metal needle suspended in the bottle to a friction machine (Fig.5). This is the condensation of vapour due to point discharge of electricity. The book was not allowed to be published, but only distributed from hand to hand. In Europe, the problem of condensation nuclei was studied in various countries (Coulier, 1875; Aitken, 1880–81,

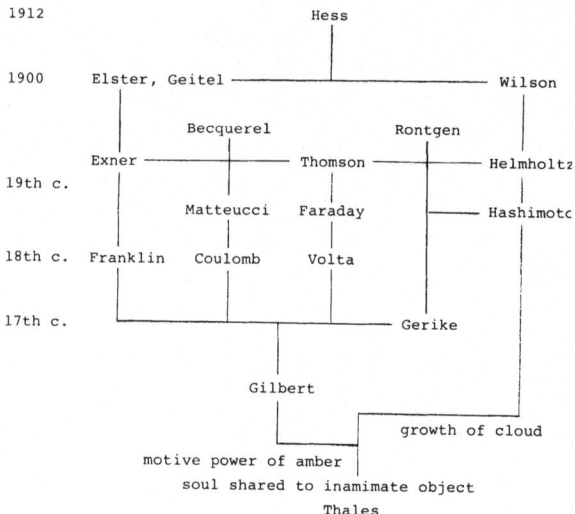

```
1912                                    Hess

1900           Elster, Geitel ─────────────┼───────────── Wilson

                      Becquerel              Rontgen
               Exner ─────────── Thomson ──────── Helmholtz
19th c.                │               │      │
                   Matteucci      Faraday    └── Hashimoto
18th c. Franklin   Coulomb         Volta

17th c.        └───────────────┬───────── Gerike

                        Gilbert
                                          growth of cloud
       motive power of amber   │
       soul shared to inamimate object
                      Thales
```

Fig. 4

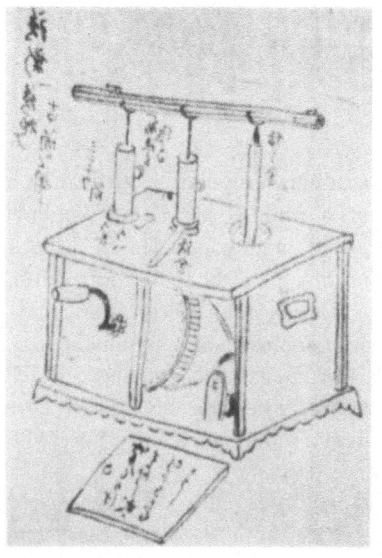

Fig. 5

1890, 1892; Kiessling, 1884a,b) since about 1875. Dust was known to be one of them, and condensation of vapour in dust free air became a current problem. In 1886, von Helmholtz found independently that a point discharge of electricity accelerated vapour condensation, and he assumed gaseous ions as the explanation (Helmholtz, 1886).

In 1893, Thomson, the Cavendish Professor of Cambridge University, supported this interpretation from his consideration of surface tension. Therefore, this idea might be latent in the motive of Wilson's study on the growth of a cloud at the Cavendish

Laboratory. By making a primitive expansion chamber with a glass bottle (Wilson, 1897) (Fig. 6), he found condensation nuclei in dust free air in 1895 which increased by irradiation of X rays in 1896. He concluded in 1897 that the natural condensation nuclei in a dust free vessel were the same thing as those produced by X rays or uranium, i.e. gaseous ions, because the critical expansion ratio was the same. In 1900, he found the existence of the natural condensation nuclei even in a sweeping electric field and tried to make a continuously operating cloud chamber to observe the steady production of ions. Probably due to its technical difficulty, he changed his method to observe the dark current in a closed flask (Fig.7). He carried his instrument to Peebles to avoid the radioactive contamination of Cavendish Lab., but ion production of 20 pairs per cc per sec still remained (Wilson, 1900).

Cosmic rays were first imagined by the inventor of cloud chamber, C.T.R.Wilson. To explain the above described residual ionization, he imagined a strange radiation of extra-terrestrial origin. Modifying his instrument into a portable style (Fig.8), he tried an underground experiment as described by R.Steinmaurer in this book. He did not know that cosmic rays were responsible for only one tenths of the residual ionization. A similar failure took place also in 1906. Campbell and Wood suggested a connection between the daily variation of ionization in a closed vessel and that in the Earth's electric field, and O.W.Richardson (Richardson, 1906) published an idea for its explanation by assuming penetrating radiations from the sun, though the maximum

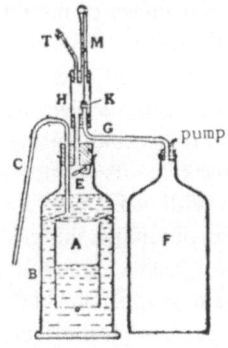

Fig. 6

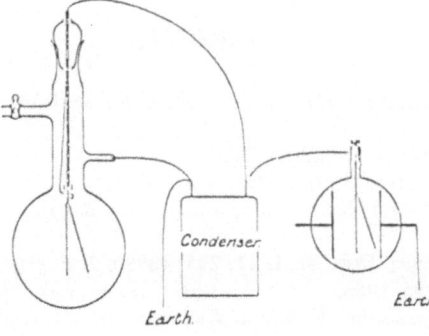

Fig. 7

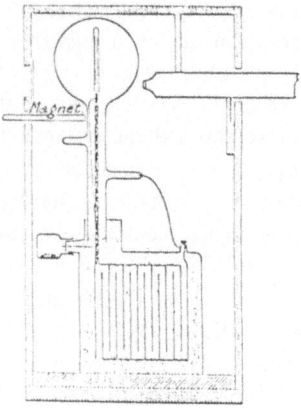

Fig. 8

was at night. Wilson also published a meteorological argument which favored Richardson's explanation, and Hess tried one balloon flight during the solar eclipse in April 1912. The work since 1900 on the approach to Hess' success are described in this book by Steinmaurer. The charactor of the lineage since Thales of Miletus was revealed evidently by Hess' discovery of a stranger from space, who brought presents known by C.D.Anderson's double discovery, positron and mesotron, and also many other gems of strangeness and charm to improve our image of nature.

I visited Professor V.F.Hess at Fordham University, New York, on 16 August 1952, when he was 69 years old. He answered my questions with a tender heart. His field of research before cosmic rays was atmospheric electricity. He was F.Exner's student and Th.Wulf's Junior. He discovered cosmic rays during his study of natural ionization. When I saw him, he had come back to his former field, atmospheric electricity, and was studying natural ionization with an open ion chamber at sea to avoid the radioactivity of earth. He was also observing the electrical conductivity of the breath from the lungs of man and found higher conductivity for those working on nuclear experiments. In his laboratory, there were some shielding plates of iron, less radioactive than lead, used for his study of the biological effect of cosmic rays. He showed me an old instrument, which was the ion chamber used in his discovery of cosmic rays, and allowed me to take a photograph of him looking through its telescope (Fig. 9*).

REFERENCES

Aitken: 1880-81, *Trans. Roy. Soc. Edin.* **30**, 337; 1890, *ibid* 35, I; 1980, *Proc. Roy. Soc. Lond.* **51**, 408.
Coulier: 1875, *J. Pharmacie et Chimie* **22**, 165, 254.
Coulomb, C.A.: 1785, *Mem. Akad. Paris*, p.612, p.616.
Egbert Buys: 1771, *Nieuw en Volkomen Woordenboek van Konsten en Weetenschappen* 10 vols. (according to H. Saigusa)
Elster, J. and Geitel, H.: 1899a, *Phys. Zs.* **1**, 11, 245; 1899b, *Terr. Mag. Atmos. Elec.* **4**, 213.
Exner, F.: 1887, *Wien Ber.* **95**, 1088.
Hashimoto, D.: 1813, *Orandashisei Erekiteru, Kyurigen*; revised edition by Saigusa, H.: 1942,

*<Ed.> Figure 9 is displaced to page 3 (Pictorial Prologue).

Nihon Kagaku Koten Zensho, **6**, 571, Press Asahi.

Helmholtz, V.: 1886, *Wied. Ann.* **27**, 509.

Kiessling: 1884a, *Hamburg. Abhandl.* **8**; 1884b, *Götting. Nachr.* **122**, 226.

Kirk, G.S. and Raven, J.E.: 1957, 1975, *The Presocratic Philosophers*, Camb. Univ. Press, p.94.

Linss, W.: 1887, *Meteoro. Zeits.* **4**, 345, 352.

Matteucci, M.: 1849, *Annal. Chim. Phys. (3)* **27**, 133; 1850, *ibid* **28**, 385.

Miehlnickel, E.: 1938, *Höhenstrahlung*, Theodor Steinkopff, Leipzig.

Millikan, R.A.: 1935, *Electrons (+ and —), Protons, Photons, Neutrons, and Cosmic Rays*, Univ. Chicago Press, p.l.

Nishina, Y., Sekido, Y., Tekeuchi, M. and Ichimiya, T.: 1941, *Uchusen*, Iwanami Koza Butsurigaku XI. B, Iwanami Shoten, Tokyo.

Richardson, O.W.: 1906, *Nature* **74**, 55; 1921, *The Emission of Electricity from Hot Bodies*.

Sekido, Y.: 1944, *Ushusen*, Kawade Shobo, Tokyo.

Thomson, J.J.: *Reflections and Recollections*.

Wilson, C.T.R.: 1897, *Philos. Transa.*, 189, 265; 1900, *Proc. Camb. Philos. Soc.*, **11**, 32.

ERINNERUNGEN AN V. F. HESS, DEN ENTDECKER DER KOSMISCHEN STRAHLUNG, UND AN DIE ERSTEN JAHRE DES BETRIEBES DES HAFELEKAR-LABORS

Rudolf STEINMAURER*

Um zu verstehen, wie es zur Entdeckung der Kosmischen Strahlung durch Hess im Jahre 1912 kommen konnte, müssen wir uns in die Zeit um 1900 zurückversetzen, als man die Leitfähigkeit der Atmosphäre und ihre Ursachen zu studieren begann.

Geitel, Elster u. Geitel und Wilson hatten fast gleichzeitig festgestellt, daß staubfreie Luft auch in geschlossenen Gefäßen stets schwach leitet. Als Ursache nahmen Elster u. Geitel Spuren radioaktiver Stoffe in der Luft selbst oder in den Gefäßwänden an. Wilson erwog als ionisierendes Agens eine Strahlung, deren Quelle außerhalb der Atmosphäre liegen könnte (Wilson, 1901). Er schreibt:

" ... whether the continuous production of ions in dust-free air could be explained as being due to radiation from sources outside our atmosphere, possibly radiation like Röntgen rays or like cathode rays, but of enormously greater penetrating power".

Da, wenn von außerhalb der Atmosphäre kommend, die Wirkung der Strahlung unter "many feet of solid rock" schwächer sein müßte, transportierte er die Apparatur in den Caledonian Railwaytunnel bei Peebles. Da er aber dort nicht die erwartete Abnahme fand, sondern eine innerhalb der Fehlergrenzen liegende Zunahme, schloß er:

"It is unlikely, therefore, that the ionization is due to radiation which has traversed our atmosphere; it seems to be, as Elster u. Geitel concludes, a property of the air itself".

Daß die Ionisierung durch eine von außen in das Gefäß dringende Strahlung bewirkt wird, zeigten die Panzerungsversuche von Rutherford und Cooke. Als Quelle dieser durchdringenden Strahlung wurden bald die radioaktiven Substanzen im Erdboden ("Erdstrahlung") und in der Luft ("Luftstrahlung") erkannt.

Mit dieser Deutung eines radioaktiven Ursprunges der gesamten "Reststrahlung" gab man sich mehrere Jahre lang zufrieden. Ihre Unzulänglichkeit wurde aber offenbar, nachdem man mit verbesserten Apparaten Messungen über See und in größeren Höhen in der Atmosphäre ausgeführt hatte.

So fanden Mc Lennan und ebenso Pacini bei Vergleichsmessungen über See und Land einen unerwartet hohen Wert über See, trotz der geringen Radioactivität des Meerwassers. Pacini schätzte diesen durchdringenden, nicht radioaktiven Anteil zu etwa 2 I (1 I = 1 Jonenpaar pro cm³ und Sekunde).

Daß nicht die gesamte beobachtete durchdringende Strahlung von den radioaktiven Stoffen im Erdboden stammen kann, folgt auch aus Messungen der Abnahme der

* University of Innsbruck (emeritus), Innsbruck, Austria.
<Ed.> The manuscript of this article written in German was not translated into English according to the suggestion of editors who hope that the readers might be invited into the world of cosmic ray classics because all of the very early papers on cosmic rays were published in German.

Y. Sekido and H. Elliot (eds.), Early History of Cosmic Ray Studies, 17–31.

V.F. Hess in 1936–37, on the occasion of Nobel prize.

Ionisation mit der Höhe. So fand Wulf nach Abzug der Wandstrahlung seiner Apparatur auf dem Boden 6 I, auf dem Eiffelturm in 300m 3,5 I, einen Wert, der zu groß war, denn nach Eve werden Gamma-Strahlen durch 300m Luft fast völlig absorbiert. Wulf schloß:

"... Die bisher gemachten Versuche verlangen daher außer der Erdrinde noch eine andere Quelle für die Gamma–Strahlung in den höheren Luftschichten oder eine wesentlich schwächere Absorption durch die Luft, als bisher angenommen".

Messungen im Freiballon wurden von Bergwitz und von Gockel (bis 4500m) durchgeführt. Gockel berichtet, daß er zwar eine Verminderung der Strahlung mit der Höhe feststellen konnte, aber

"lange nicht in dem Maße, wie sie zu erwarten gewesen wäre, wenn die Strahlung ihrer Hauptsache nach vom Erdboden ausgeht".

So stand das Problem, als Hess sich einschaltete. Hess, damals 1. Assistent am neugegründeten Institut für Radiumforschung in Wien, war mit der radioaktiven und der luftelektrischen Meßtechnik bestens vertraut. In einem Vortrag vor der 83. Naturforscherversammlung in Karlsruhe im Jahre 1911 erklärte er (Hess, 1911):

"Die zu geringe Abnahme der Ionisation mit der Höhe in einem geschlossenen Gefäß könnte zweierlei Ursachen haben:"

"... erstens kann außer den radioaktiven Substanzen der Erde ein amderer, uns noch unbekannter Ionisator in der Atmosphäre wirksam sein" oder es

"kann die Absorption der Gamma-Strahlen in Luft vielleicht viel langsamer erfolgen als bisher angenommen wurde".

Hess schildert seine Untersuchungen, die zur Entscheidung zwischen diesen beiden Möglichkeiten und schließlich zur Entdeckung der Kosmischen Strahlung führten, wie folgt (Hess, 1912, 1940):

"At that time, in the spring of 1911, after reading an account of Father Wulf's Eiffel-Tower experiments, I was inclined to believe that a hitherto unknown source of ionization may have been in evidence in all these experiments; and I decided to attack the problem by direct experiments of my own.

It seemed to me necessary to measure accurately the absorption of gamma rays from radium in air in order to find out how far above the ground gamma rays could act as an ionizing agency. I made this measurement in Vienna, using a large quantity of radium (1500 milligrams) as source of gamma rays and observing the ionization in a closed vessel placed outdoors at various distances up to 90 meters from the source. The coefficient of the absorption in air calculated from these measurements was of the expected magnitude; and from this it could be said with certainty that gamma rays from the ground are almost completely absorbed at 500 meters above the ground.

The next step was the construction of an air-tight ionization apparatus which could be used during balloon flights and fitted with a sensitive electrometric system which was not influenced by the large fluctuations of temperature occurring in the flights. I used a modification of Th. Wulf's apparatus with walls of zink, thick enough to withstand the excess pressure of one atmosphere and a temperature compensation for the fibre electrometer. Furthermore, I found it very important always to use two or three of the instruments simultaneously in order to avoid errors from instrumental defects. With such instruments, I made ten balloon ascents: two in 1911, seven in 1912 and one in 1913. Five of them were carried out at night, and some of them continued during the following morning. One flight was made during a solar eclipse, in April 1912.

By taking successive readings of the ionization with two or three instruments at a time, much more reliable data were obtained. I found that at 500 meters above the ground the ionization was, on the average, about 2 I lower than on the ground and that, from about 1800 meters upwards, an increase of ionization is undoubtedly in evidence. At 1500 meters, the ionization increased to the same value as had been found on the ground. At 3500 meters, the increase amounted to no less than 4 I, at 5000 meters to 16 I above the ground value. No difference between day and night observations was noticed.

An explanation of the increase of ionization with increasing altitude on account of the action of radioactive substances was impossible. I calculated that the known quantities of radium emanation (radon) and other substances in the atmosphere could not produce more than one twentieth of even the small effect observed at an altitude of one to two kilometers.

The only possible way to interpret my experimental findings was to conclude to the existence of a hitherto unknown and very penetrating radiation, coming mainly from above and being most probably of extra-terrestrial (cosmic) origin. The extraordinary penetrating power of this radiation could be derived from the fact that its ionizing action was even noticeable at sea level, after having penetrated the whole atmosphere of which the mass is equivalent to a layer of 76 centimeters of mercury or 10.5 meters of water".

Soweit der Bericht. Schon auf Grund der ersten Fahrten hatte Hess die Existenz einer derartigen Strahlung vermutet, aber Gewißheit brachte erst die Hochfahrt am 7. August 1912. Dieser Tag kann also als Entdeckungstag der Kosmischen Strahlung angesehen werden.

Zur Gegenüberstellung mit den heutigen Forschungsmethoden, die sich der Raketen und der künstlichen Satelliten bedienen, sei Hess wörtlich in seiner Schilderung der Fahrt vom 7. August 1912 wiedergegeben (Hess, 1913):

"Es erübrigte jetzt nur noch, einmal eine Hochfahrt zu machen, um das Verhalten der Gamma-Strahlung in Höhen von mehr als 3000 m zu erforschen. Zu diesem Zwecke habe ich am 7. August eine Fahrt mit dem 1.680 m³ fassenden Ballon "Böhmen" des deutschen Luftfahrtvereines in Böhmen von Aussig a. d. Elbe aus unternommen. Die Füllung des Ballons erfolgte bereits während der Nacht. Es wurde sehr reiner Wasserstoff zur Füllung benützt, der pro Kubikmeter eine Tragfähigkeit von 1,1 kg aufwies. Es war also Hoffnung vorhanden, eine Höhe von 6000 m zu erreichen.

An der Fahrt nahmen teil: Herr Hauptmann W. Hoffory als Führer, Herr E. Wolf als meteorologischer Beobachter und meine Wenigkeit. Um 6 Uhr 12 Minuten früh wurden wir vom Fahrwart, Herrn Ing. Walter Mitscherlich, hochgelassen. Die Platzfrage gestaltete sich recht schwierig, denn für drei Personen, drei große Sauerstoffzylinder, Sitzbank, Instrumentenkorb und Handgepäck schien der an sich bequeme Ballonkorb doch etwas zu eng. Meine drei Strahlungsapparate wurden wie gewöhnlich am Korbrande an kleinen, mit Stellschrauben versehenen Konsolen montiert. An Ballast nahmen wir 52 Sack mit (etwa 800 kg). Ein Teil der Säcke war so gehängt, dass ihre Entleerung durch Abschneiden eines Bindfadens bewirkt werden konnte, was in grösseren Höhen zur Vermeidung jeder körperlichen Anstrengung wichtig ist. Nach Abgabe von zehn Sack Ballast waren wir in 1500 m Höhe. Kurz vorher begann ich mit meinen Beobachtungen. Unsere Fahrtrichtung war bisher westlich, dann aber nördlich, so dass wir um 7.30 Uhr die deutsche Grenze bei Peterswalde übersetzten. Um 8 Uhr schwebten wir über Struppen bei Pirna mit prächtiger Aussicht auf die Festung

Route des Entdeckungsfluges der kosmischen Strahlung.

Hess bei Ballonlandung (1912).

Königsstein und die sächsische Schweiz. Um 8.30 Uhr waren 3000 m erreicht – nach Verbrauch von 20 Sack Ballast. Als ich bis 3500 m genügend Messungen gemacht hatte, trieben wir den Ballon wieder rascher aufwärts. Um 9.15 Uhr wurden 4000 m erreicht; wir standen über Elstra im östlichen Sachsen. Da ich mich etwas müde zu fühlen begann, griff ich schon zum Sauerstoffinhalationsapparat, um mich für die immerhin anstrengenden Ablesungen munter zu erhalten. Die Kälte war auch schon merklich. In 4200 m massen wir 8 1/2 Grad unter Null, was umso fühlbarer war, als die Sonne durch einen dünnen Wolkenschleier, der in enormer Höhe schwebte, nur sehr ge-schwächt durchschimmerte. In 4800 m begann auch Herr Wolf mit der Sauerstoff-atmung, die sowohl bei ihm wie bei mir sehr belebend wirkte und die hohe Pulsfrequenz herabminderte. Um 10.45 Uhr hatten wir 5350 m erreicht. Trotz Sauer-stoff fühlte ich mich so schwach, dass ich nur noch mit Anstrengung an zwei Apparaten die Ablesungen ausführen konnte, die dritte Ablesung misslang. So ent-schloss ich mich, obwohl wir noch zwölf Sack Ballast hatten, herunterzugehen, und bat Hauptmann Hoffory, Ventil zu ziehen. Bei dem Abstiege fühlten wir uns noch recht matt bis 4000 m, dann aber erholten wir uns überraschend schnell. Hauptman Hoffory hatte die Höhe von 5350 m ohne Sauerstoffatmung ausgehalten. Bei mir hatte wohl die vorausgegangene ungenügende Nachtruhe (ich war um 1/2 4 Uhr aufgestanden) und eine Magenindisposition die Widerstandskraft gegen die Höhenkrankheit so sehr vermindert.

Wir liessen den Ballon, der mehrmals von selbst zu fallen aufhörte, durch mehrfaches Ventilziehen bis zirka 1000 m herabgehen, dann orientierten wir uns über die Terrain- und Windverhältnisse und landeten bei schwachem Bodenwinde sehr glatt um

12.15 Uhr mittags auf einer sandigen Wiese bei Pieskow (Brandenburg), etwa 50 km östlich von Berlin. Nachdem uns ein Gutsbesitzer in freundlicher Weise Landungshilfe zur Verfügung gestellt und uns gastfreundlich aufgenommen hatte, fuhren wir gegen 4 Uhr nach Berlin, von wo wir mit dem Nachtschnellzuge nach Wien zurückkehrten.

Mit dem wissenschaftlichen Ergebnis dieser Fahrt konnte ich sehr zufrieden sein; es war mir gelungen, mit drei Apparaten unabhängig den Verlauf der durchdringenden Strahlung bis über 5000 m zu verfolgen. Nachdem ich schon bei den acht vorhergehenden, von Wien aus unternommenen Fahrten gefunden hatte, dass die Gesamtstrahlung in Höhen von 1000 bis 2000 m ebenso gross ist wie am Boden, also grösser als in 500 und 1000 m ist, ergab sich in den Höhen von 2000 m aufwärts eine noch viel auffallendere Steigerung der Gamma-Strahlung mit der Höhe. In 4000 m war die Gamma-Strahlung schon um die Hälfte stärker als am Boden, im 5000 m mehr als doppelt so stark. Das war ein Ergebnis, welches vollkommen neue Gesichtspunkte schuf: es war der Beweis erbracht, dass in 5000 m Höhe eine viel stärkere Gamma-Strahlung wirkt als auf der Erde.

Die Strahlung der Erde ist in diesen Höhen längst unwirksam. Wenn man an der Ansicht festhält, dass nur die bekannten radioaktiven Produkte in der Atmosphäre eine Gamma-Strahlung erregen, so müsste man zur Erklärung der von mir beobachteten Strahlungserhöhung in 5000 m annehmen, dass zufällig gerade in dieser Höhe eine lokale Anhäufung radioaktiver Materie stattgefunden habe. Dies ist aber so unwahrscheinlich, dass man eher der Ansicht zuneigen wird, dass ein Teil der beobachteten Gamma-Strahlung von oben her in die Atmosphäre eindringt – also ausserterrestrischen Ursprunges ist. Dass ein Teil der Gamma-Strahlung von der Sonne kommt, ist unwahrscheinlich, da ich bei den von Wien aus unternommenen Fahrten in der Nacht und während der Sonnenfinsternis am 17. April 1912 keine Verminderung der Gamma-Strahlung fand".

Eine volle Bestätigung der Hess'schen Messungen brachten die Hochflüge Kolhörster's 1913/14. Bis 5000 m stimmten die Ionisationswerte mit den von Hess gemessenen

Hess vor dem Hafelekar-Labor am Tage nachdem er die Mitteilung von der Zuerkennung des Nobelpreises erhalten hatte (November 1936).

überein, dann folgte ein rapider Anstieg, bei 9000 m war die Ionisation auf das etwa vierzigfache des Bodenwertes gestiegen. Der für die durchdringende Strahlung berechnete Absorptionskoeffizient war etwa zehnmal kleiner als der der RaC Gamma-Strahlung.

Hess' Entdeckung fand zunächst nur in einem engen Kreis Beachtung. Die Zeit war für sie noch nicht reif. Ihre große Bedeutung für die Entwicklung der Physik konnte damals weder von Hess noch von seinen Zeitgenossen erkannt werden. Niemand hätte zu dieser Zeit auch nur geahnt, daß die weitere Erforschung der "durchdringenden Strahlung" zur Hochenergie- und Elementarteilchenphysik führen würde. So ist es auch verständlich, daß Hess erst ein Vierteljahrhundert später, 1936, den Nobelpreis erhielt.

Es war naheliegend, die weiteren Studien der Strahlung in möglichst großen Höhen auszuführen. So richtete Hess im Herbst 1913 eine Dauerbeobachtung auf dem Obir (2044 m) in Kärnten ein, um zu untersuchen, ob die Strahlung Schwankungen unterworfen ist und welcher Art diese sind. Der Beginn des ersten Weltkrieges setzte aber den Arbeiten ein Ende, und die Strahlung geriet fast in Vergessenheit.

Erst durch Nernst's Buch "Das Weltgebäude im Lichte der neueren Forschung" (Berlin, 1921), in dem er die Idee aussprach, die durchdringende Strahlung könnte in gewissen Sterntypen beim Zerfall transuranischer Kerne emittiert werden, wurde das Interesse an der Strahlung neu erweckt. Gleichzeitig wurden aber auch Bedenken an ihrer Existenz geäußert. Man argwöhnte, an der Ballonhülle hätten sich radioaktive Stoffe angelagert, ein radioaktives Gas oder radioaktive Zerfallsprodukte hätten sich in der Tropopause angesammelt, auch eine Auslösung der Strahlung in den höchsten Schichten durch solare Elektronen wurde erwogen – alles Einwände, die entkräftet werden konnten. Behounek und G. Hoffmann zweifelten auf Grund von Absorptionsmessungen, Millikan und Mitarbeiter auf Grund von Registrierballonaufstiegen bis 15 km an der Existenz einer Strahlung angegebener Eigenschaften. Weitere Versuche veranlaßten jedoch die Genannten, ihre Ansichten zu revidieren. Im Zuge der damit verbundenen Auseinandersetzungen (Bergwitz et al., 1928) ergaben sich auch Prioritätsstreitigkeiten, die erst 1932 durch die Zuerkennung des Abbe-Preises und 1936 des Nobelpreises, den Hess gemeinsam mit C.D. Anderson erhalten hatte, beendet wurden. Spät kam es auch zu einer endgültigen Namensgebung der Strahlung. Von einigen Autoren war "Hess' sche Strahlung", in den USA auch "Millikan-Strahlung" vorgeschlagen worden, Hess schlug 1926 "Ultragammastrahlung" vor. Später, nachdem ihre Teilchennatur erkannt war, "Ultrastrahlung", auch "Kosmische Ultrastrahlung", und etwa 1940 schloss Hess sich dem von Millikan stammenden Namen "Kosmische Strahlung" an. Im deutschen Sprachraum wird auch oft der von Kolhörster stammende, von Hess aber abgelehnte Name "Höhenstrahlung" verwendet.

Nachdem die Existenz der Strahlung, die man bis zur Entdeckung ihrer Teilchennatur durch Bothe und Kolhörster (1929) für eine ultraharte Gammastrahlung gehalten hatte, nun gesichert war, begann man, nach ihrem Ursprung zu suchen. Angeregt durch Nernst's Idee, setzte eine Suche nach Intensitätsmaxima ein, die der Kulmination gewisser Himmelsbereiche entsprechen könnten. So wurde unter anderem in Gletscherspalten auf dem Jungfraujoch, auf dem Mönchsgipfel (Kolhörster und Salis), auf der Zugspitze (Büttner) in mehrtägigen Meßreihen nach einer Sternzeitperiode gesucht. Die dabei gefundenen Anzeichen für eine solche konnten jedoch von anderen (Hoffmann, Steinke, Millikan u.a.) nicht bestätigt werden.

Hier trat nun Hess, inzwischen als Ordinarius an die Grazer Universität berufen, wieder auf den Plan. Mit Unterstützung verschiedener Institutionen konnte er mehrere Strahlungsmeßapparate vom verbesserten Kolhörster'schen Typ (4,3 1, Zweischlingen-elektrometer) sowie Panzermaterial anschaffen. Nach Vorversuchen in Graz, im Innsbrucker Mittelgebirge und auf dem Patscherkofel organisierte er 1927 eine einmonatige Registrierung auf dem Sonnblick (3100 m, in den Hohen Tauern) mit O. Mathias als Beobachter.

Dies war die Zeit, als der Verfasser in Hess' Institut eintrat. Da die Registrierungen auf dem Sonnblick nur unregelmäßige, den mittleren Fehler kaum übersteigende Strahlungsschwankungen gezeigt hatten, entschloß sich Hess, auf dem Sonnblick-Observatorium eine ganzjährige Registrierung mit drei parallel laufenden Apparaten nach einer verbesserten Methode einzurichten. Der Verfasser wurde beauftragt, mit A. Reitz mehrere Wochen lang rund um die Uhr die Strahlung zu registrieren und den Beobachter am Observatorium für die Fortsetzung der Beobachtungen einzuschulen. Der Restgang der Apparate war vorher in einer Kalksteinhöhle nördlich von Graz unter 70 m Gestein hinter 7 cm Eisen bestimmt worden.

Beobachtungen im Hochgebirge waren zu jener Zeit, da es noch keine Autostraßen in die Hochtäler und keine Bergbahnen gab, mit erheblichen Mühen verbunden. Ein Bild davon gibt der nachfolgende (gekürzte Bericht des Verfassers (Steinmaurer, 1930)):

"Mein Kollege, Herr A. Reitz, und ich traten am 29. Juni 1929 die Reise auf den Sonnblick an. Wir wählten, wie alle wissenschaftlichen Expeditionen, des schweren Gepäcks halber den Weg über Taxenbach, Rauris, Kolm — Saigurn. Unser Gepäck wog etwa 200 kg und bestand aus den drei Strahlungsapparaten, die wir auf der Eisenbahn im Abteil mitführten, einer großen Kiste, welche die Registrieruhr enthielt, einem Reisekorb, in dem die drei Registrierapparate, ein Akkumulator und verschiedene Reserveteile verpackt waren, ferner eine Kiste mit Anodenbatterien, dem Ausmeßtisch und nicht an letzter Stelle einer Kiste mit Konserven, Keks, Marmelade, Zwieback usw., Dingen, die es uns gestatten sollten, dem dräuend unbekannten Speisezettel der nächsten Wochen gegenüber ein wenig unsere Selbständigkeit zu bewahren.

Der etwa 400 kg schwere Eisenpanzer — weitere 320 kg Panzermaterial waren noch von Dr. Mathias' Beobachtungen her auf dem Observatorium verblieben — war bereits im Laufe des Juni auf den Sonnblick geschafft worden. Zum Transport des Materials von Taxenbach nach Kolm-Saigurn wurde ein Pferdefuhrwerk gemietet, die 1500 m von dort zum Observatorium mußten von Trägern — jeder trug eine Last von 40 bis 60 kg — bewältigt werden. Diese Transporte waren natürlich mit hohen Kosten verbunden, obgleich die Trägerlöhne gegenüber 1927 um 20% erniedrigt waren, machten die Transportkosten allein über 25% der gesamten Expeditionskosten aus.

Der Aufstieg war herrlich. Sonne über einigen Zentimetern Neuschnee! Auf dem Gipfel eine Fernsicht vom Traunstein bis zu den Dolomiten! Das Gepäck war glücklich angelangt, nur einer der Strahlungsapparate hatte dem inneren Überdruck nicht standgehalten und war ausgelaufen. Das bedeutete aber keinen großen Schaden, denn durch Anbringung einer Korrektur können die Messungen auf Normaldruck reduziert werden ...

Die Apparate wurden auf Betonsockeln in der sogenannten Pendelhütte aufgestellt, in der 1911 gravimetrische Messungen ausgeführt worden waren. Am 3. Juli war die Aufstellung beendet. Da die Apparate alle vier Stunden neuer Aufladung bedurften, hatten wir einen Tag- und Nachtdienst eingerichtet. Der "Tagdienst" mußte die Apparate zwischen 3 und 4 Uhr früh zum erstenmal und dann bis 20 Uhr nach Ablauf von je vier Stunden laden und überprüfen. Der "Nachtdienst" wechseite um Mitternacht die Registrierstreifen und unterzog die Apparatur einer Generalrevision ..."

Auch diese Beobachtungen brachten keine Bestätigung einer Sternzeitperiode, auch nicht nach Eliminierung des den Schwankungen überlagerten Luftdruck-Einflusses durch Reduktion aller Meßwerte auf Normaldruck. Doch wurden Andeutungen eines Tagesganges und ausgeprägte jajreszeitliche Schwankungen festgestellt. Hess entschloß sich daher, die Messungen fortzusetzen und zwar im Hinblick auf die Kleinheit der Schwankungen mit einer Apparatur höchster Präzision, ähnlich der von Hoffmann gebauten. Für die Aufstellung einer solchen viel größeren und schwereren Apparatur war der Sonnblick wegen seiner schweren Zugänglichkeit ungeeignet. Günstiger erschien die leichter erreichbare Zugspitze (2963 m). Doch die Wahl fiel auf das Hafelekar bei Innsbruck, das von der Universität mit der Seilbahn in weniger als einer Stunde erreichbar ist. Das österreichische Bundesministerium für Unterricht hatte an der Universität Innsbruck ein "Institut für Strahlenforschung" errichtet, an das Hess als Vorstand berufen wurde.

So übersiedelte Hess 1931 nach Innsbruck. Das Innsbrucker Institut war sehr bescheiden. Es bestand aus drei Zimmern, jegliche apparative Grundausstattung fehlte, und außer einem Assistenten (der Verfasser), einem wissenschaftlichen Mitarbeiter (J. Priebsch) und einem an zwei Tagen der Woche zur Verfügung stehenden Mechaniker war kein Personal vorhanden. Die Kolhörster'schen Strahlungsapparate mit Panzermaterial, einige Geräte für luftelektrische Messungen hatte Hess aus Graz mitnehmen können.

Hess kam mit großen Plänen. Sein Wunsch war die Errichtung eines Gebäudes auf dem Hafelekar, das außer der "Station für Ultrastrahlenforschung" auch einem meteorologischen Observatorium, einer kleinen Sternwarte und einem Labor für alpine Forschungsarbeiten Raum geboten hätte. Da aber die Mittel nicht reichten, wurde im "Unterkunftshaus Hafelekar" ein 3 mal 4,5 m^2 großes Zimmer als Labor eingerichtet. Elektrischer Strom war vorhanden, Wasser mußte im Winter von der Bergstation der Seilbahn geholt werden.

Durch die finanzielle Hilfe der Österrichischen Akademie der Wissenschaften, der Preussischen Akademie der Wissenschaften, der Österreichisch-deutschen Wissenschaftshilfe und des Sonnblick-Vereines war es Hess möglich, eine von E. Steinke, einem Mitarbeiter G. Hoffmann's, konstruierte "Steinke Standard-Apparatur" zu beschaffen. Diese bestand aus einer 22,6 1 fassenden, mit 10 bar CO_2 gefüllten Ionisationskammer, die nach der Auflademethode mit Ladungskompensation über einen Influenzierungskondensator arbeitete. Registriert wurden stündlich die Ausschläge eines Lindemann-Elektrometers. Die Apparatur lief fünf Tage wartungsfrei. Gegen die Umgebungsstrahlung schützte ein 1,5 t schwerer, allseitig 10 cm starker Bleipanzer ("Vollpanzer"), der oben geöffnet werden konnte ("Halbpanzer").

Schon 1913 hatte Hess die Idee von Simultanbeobachtungen mit gleichartigen Apparaten geäußert. Bei einer Zusammenkunft mit A. Corlin, W. Kolhörster und E.

Die "Station für Ultrastrahlenforschung" auf dem Hafelekar bei Innsbruck (2300 m), 1960, vor
dem späteren Ausbau.

Steinke in Berlin im Dezember 1930 kam er darauf zurück, Durch die Entdeckung der
Teilchennatur der Strahlung waren auch Beobachtungen in verschiedenen geogra-
phischen Breiten aktuell geworden, Steinke erklärte sich bereit, mehrere Exemplare
seiner Standard-Apparatur herstellen zu lassen. In Aussicht genommen waren Stationen
in Åbisko (Corlin), Königsberg (Steinke), Dublin (Nolan u. O'Brolchain), Hafelekar
(Hess), Bandung (Clay), und Kapstadt (Schonland).

Im Juli 1931 wurden die ersten Apparate von Corlin, O'Brolchain und dem Verfasser
in Königsberg übermommen. Ende August begannen die Registrierungen auf dem
Hafelekar.

Einer der ersten Besucher war A. Piccard, wenige Wochen nach der Landung auf dem
Gurgler Ferner, nach seinem ersten Stratosphärenflug. Er ließ seinen bei dem Stratosphärenflug benützten Strahlungsapparat durch seinen Mitarbeiter Kipfer mit den
Hess'schen Apparaten vergleichen. Auch andere prominente Wissenschaftler haben sich
im Gästebuch der Station eingetragen.

Das Hauptergebnis der Registrierungen mit den Steinke-Apparaten in den ersten
Jahren war der Nachweis der Existenz einer regelmäßigen täglichen Schwankung nach
Ortszeit (Hess *et al.,* 1934, Hess u. Graziadei, 1936), die bei den Vollpanzer- und
Halbpanzermessungen konform auftritt. Nach rechnerischen Eiiminierung des Einflusses
der unperiodischen Schwankungen sowie der Tagesperiode des Luftdruckes auf die
Meßwerte ergab sich in beiden Jahren ein Maximum der Strahlung zur Mittagsstunde,
ein Minimum zwischen 21 Uhr und 3 Uhr. Ob der Mittagsanstieg, so wie von Hess
früher angenommen, durch eine solare Komponente der Strahlung oder durch indirek-
ten Einfluß der Sonne erzeugt ist, wird offen gelassen. Außer dem täglichen Gang
werden auch unregelmäßige, sich über Tage erstreckende Schwankungen ("Schwankun-

Three pioneers of Cosmic Ray research von links nach rechts: Hess, Steinke, E. Regener
Regener demonstriert sein Ballonelektrometer
(Immenstaad/Bodensee, August 1932).

gen zweiter Art") ebenso bestätigt wie plötzliche kurzzeitige Anstiege der Ionisation, die sogenannten "Hoffmann'schen Stöße."

Im Jahre 1933 konnte Hess durch eine Widmug der Rockefeller Foundation seine Forschungsstation erweitern. Drei Räume des Unterkunftshauses wurden in das Labor einbezogen und zwei weitere "Steinke-Standardapparate" angeschafft. Nun konnten zwei Apparate auf dem Hafelekar parallel laufen, einer im Voll-, der andere im Halbpanzer, während ein dritter in Innsbruck betrieben wurde.

Hess' Absicht war, die zeitlichen Variationen der Strahlung dauernd zu verfolgen, das gewonnene umfangreiche Beobachtungsmaterial in Bezug auf periodische Schwankungen nach Orts- und Sternzeit zu analysieren und vorhandene unperiodische Schwankungen auf einen Zusammenhang mit Änderungen der Außentemperatur, des erdmagnetischen Feldes und der Aktivität der Sonne zu untersuchen. Von den Ergebnissen der Arbeiten der Jahre 1934 bis 37 sei kurz folgendes mitgeteilt:

Die Einflüsse des Luftdruckes und der Außentemperatur ("Luftdruck- und Temperatureffekt") wurden von Priebsch (Priebsch u. Baldauf, 1936) nach der Methode der multiplen Korrelation berechnet. Gegenüber Luftdruck und Temperatur verläuft die Strahlung antiparallel. Die gefundene negative Temperaturabhängigkeit wurde zu deuten versucht als Folge der Änderung der Luftdichte und damit der Streubedingungen. Die korrekte Erklärung konnte damals nicht gegeben werden, da die Entstehung der Muonenkomponente noch unbekannt war.

Auch nach Berücksichtigung der Einflüsse von Luftdruck und Außentemperatur verblieb ein täglicher Gang nach Ortszeit mit einem Mittagsmaximum. Als Ursache des Tagesganges wurde eine Beeinflussung der Strahlung in den höchsten Schichten angenommen, wobei auch auf den antiparallelen Gang mit der Horizontalintensität des erdmagnetischen Feldes hingewiesen wurde. Auch für den Tagesgang konnte die richtige Erklärung noch nicht gegeben werden.

Ein regelmäßiger Gang nach Sternzeit wurde nicht gefunden. Nur ein Maximum um 20.40 Uhr war in den Meßreihen angedeutet, das nach Compton' und Getting's Berechnung, durch die Rotation der Milchstraße hervorgerufen, ein Hinweis für einen extragalaktischen Ursprung der Strahlung wäre.

Bei der Suche nach einem Zusammenhang zwischen Strahlung und Sonnenaktivität konnte eine 27,2 tägige Periode nachgewiesen werden, entsprechend der Umdrehungs-dauer der Sonnenfleckenzone (Graziadei, 1936).

Die Variationen des erdmagnetischen Feldes und der Strahlung zeigten teils parallelen, teils antiparallelen Verlauf, letzteren vor allem in den Tages- und Jahresgängen. Ein ausgeprägter Parallellauf wurde bei magnetischen Stürmen gefunden, worauf schon Forbush verwiesen hatte. Bei den Stürmen Ende April 1937 und am 25-26. Jänner 1938 wurde dieser Effekt weltweit beobachtet (Hess *et al.*, 1938). Mitte der Dreißigerjahre erlaubten es die Mittel, auch Zählrohr und Nebelkammer einzusetzen. Nach Einschu-lung bei Regener in Stuttgart konnte Priebsch (Priebsch, 1936) eine Dauerregistrierung der Zahl der stündlichen Dreieckskoinzidenzen mit einem selbstgebauten Apparat beginnen, wobei sich im allgemeinen ein guter Gleichlauf mit den Ionisationskammern ergab. Daneben wurden Sekundäreffekte der Strahlung studiert, so die Rossi-Kurve bis 36 cm Blei aufgenommen.

Über Einladung von Hess führte im Sommer 1934 F. Rieder (Wien) Versuche mit einer Nebelkammer durch. Es war eine kleine, von Hand zu bedienende Kammer nach Philipp und Dörffel von 12 cm Durchmesser. Umgebende Spulen erzeugten ein Magnetfeld von 1500 Gauss. Mit einer "Contax"-Kamera wurden die Spuren foto-graphiert. Bei einem Teil der Versuche lag quer durch die Kammer ein Bleistreifen. Gefunden wurden Elektronen, Schauer und einige schwere Teilchen. Eine Statistik über Ladungssinn und Energie wurde erstellt.

Auch mit Kernspurplatten wurde zu dieser Zeit experimentiert. Die beiden Wiener Physikerinnen M. Blau und H. Wambacher (Blau u.Wambacher, 1937) sandten Platten, die unter verschiedenen Bedingungen mehrere Wochen auf dem Hafelekar exponiert wurden. Die dann in Wien entwickelten und durchmusterten Aufnahmen brachten eine Über-raschung. Es wurden zwei Gruppen von Bahnspuren gefunden: Neben langen Spuren, die man Protonen von bisher nie beobachteter Energie zuordnen konnte, wurden kurze Spuren entdeckt, die von einem Zentrum sternförmig ausgingen. Verfasser erinnert sich noch, wie Frau Blau, 1937 auf Sommeraufenthalt in Tirol, zu ihm vor der Talstation der Innsbrucker Hungerburgbahn sagte: "Eben hat mir Frau Wambacher aus Wien geschrieben, sie habe auf den Platten, die am Hafelekar exponiert waren, merkwürdige, sternförmige Teilchenbahnen gefunden, die bestimmt nicht von radioaktiven Verun-reinigungen herrühren können." Dies war der erste Nachweis eines Zertrümmerungs-sterns.

Von all den Untersuchungen, die in den Dreißigerjahren im Hafelekar-Labor ausge-führt wurden, seien ferner die Versuche über die biologischen Wirkungen der Strahlung hervorgehoben. Sie wurden vom Schweizer Arzt J. Eugster als Gast an Taufliegen, Kaninchen, Bakterien und Samen ausgeführt. Die Ergebnisse sind in dem Buch von Eugster-Hess "Die Weltraumstrahlung und ihre biologische Wirkung" (Zürich, 1940) niedergelegt.

Weiters wurden auf Hess' Anregung auf dem Hafelekar Zählungen der Groß- und Kleinionen, der Staub- und Kondensationskerne, des Emanationsgehaltes der Freiluft, stets im Vergleich mit Messungen in Innsbruck, sowie eine Registrierung des luftelek-trischen Potentialgefälles ausgeführt.

Hess verließ Innsbruck im Mai 1937 und übernahm als Nachfolger H. Benndorf's das

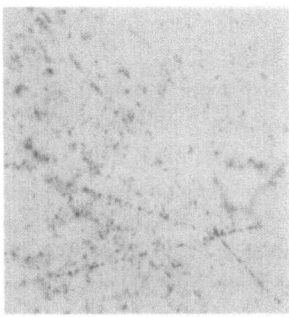

Auf Kernemulsionsplatten, die einige Wochen in 2300 m Hohe (Hafelekar) gelagert waren, fand Hertha Wambacher (Wien) 1937 sternartige Teilchenbahnen, die als Spuren einer Kernzertrümmerung durch die kosmischen Strahlung gedeutet wurden.

Grazer Physikalische Institut. Das Hafelekar-Labor, vor allem die Registrierung der Strahlungsintensität, wurde vom Verfasser weitergeführt.

Zum Schluß noch eine kurze Würdigung der Persönlichkeit Hess'. Victor Franz Hess wurde am 24. Juni 1883 in Waldstein geboren, einem kleinen Ort nördlich von Graz (Österreich). Nach seiner Promotion "Sub auspiciis Imperatoris" 1906 hatte er die Absicht, bei Drude in Berlin ein Thema aus dem Gebiet der Optik zu bearbeiten. Ein Ausbildungsstipendium war ihm bereits bewilligt. Aber Drude's plötzlicher Tod vereitelte diesen Plan. So entschloß sich Hess, sich in Wien weiterzubilden, wo er, zuerst am 2. Physikalischen Institut, dann am neugegründeten Institut für Radiumforschung, luftelektrische und radioaktive Studien betrieb, in deren Verfolgung er die Kosmische Strahlung entdeckte. In seiner Selbstbiographie bezeichnet er die Zeit am Radiuminstitut als die glücklichste seines Lebens. Dem Wiener Physikerkreis gehörten damals neben Benndorf, Kohlrausch, Hasenöhrl, Pribram, Schweidler auch die nachmaligen Nobelpreisträger Hevesy, Paneth, Schrödinger an. Im ersten Welt-Krieg war Hess Leiter der Röntgenabteilung eines Reservelazarettes. Daneben führte er mit R.W. Lawson, einem Engländer, der in Wien vom Kriegsausbruch überrascht worden war, eine international anerkannte Präzisionsbestimmung der Zahl der von 1 Gramm Radium sekundlich ausgesandten Alphateilchen aus und wendete als erster die Rutherford-Geiger'sche Zählkammer auf die Zählung von Gammastrahlen an.

1920 wurde Hess an die Universität Graz berufen. Fast gleichzeitig erhielt er eine Berufung als Chefphysiker der Radium Corporation in Orange N.Y. und als "Consulting Physicist" im US Bureau of Mines im Ministerium des Inneren. Arbeiten über medizinische Anwendungen des Radiums und Methoden zur Bestimmung des Radiumgehaltes von Erzen und zur Reinigung der Radiumemanation stammen aus dieser Zeit.

Nach zwei Jahren kehrte Hess nach Graz zurück. Da ihm aber dort keine Radiumpräparate und nur sehr beschränkte Mittel zur Verfügung standen (sein Labor bestand aus zwei Zimmern, er hatte weder einen eigenen Assistenten noch einen eigenen Mechaniker), setzte er seine in Wien begonnenen, wenig apparativen Aufwand erfordernden luftelektrischen Arbeiten fort und war literarisch tätig. Das Buch "Die elektrische Leitfähigkeit der Atmosphäre und ihre Ursachen" (Vieweg, 1926) und die

Monographie "Ionisierungsbilanz der Atmosphäre" (in *Ergebnisse der Kosmischen Physik*, 2 (1934) einstanden in dieser Zeit.

1931 wurde Hess an die Universität Innsbruck berufen. Besonders während der ersten Jahre fühlte er sich in Innsbruck sehr wohl. Sein Werk gedieh und fand Anerkennung. Getrübt wurden die Innsbrucker Jahre durch zwei Operationen, die sich als Folge einer noch in Wien erlittenen Radiumverbrennung als notwendig erwiesen hatten: die Amputation des linken Daumens und eine Kehlkopfoperation, von der eine ständige Heiserkeit zurückblieb. Seiner neuen Tätigkeit an der Grazer Universität, an die er im Frühjahr 1937 berufen wurde, konnte er sich nur ein Jahr lang widmen. Bald nach der Besetzung Österreichs durch das nationalsozialistische Deutschland wurde Hess zuerst in den Ruhestand versetzt und im September 1938 fristlos ohne Pension aus politischen Gründen entlassen und gezwungen, den Nobelpreis, den er in Schweden investiert hatte, einzuberufen und gegen deutsche Reichsschatzscheine umzutauschen. Als kosmopolitisch denkender Wissenschaftler und aktiver Katholik hatte Hess aus seiner Ablehnung des Nationalsozialismus nie ein Hehl gemacht. Er emigrierte nach den USA, wo er sogleich an der Fordham Universität in New York ein neues Betätigungsfeld fand.

Persönlich war Hess von großer Liebenswürdigkeit, Konzilianz und Aufgeschlossenheit, durchaus nicht der Typus eines unnahbaren "Olympiers", wennglich er auch eine gewisse Respektierung wünschte. Seine Gattin schuf ihm jene Atmosphäre, die er für seine wissenschaftliche Arbeit, in der er ganz aufging, benötigte. Er war ein fleißiger, flinker, ordnungsliebender, stetiger Arbeiter. Stress wußte er zu vermeiden. Seiner Arbeit zuliebe verzichtete er auf regen gesellschaftlichen Verkehr, auch Hobby pflegte er keines. Sein ständiger Begleiter, auch im Institut, war ein brauner Dackel. Hess war von kräftiger, voller Statur, aber kein Alpinist oder Schisportler, doch ein großer Naturfreund und Bewunderer der Tiroler Bergwelt. Neben seiner wissenschaftlichen

V.F. Hess, 1960. – Auf dem Tisch ein Strahlungsapparat (nach Kolhörster), wie er in der Zeit von 1927 bis 1931 von Hess u. Mitarbeitern in Graz, auf dem Sonnblick und in Innsbruck verwendet wurde. Daneben eine Zamboni-Säule zur Auflsdung des Elektrometer-Systems.

Tätigkeit widmete er sich mit der ihm eigenen Gründlichkeit auch der Lehre, nach dem Grundsatz, daß Lehre und Forschung Hand in Hand gehen müßten. Obgleich ein strenger, aber gerechter Prüfer, war er bei den Studenten beliebt. Seine besondere Fürsorge galt den Dissertanten und seinen Mitarbeitern, zu denen er ein fast väterliches Verhältnis hatte. Er stellte hohe Anforderungen, ging aber selbst mit gutem Beispiel voran. Leistungen wurden gewürdigt und durch Förderung belohnt. Mit Tadel war er sparsam, heftig wurde er nie. Mit Fachkollegen pflegte er regen Kontakt, über divergierende Ansichten wurde sachlich diskutiert.

Die Entlassung im Jahre 1938 traf Hess sehr schwer. Nach Kriegsende kam er noch einige Male nach Österreich, besuchte sein Hafelekar-Labor, aber zur Annahme des Angebotes, auf seine alte Lehrkanzel zurückzukehren, konnte er sich nicht entschließen. Hess starb am 17 Dezember 1964 in New York.

Der Verfasser dankt Herrn Prof. F. Fliri für die Ausarbeitung der Flugroute Prof. Hess', Herrn Prof. J. Priebsch für Beratung und Durchsicht des Manuskripts, sowie Herrn Prof. D. Burkert für Überlassung von Lichtbildern.

REFERENZEN

Bergwitz, K., Hess, V.F., Kolhörster, W. u. Schweidler, E.: 1928, *Physikal. Zs.*, **29**, 327.

Blau, M. u. Wambacher, H.: 1937, *Nature* **140**, 585; 1937, *Sitzungsber. Akad. Wiss. Wien*, *IIa* **146**, 623.

Graziadei, H.Th.: 1936, *Sitzungsber. Akad. Wiss. Wien*, *IIa* **145**, 495.

Hess, V.F.: 1911, *Physikal. Zs.* **12**, 998.

Hess, V.F.: 1912, *Physikal. Zs.* **13**, 1084 ; 1912, *Sitzungsber, Akad. Wiss. Wien*, *IIa* **121**, 2001.

Hess, V.F.: 1913, Jahrbuch d. Österreich. Aeroclubs 1912, 190.

Hess, V.F.: 1940, *Thought (Fordham Quaterly)* **XV**, 225.

Hess, V.F., Graziadei, H.Th. u. Steinmaurer, R.: 1934, *Sitzungsber. Akad. Wiss. Wien*, *IIa* **143**, 313.

Hess, V.F., u. Graziade, H.Th.: 1936, *Terr. Mag.*, **41**, 11.

Hess, V.F., Demmelmair, A. u. Steinmaurer, R.: 1938, *Terr. Mag.* **43**, 7; 1938, *Sitzungsber. Akad. Wiss. Wien*, *IIa* **147**, 89.

Hess, V.F., Steinmaurer, R. u. Demmelmair, A.: 1938, *Nature*, **141**, 686.

Priebsch, J.A.: 1936, *Sitzungsber. Akad. Wiss. Wien*, *IIa* **145**, 101.

Priebsch, J.A., u. Baldauf, W.: 1936, *Sitzungsber. Akad. Wiss. Wien*, *IIa* **145**, 583.

Steinmaurer, R.: 1930, 38. Jahresbericht d. Sonnblick-Vereines, 1929.

Steinmaurer, R.: 1930, *Sitzungsber. Akad. Wiss. Wien*, *IIa* **139**, 281.

Wilson, C.T.R.: 1901, *Proc. Roy. Soc. London*, *A* **68**, 151.

EARLY PROPAGATION OF COSMIC RAY STUDIES

OLD PHOTOGRAPHS

Dr. Werner Kolhörster im Jahre 1912.

Dr. Werner Kolhörster (1887–1946), 1935 Professor of Physics at the University of Berlin, photographed 1912, the year of his first balloon flights. (Courtesy of Mrs. Editha Kolhörster)

Werner Kolhörster, after his famous balloon flights by which he had confirmed the increase of the ionization strengh with altitude, discovered by Victor Hess, and by which he had continued the measurements up to 9000 m, devoted himself with the passion of the strongly motivated young scientist to the further exploration of the penetrating "Höhenstrahlung". In order to broaden his knowledge in the relevant field of radioactivity he returned to the "Reichsanstalt" as a permanent guest where he could now continue his work on Cosmic Rays. 1930 in rewarding his merits the Prussian Academy of Sciences granted him the funds to build the "Höhenstrahlen-Laboratorium" (Cosmic Ray Laboratory) at Potsdam. This was the first special Institute for Cosmic Ray Research.

<Ed> Displaced from p. 39, Fig. 1, contributed by G. Pfotzer.

33

Y. Sekido and H. Elliot (eds.), Early History of Cosmic Ray Studies, 33–38.
© 1985 by D. Reidel Publishing Company.

(b)

(a)

a) Dr. Kolhörster and Dr. von Salis with Kolhörster's portable ionization chamber which they used for their early experiments in 1923 for the absorption measurements in the ice-holes of the alpes. b) Im Eisstollen an der Station Eigergletscher. Professor Maurer, Dr. v. Salis, Dr. Kolhörster. Measurements with Kolhörster's instruments in the ice-hole at the Eiger glacier. To the right Dr. Kolhörster taking visual readings of an electrometer. a & b) Courtesy of Mrs. Editha Kolhörster. Instruments used by Kolhörster and von Salis and relevant pictures as a remembrance of these early measurements are still exhibited in the "Jungfraujoch Cosmic Ray Station" in Switzerland.

<Ed> Contributed by G. Pfotzer. See p. 75.

34

R. A. Millikan and G. Harvey Cameron 1925.

Holding ion chambers used to make measurements at Pike's Peak. The lead shields were used to reduce local radiation. (Courtesy of the Archives, California Institute of Technology for caption also)

⟨Ed⟩ Contributed by H. V. Neher. See p.91. R. A. Millikan proposed the name "Cosmic Rays" just after this experiment.

1924 — "Stone age" of the nuclear physics. The first observations of the Compton-effect of γ-rays with a cloud-chamber in a magnetic field (about 1000 gauss).

<Ed> Contributed by D. V. Skobeltzyn. See p. 47. D. V. Skobeltzyn in the photo was 32 years old.

The arrangement, used since 1925—1926, with a magnetic field of 1500-2000 gauss. Most of the cosmic-ray tracks were observed in this field.

<Ed> Contributed by D. V. Skobeltzyn. See p. 47.

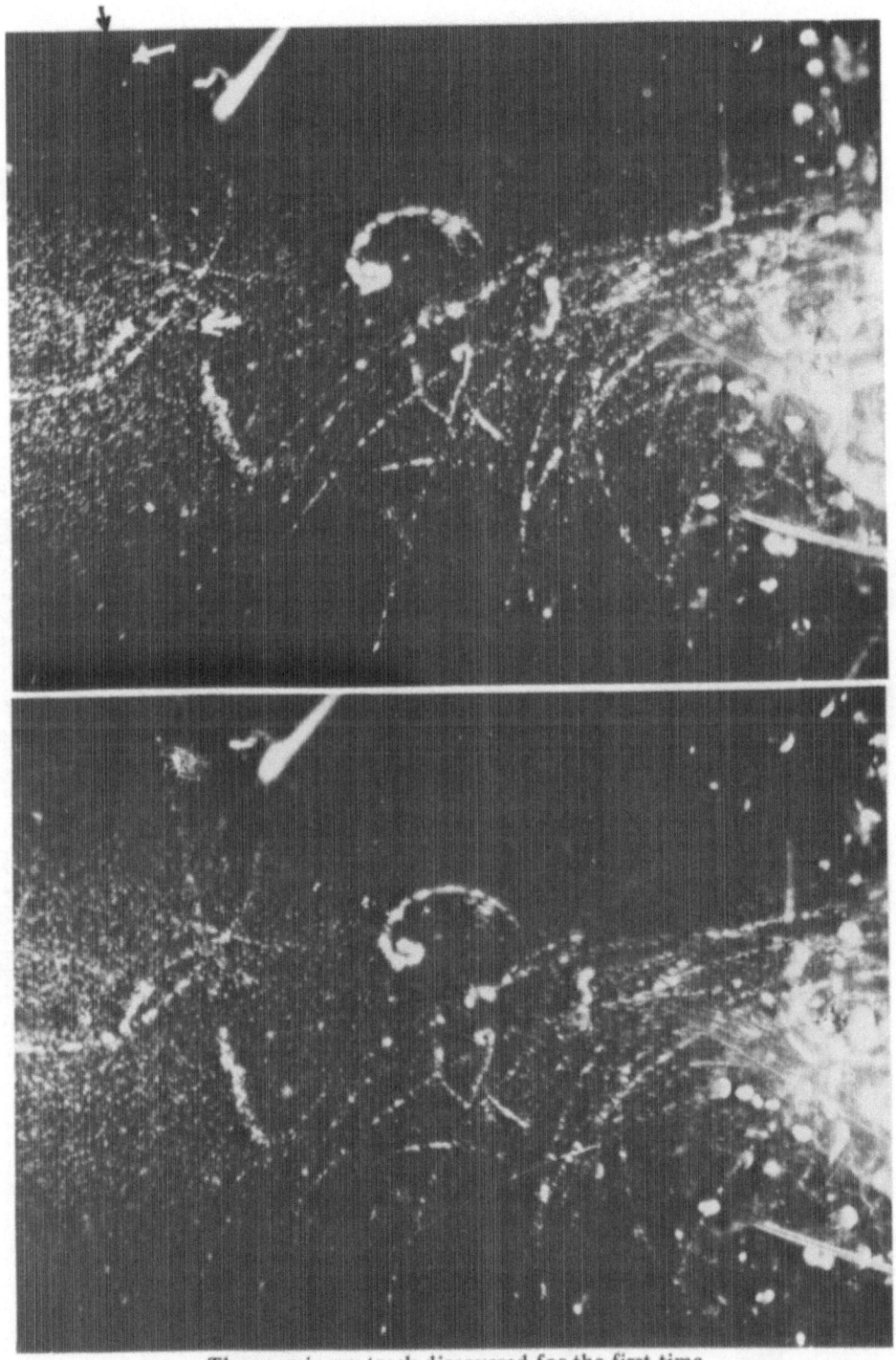

The cosmic-ray track discovered for the first time.

<Ed> Contributed by D. V. Skobeltzyn. See p. 47. This stereoscopic pair of photos was published in *Zs. f. Physik*, **43**, 363, Fig. 12 in 1927. The track is indicated by a black and two white arrows. H = 1030.

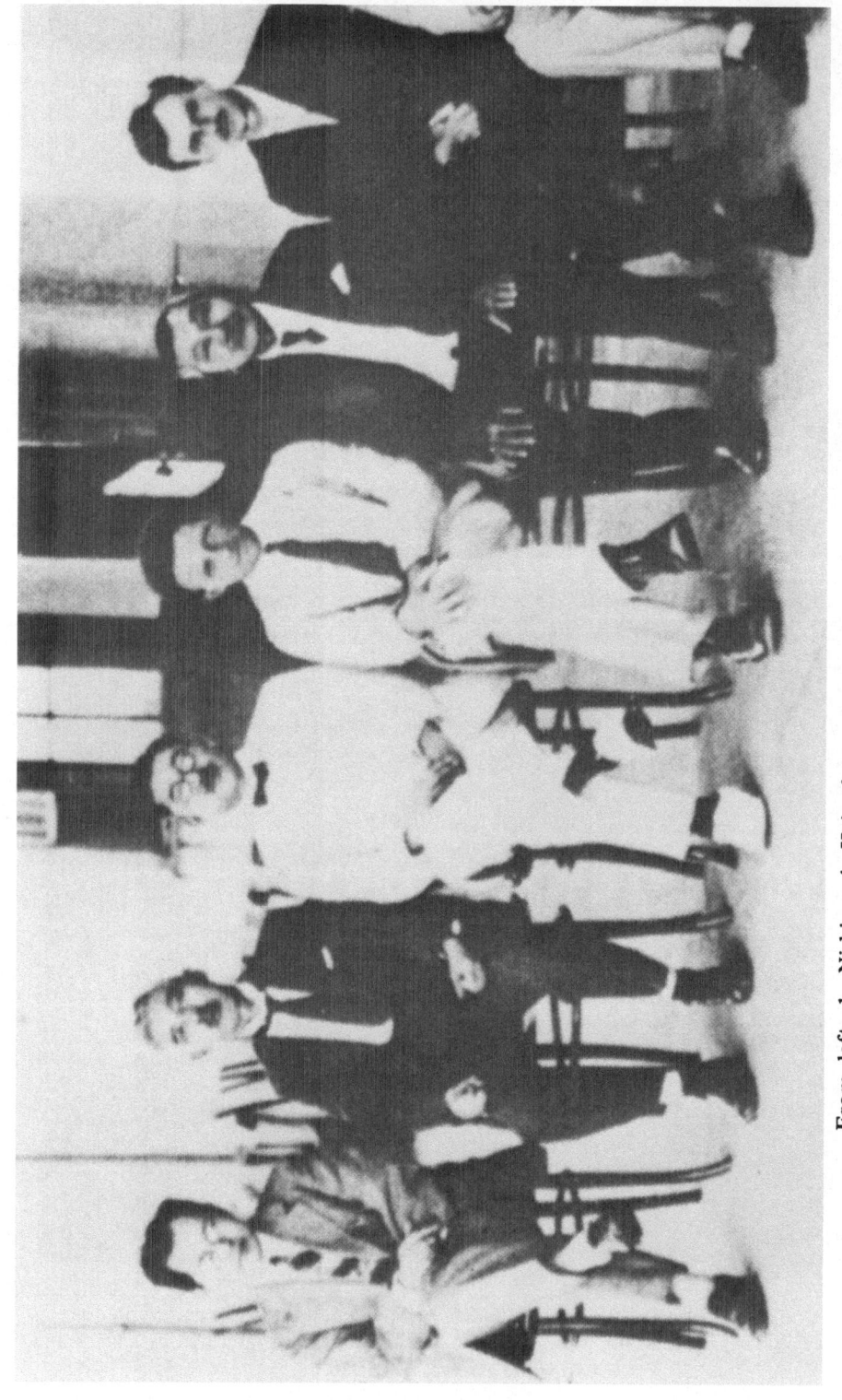

From left, 1: Nishina, 4: Heisenberg, 5: Nagaoka, 6: Dirac, at IPCR, Tokyo (1929).

<Ed> Displaced from p. 188, Fig. 2, contributed by Y. Sekido.

38

EARLY EVOLUTION OF COINCIDENCE COUNTING
A FUNDAMENTAL METHOD IN COSMIC RAY PHYSICS

Georg PFOTZER*

1. Remarks on the Scientists Involved

The roots of the coincidence method introduced in Cosmic Ray physics by Bothe and Kolhörster (1929) can be traced back to Professor Geiger's laboratory for radioactivity in the former "Physikalisch-Technische-Reichsanstalt" in Berlin.

Werner Kolhörster (Fig.1)**after his famous balloon flights by which he had confirmed the increase of the ionization strength with altitude, discovered by Victor Hess, and by which he had continued the measurements up to 9000 m, devoted himself with the passion of the strongly motivated young scientist to the further exploration of the penetrating "Höhenstrahlung". In order to broaden his knowledge in the relevant field of radioactivity he entered 1914 Geiger's laboratory. The outbreak of World War I interrupted, however, his career in this newly disclosed field. After the war and a short intermezzo as teacher, he returned to the "Reichsanstalt" as a permanent guest where he could now continue his work on Cosmic Rays. 1930 in rewarding his merits the Prussian Academy of Sciences granted him the funds to build the "Höhenstrahlen-Laboratorium" (Cosmic Ray Laboratory) at Potsdam. This was the first special Institute for Cosmic Ray Research. ((Ed.) Photo of Kolhörster is in "Early Propagation of Cosmic Ray Studies" of this book.)

Walter Bothe, Nobel prize winner in 1954, joined Geiger's laboratory in 1920. He had taken his doctor's degree, 23 years old, as a student of Max Planck in 1914, shortly before the outbreak of World War I. Referring to an extended necrology by his former associate, Professor Fleischmann (1957), the following story about his fate during the war seems to be worthy of being briefly related here: Drafted by the army which operated against the Russians, he was taken prisoner in 1915 and stationed in Siberia. There he learned the Russian language and occupied himself with extended studies on the problems treated in his thesis on "The molecular theory of diffraction, reflection, scattering and extinction". During the Russian revolution the prisoners of war had been left more or less on their own resources. In that situation Bothe, making use of his knowledge in chemistry, succeeded in starting the production of matches in a small factory in order to secure his and his comrades maintenance. He returned only in 1920, and on the way back he married at Moscow Barbara Belowa, whose acquaintance he had made in Berlin before the war and with whom he could correspond during his captivity. In Geiger's laboratory he was then introduced to the experimental state of the art in radioactivity.

Hans Geiger (see e.g. Stuhlinger, 1946) had already made himself a name through his work in Lord Rutherford's laboratory between 1906 and 1912, particularly by the fundamental experiments on the atomic structure, carried out jointly with Marsden, and on several rules of alpha-decay. In connection with this account, it is remarkable

* Max-Planck-Institut für Aeronomie (emeritus), Katlenburg-Lindau, W-Germany.
**(Ed.) Figure 1 is displaced to page 33 (Old Photographs).

Y. Sekido and H. Elliot (eds.), Early History of Cosmic Ray Studies, 39–44.
© *1985 by D. Reidel Publishing Company.*

Fig. 2. Professor Walter Bothe (1891–1957) (right), Nobel prize winner 1954, (coincidence method and discovery of artificial nuclear gamma radiation) and Professor Erich Regener (1881–1955) at a meeting in 1937. (Courtesy of Max-Planck-Institut für Kernphysik, Heidelberg).

Fig. 3. Professor Hans Geiger (1882–1945), photograph taken in 1930, when he was Director of the Institute for Physics at the University of Tübingen.

that he has started, jointly with Rutherford, first attempts of an electrical counting of alpha-particles, onward from 1908. 1913, when he was already head of the newly founded laboratory in the Reichsanstalt, he succeeded in proving that individual alpha or beta particles, which were passing a suitably strong field of a point electrode, initiate reliably short discharges, which can be used for electric counting of these particles (Bothe, 1942). This finding resulted in the construction of Geiger's point-counter (Geiger, 1913, 1914, 1933). It was the precursor of the tube counter invented by Geiger and Mueller (1928) at the University of Kiel, where Geiger had been offered a chair in 1925.

Geiger's experimental skill was outstanding. This is testified by Bothe in a most interesting paper on "Geiger's counting methods" dedicated to Professor Geiger's 60th birthday (Bothe, 1942). It reads, freely translated from German: "One of the author's lasting impression from the beginning of his work in Geiger's laboratory was, how Geiger with effortless ease set up an arrangement for counting alpha and beta particles composed of a brass tube, pieces of Wollastone wire, ebonit, sticking-wax, a sewing-needle, an electrostatic machine and finally a touch of Indian-ink on a piece of paper", serving as quenching resistor.

After this prelude on personalities I should like to turn now to the evolution of the coincidence method itself.

2. The Approach to the Coincidence Method

Cloud chambers had already given the first precise evidence on the spatial passage of particles and the branching of their tracks, when electrical counters came into play. The fast response of Geiger's counters permitted to determine rather exactly the time of an ionizing particle's passage through the sensitive region. This opened the door for a new counting mode of outstanding usefulness, the "Coincidence-Method". Although its number of applications is nearly unlimited, it serves basically only two purposes: Firstly, to infer the coherence of different particles with respect to their common origin from the coincident response of suitably placed counters; Secondly, to detect an ionizing particle which is passing at least two counters.

It was the merit of Bothe of having recognized the far-reaching importance of this method. His first coincidence work was carried out jointly with Geiger already in 1925 (Bothe and Geiger, 1925). It was related to the Compton effect and aimed at the decision whether the conservation of energy and momentum is also valid for the individual elementary process — as was claimed by Compton and Debye — or whether it is true only for the statistical average after a new theory by Bohr *et al.* (1924). Bothe and Geiger, using two point-counters, could show that the scattered photon and the recoiled electron were coincidentally detected by the two counters in time limits of less than $100\,\mu s$. This allowed to conclude that Compton's and Debye's view was correct. The coincidences were recorded on a moving film as shown in Fig. 4. A little detail may be mentioned (Fleischmann, 1957): At the beginning of the experiment many countings occurred which exhibited time differences between the responses of the photon and the electron counter of some 10^{-3} seconds. Geiger recognized that this was due to the slow motion of the ions into the high field region of the point, what influenced statistically the accuracy of the timing. Bothe's part was to find the remedy by

Fig. 4. Recording of coincidences on a moving film in the early experiments, via excursions of electrometer-fibres. The dense vertical lines are time signals in intervals of 10^{-3} seconds. By using the white dashes marking the steep part of the pulse, the exactitude of a coincidence could be improved to within 10^{-4} seconds. (Bothe and Geiger, 1925).

the superposition of a homogeneous electric field on that of the point by letting protrude the tip on a metallic archtorus (see e.g. Geiger, 1933). Thereafter the time differences of the coincident pulses remained in the above-mentioned limits of 100 μs.

In the course of another investigation, Bothe (1926) aimed at the decision, whether in the elementary process of X-ray fluorescence the radiation is emitted as a spherical electromagnetic wave or as a quantum propagating distinctly in a certain direction. In the first case coincidences were to be expected, if the emitter was placed between the sensitive regions of two point counters, in the latter case no coincidences should have occurred. The correct solution of no coincidences could, however, only be ensured after a background of really observed coincidences could be shown as due to those single beta particles of a radioactive contamination, which passed both counters.

This experience then led Kolhörster (1928) to place two of the newly by Geiger and Mueller (1928) introduced tube counters side by side into a beam of gamma rays in order to detect the recoil electrons by coincidences. As the recoil electrons with the highest energies are scattered practically in the forward direction, he considered such an arrangement as suitable to determine roughly the directional distribution of gamma rays. The following notice on the same page by Bothe and Kolhörster (1928) referred to measurements of the absorption of the secondary electrons by placing lead plates of increasing thicknesses between the counters.

This was the last step before the coincidence method was introduced in Cosmic Ray physics (Bothe and Kolhörster, 1929).

3. The First Coincidence Counting in Cosmic Ray Research –
An Experiment with a Golden Absorber by Bothe and Kolhörster (1929)

Bothe and Kolhörster started their famous experiment on the ground that the tracks of fast "electrons" which Skobelzyn (1927, 1929) had seen in the cloud chamber, are in fact related to the "Höhenstrahlung", as he had assumed. Furthermore, it seemed clear that charged particles were the immediate cause of the commonly measured ionization. The real problem was therefore whether these particles are of secondary origin, or whether they had to be considered as the "Höhenstrahlung" proper. They attempted therefore to decide this question by a search for coincidences in the highly filtered Cosmic Ray beam. They placed two counters one above the other inside of a housing

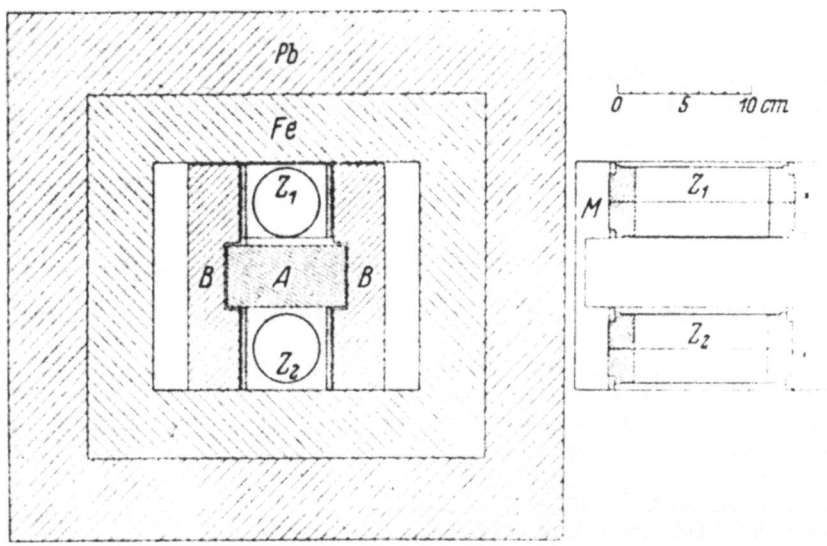

Fig. 5. Bothe and Kolhörster's crucial experiment on the corpuscular nature of the strongly penetrating radiation by measuring coincidences between Z_1 and Z_2 separated by a gold-block A of 4.1 cm thickness (weight $\approx$ 12 kg).

(Bothe and Kolhörster, 1929)

shielded on all sides by 5 cm iron and 6 cm lead (Fig. 5). The spacing between the counters amounted to 4.5 cm in order that absorbing material up to that thickness could be piled up. The first measurement with this arrangement was carried out in the basement of the Reichsanstalt. This increased the total shielding against the vertical beam through the concrete ceilings by an equivalent of 2 m water. Under these conditions they obtained systematic coincidences in excess of what accidentally could be expected. To their own surprise the coincidence rate did not change in the limits of the standard error if up to 4 cm of lead was placed between the counters. In order to increase the absorbing equivalent without an inconvenient change of the whole arrangement, they interposed a block of gold, 4.1 cm thick with an area of 8.9 x 17.5 cm (12.3 kg) just available for a short time. Although this increased the absorber to about 7 cm equivalent of lead, no clearly measurable decrease of the coincidence rate was obtained. In order to obtain clear reactions on the interposing of 4.1 cm gold, they placed now the whole arrangement immediately below on open window in the roof of the Reichsanstalt and removed in addition the upper shield of the housing. Only thereafter a reduction of the counting rate by 24% was achieved by the gold block. This corresponded to a mass absorption coefficient of the particulate radiation of $\mu/\rho = (3.5 \pm 5) \cdot 10^{-3}$ cm^2/g, as also measured commonly under the same conditions for what was assumed to be the "penetrating extraterrestrial radiation". By using also the coincidence rate obtained in the basement (total absorber, equivalent to 430 g/cm^2) they had now three points for drawing the nowadays well-known absorption curve.

Omitting the lengthy and careful discussion of the results by Bothe and Kolhörster, it remains to cite their main conclusion that "a corpuscular radiation was detected by which all hitherto observed effects of the "Höhenstrahlung" can be explained. It is unlikely that a gamma-radiation of a comparable penetrating power exists and there is, at least presently, no reason for such an assumption." Besides this very important result which was achieved by the first application of the coincidence method in Cosmic Ray physics, the method itself proved to be an indispensible tool in elementary particle physics, up to the present day.

REFERENCES

Bohr, N., Kramers, H. A. and Slater, J. C.: 1924, *Z. Physik* **24**, 69.

Bothe, W.: 1926, *Z. Physik.* **37**, 547; 1942, *Naturwiss.* **30**, 593.

Bothe, W. and Geiger, H.: 1925, *Zs. Phys.* **32**, 639.

Bothe, W. and Kolhörster, W.: 1928, *Naturwiss.* **16**, 1045; 1929, *Z. Physik.* **56**, 751.

Fleischmann, R.: 1957, *Naturwiss.* **44**, 457.

Geiger, H.: 1913, *Verh. d. D. Phys. Ges.* **15**, 534; 1914, *Ann. Phys.* **44**, 813; 1933, *Hdb. d. Physik.* **22/2**, 159.

Geiger, H. and Mueller, W.: 1928, *Naturwiss.* **16**, 617.

Kolhörster, W.: 1928, *Naturwiss.* **16**, 1044.

Skobelzyn, D.: 1927, *Z. Physik.* **43**, 354; 1929, *Z. Physik.* **54**, 686.

Stuhlinger, E.: 1946, *Z. Naturf.* **1**, 50.

PART 2   MEMORY OF REMOTE DAYS

THE EARLY STAGE OF COSMIC RAY PARTICLE RESEARCH

Dimitry V. SKOBELTZYN*

What follows is a brief account of "personal recollections" pertaining to the pre-history of cosmic-ray research, written by probably the oldest living participant of the scientific events of that period.

In 1927, I published photographs of secondary electron tracks produced by a beam of γ rays in a Wilson cloud chamber placed in a magnetic field (Skobeltzyn, 1927). On two of these photographs, the tracks of one or two unusually high-energy particles, not connected with the γ ray beam, were discovered. Subsequent observations revealed the relatively frequent appearance of similar tracks and a very striking peculiarity: the appearance of groups of simultaneous correlated tracks. ((Ed.) For a very interesting discussion of Skobeltzyn's first observations and reactions to them, see Norwood Russell Hanson, *The Concept of the Positron* (Cambridge University Press, 1963), especially pp. 136–139 and 216–217)

I began research on gamma-rays (and subsequently cosmic rays) in late 1923 in the laboratory of my father — a professor of physics at the Leningrad Polytechnical Institute. It started spontaneously under the impact of a very important discovery of that time, the Compton effect, and a fortunate idea of my own; namely, to investigate in a Wilson chamber the recoil-electron tracks of γ rays. My first photographs were taken without a magnetic field present.

The use of a magnetic field combined with a cloud chamber, which turned out to be so fruitful, was not an invention. I used such a field (of 1500–2000 gauss) as an auxiliary means to deflect the tracks of secondary β rays produced by γ rays in the walls of the Wilson chamber. The spurious background of such tracks hindered observation of the recoil-electron tracks produced in the gas of the chamber.

At that time, I was not interested in cosmic rays. I was, however, aware of the work being done in this field. One of my older university colleagues, Leo Mysovski, performed many important experiments in studying cosmic rays, called at that time "Höhenstrahlung". To him and his co-worker Tuvim belongs, for instance, credit for discovery of the "barometric effect" exhibited by this radiation. The highlights in this field of research at that time were the very important results of the experiments by Robert Andrews Millikan and G. Harvey Cameron concerning the absorption of cosmic rays in mountain lakes and Millikan's (unfounded) hypothesis on the nature and origin of cosmic rays as very hard "ultra γ rays" produced as a product of synthesis of various light nuclei, such as He, O, Si, etc.

It appears that before my Wilson photographs were published nobody tried to observe the secondary β particles of such hypothetical ultra γ rays. In this connection, I shall quote a paper by a renowned and very competent experimenter, Walter Bothe, which appeared in early 1923 (Bothe, 1923). One finds the following astonishing

* P.N. Lebedev Physical Institute, Acad. Sci., Moscow, U. S. S. R.

Y. Sekido and H. Elliot (eds.), Early History of Cosmic Ray Studies, 47–52.
© *1985 by D. Reidel Publishing Company.*

statement: "Man muss offenbar schliessen, dass ein β-Strahl, dessen Geschwindigkeit der Lichtgeschwindigkeit nahekommt, nach der Wilsonschen Methode nicht oder nur sehr schwer zu erkenne ist". ((Ed.) "One must apparently conclude that a β ray whose speed approaches the speed of light, is not recognizable by the Wilson method, or only with great difficulty.")

Luckily for me, I could not be impressed by such a pessimistic pronouncement because before I began my observations with the Wilson chamber I had the opportunity to study thoroughly two fundamental papers by Niels Bohr on the theory of ionisation produced by fast β particles (Bohr, 1913, 1915). Bothe probably overlooked this important contribution of Bohr, which seems to have been over-shadowed by his papers containing the famous quantum postulates, that appeared practically simultaneously.

After my fitst casual observation (Skobeltzyn, 1927) on the appearance of cosmic-ray tracks in a Wilson chamber, (The term *cosmic rays* (which had not yet been adopted) was not mentioned by me on that occasion, nor even *Höhenstrahlung*.) a full report on the corresponding results of my work was published in 1929 (Skobeltzyn, 1929, 1932). But most of the pertinent facts and photographs were presented by me earlier in the course of a discussion at an informal conference "On γ and β Ray Problems", held in Cambridge from July 23 to 27, 1928, under the patronage of Ernest Rutherford. One session of the conference was scheduled for a discussion on β ray problems. (It seems that no paper was read at that session.) During this discussion, I demonstrated a collection of photographs of cosmic ray tracks and, I dare say, it produced some impression on the audience. Incidentally, I began with a comment on the Bohr theory of the ionisation produced by very fast β-particles.

Immediately after my impromptu remarks, Hans Geiger took the floor to announce that Bothe and Werner Kolhörster were working on a method to register cosmic rays by the coincidence of pulses in two wire-counters, and that they hoped to be able to study the penetrating power of the rays by this method.

In connection with this communication of Prof. Geiger on July 25 (evoked, I believe, by my own presentation), I call attention to the following dates. The announcement by Geiger and Müller of their invention of wire counters was published on the 7th of the same month (July, 1928) (Geiger and Müller, 1928). A very brief communication of Bothe and Kolhörster, informing that as a result of their coincidence observations with recently invented wire counters there were observed ionising particles penetrating 1 cm of lead, is dated November 2, 1928) (Bothe and Kolhörster, 1928). A detailed communication of my results, published in Zeitschrift für Physik, is dated February 5, 1929 and that of Bothe and Kolhörster is dated May, 1929 (Skobeltzyn, 1929, 1932; Bothe and Kolhörster, 1929). ((Ed.) These are dates of submission, as given in the aritcles themselves.)

It is well known what extremely important results followed from further develop-ment of the technique of using a Wilson chamber plus a magnetic field. The next step that appeared natural was to use a magnetic field of much higher strength. There were many reasons why I myself never tried to do this. In 1929–31, I was working at the Curie Laboratory in Paris. Pierre Auger, who had been working at the neighbouring Institute of Jean Perrin, asked me (probably at the beginning of 1931) whether I intended to undertake such investigations. I answered in the negative, whereupon he told me that he would try to perform that kind of experiment. Soon thereafter, he showed me his installation that was ready for operation. However, his attempt turned out to be unsuccessful and probably was dropped by him in the fall of 1931 when (as we shall see later) it was disclosed that Carl David Anderson had already obtained some thousand beautiful pictures of cosmic ray tracks in a strong magnetic field (13000 gauss).

It seems that something was wrong with the cloud chamber that Auger had constructed. I was told in the spring of 1931 that his Wilson chamber, when put into operation, showed no cosmic ray tracks whatsoever. I obtained this information from a fellow of the Curie Laboratory staff (George Fournier), who concluded from this that my observations of 1927–29 were erroneous. By that time, however, my results had already been corroborated (Mott-Smith and Locher, 1931). Until the end of my sojourn in Paris (1931), I had no further occasion to meet Auger himself, and afterwards never heard from him what had gone wrong with his cloud chamber in a strong magnetic field.

In November of 1931, when I was already in Leningrad, Millikan visited Europe and gave sensational lectures in Paris and Cambridge, showing a collection of Anderson's photographs. The main content of these lectures was published by Millikan, with Anderson as co-author, in May, 1932 (Millikan and Anderson, 1932). The tracks observed by Anderson were ascribed to high-energy protons produced by cosmic ray photons. In November and December, 1931, I received letters from Marie Curie and Frédéric Jolio-Curie from Paris, and L. Harold Gray from Cambridge, who mentioned that they attended Millikan's lectures and in a few lines informed me of the results communicated by him. Somewhat more detailed information was given in the letter of Gray, with whom at that time I was in correspondence, discussing some problems of γ ray research. He wrote (November 27): "I *think* (my emphasis) that in every case the proton tracks were more dense than those of electrons". That probably was his imagination. Millikan showed to his audience some eleven pictures of good tracks in the range of 20–80 MeV energy, say $E \simeq 50$ MeV on the average.

Now, the ionisation density (specific ionisation) depends mainly on the velocity of the particle, or on the quantity E/mc^2, where m is the proper mass of the particle. With this quantity given, the absolute value of the mass m of a particle of given charge has but little importance. It follows that the ionisation produced by a proton of 50 MeV is practically the same as that of an electron of about 25 keV. It is impossible to confuse the specific ionisation of such a slow electron with the ionisation of a fast ("relativistic") one, having several MeV or more. However, the positive tracks on pictures demonstrated by Millikan did not differ essentially from electron tracks on the same pictures, with the energies of about 50 MeV. Prof. Millikan and his audience overlooked this incosistency. After I received Joliot's letter, I wrote him straightforwardly my views on the subject and suggested that something was wrong with Millikan's photos or their interpretation.

In their note of May 1932, Millikan and Anderson repeated the interpretation Millikan had made in Europe. They asserted that the fast protons were a product of the interaction of ultra γ rays with nuclei, and even saw in the phenomenon a new confirmation of Millikan's old hypothesis on the origin of cosmic rays (ultra γ rays as a product of synthesis of certain nuclei). Progress by Anderson in the deciphering of his experimental evidence was slow. Only in September 1932 (a year after a stock of more than one thousand Wilson photos in a strong magnetic field had been obtained) did he make reference, in a brief and very cautiously drafted note, to the specific ionisation of the positively curved tracks, and concluded the existence of positively-charged particles, the mass of which "must be small compared with the mass of a proton" (Anderson, 1932). The next paper of Anderson appeared in February, 1933; it was entitled "The Positive Electron", and contained more definite statements (Anderson, 1933). In writing it, he was already aware of the results of the outstanding work of Patrick Blackett and Giuseppe Occhialini (he refers to a press report about this work) (Blackett and Occhialini, 1933).

The paper of Blackett and Occhialini was received for publication in the *Proceedings of the Royal Society of London* on February 7, even somewhat earlier than that of Anderson in the *Physical Review* (February 28). It now appears strange, perhaps, that in discussing their experiments the authors of both papers did not make any attempt to connect their results with the Dirac theory of the positron. Paul A.M. Dirac's work of 1930 was certainly known to them (to the Cambridge physicists, anyway). However, it is true that Paul A.M. Dirac himself was inclined to identify the positive particles of his theory with protons (Dirac, 1930). Anderson, in his paper, suggested some far-reaching hypotheses that appeared strange, as for instance the transformation of protons into electrons, induced by cosmic rays. (Such speculation considered now on the background of modern sub-nuclear physics may appear, perhaps, less strange?)

During some months after the discoveries of Anderson and Blackett-Occhialini, the experimental evidence obtained simultaneously by many experimenters in the field of γ ray research showed that the concepts of the Dirac theory were adequate. The conclusions drawn from this evidence were summarised by Blackett at a session of the Solvay Congress (October 22–29, 1933). The discussion that followed is certainly of historical interest. In the record of this discussion one finds, incidentally, the following characteristic remark of Rutherford: "It seems to a certain degree regrettable that we had a theory of the positive electron before the beginning of experiments ... I would be more pleased if the theory had appeared after the establishment of the experimental facts" (Structure et Propriétés des Noyeaux Atomiques, Rapports et Discussions du 7me Conseil de Physique Solvay, 1934: Translated from the French text by the author.)

I have dwelt on the facts related on the preceding pages because they can give one an idea of the psychological barriers lying in the way to discovering the first of a sequence of many generations of new particles that soon followed.

There is another important line of development that followed (with some delay) my first observation of cosmic-ray tracks. I refer to the appearance of groups of simultaneous tracks (up to four in my subsequent observations) and "showers" (up to 20 particles) on photographs of Blackett and Occhialini (1933) taken with their counter-controlled Wilson chamber (Blackett and Occhialini, 1933). This phenomenon immediately attracted the attention of physicists working in the field. But its importance in leading to perhaps the most interesting chapter of the history of this branch of science had probably not been foreseen. In fact, it was a precursor of modern high—energy physics. For a relatively long time, however, its nature remained a puzzle.

Two aborted attempts to solve this puzzle followed soon. A note by Auger and myself on this problem appeared in July 1929 (Auger and Skobelzyn, 1929). Later (early 1932), Werner Karl Heisenberg published his version (Heisenberg, 1932). How meager was the experimental evidence needed to answer the question as to the nature of the phenomenon can be seen from the fact that on 5 pages of Heisenberg's paper my observations of 1929 were quoted eleven times—quite unusual. It was unfortunate for the author to have published his paper as early as he did. During the same year (1932), the discovery of the positron was announced and a year later pictures of "showers" taken by Blackett and Occhialini appeared. Both these events changed the whole situation radically.

The suggestion of Auger and myself was that the groups of tracks could appear as a result of simultaneous production of several Compton electrons by ultra γ radiation. We even saw in this effect a reason to reject the interpretation of the nature of primary cosmic rays as high-energy β rays, an idea put forward by Bothe and Kolhörster on the basis of their measurements of the penetrating power of cosmic-ray particles (Bothe and Kolhörster, 1929). Heisenberg, on the other hand, based his deductions on the aforementioned hypothesis of Bothe and Kolhörster. According to his scheme, simultaneous tracks of an observed group were tracks of many δ rays generated by one and the same high-energy β particle. ((Ed.) δ rays are secondary electrons, usually of short range, ejected from atoms by fast charged particles.)

A long road of hard work was trodden by many theoreticians of the highest calibre (and to a lesser degree by experimenters) to arrive at an understanding of the nature of the phenomenon.

The names of Walter Heitler, Hans Albrecht Bethe, J. Robert Oppenheimer, C.F. von Weizsäcker and many others should be mentioned in this connection. The solution to the puzzle — the first version of the cascade theory of showers — came in 1937 in a paper by Homi J. Bhahba and Heitler (Bhahba and Heither, 1937).

For subsequent developments, the discovery by Auger, just before the war, of extensive atmospheric cosmic ray showers (at first called "Auger showers") was of great importance (Auger and Maze, 1939). He observed coincidences in two counters placed up to about 300 m apart (in a horizontal plane). This occurred at the time that I began my work at the P.N. Lebedev Institute in Moscow.

During the last years (1944–1945) of World War II, cosmic-ray research was resumed at the Lebedev Institute. I suggested to George Zatsepin that he begin observation of atmospheric (later known as extensive) showers and try to obtain by this method information on the highest energies of cosmic-ray particles. The first step toward this goal was to use an array of two pairs of counters separated by a distance as long as possible. By using four-fold coincidences, instead of two-fold as in Auger's experiments, one could diminish the influence of spurious background due to chance coincidences. The experiments performed by Zatsepin and co-workers were successful. They became the starting point for the development of more complicated and more refined devices in various branches of research, and they constituted a very exciting chapter in the history of modern high-energy physics.

During the following decades, work in this field was pursued by a highly qualified team of research fellows from the Lebedev Institute and Moscow University, and continues to this day.

REFERENCES

Anderson, C. D.: 1932, *Science* **76** 238.
Anderson, C.D.: 1933, *Phys. Rev.* **43** 491.
Auger, P. and Maze, R.: 1939, *Comptes Rendus* **208** 1641.
Auger, P. and Skobelzyn, D.: 1929, *Comptes Rendus* **189** 55.
Bhabha, H.J. and Heitler, W.: 1937, *Proc. Roy. Soc. A* **159** 432.
Blackett, P.M.S. and Occhialini, G.P.S.: 1933, *Proc. Roy. Soc.* **139** 699.
Bohr, N.: 1913, *Phil. Mag.* **25** 10; 1915, *Phil. Mag.* **30** *581*.
Bothe, W.: 1923, *Zeit. Physik* **12** 117.
Bothe, W. and Kolhörster, W.: 1928, *Naturwiss.* **16**, 1045.
Bothe, W. and Kolhörster, W.: 1929, *Zeit. Physik.* **56**, 751.
Dirac, P.A.M.: 1930, *Proc. Roy. Soc. A* **126**, 360.
Geiger, H. und Müller, W.: 1928, *Naturwiss.* **16**, 617.
Heisenberg, W.: 1932, *Ann. Physik* **13**, 430.
Millikan, A. and Anderson, C.D.: 1932, *Phys. Rev.* **40**, 325.
Mott-Smith, L.M. and Locher, G.L.: 1931, *Phys. Rev.* **38**, 1399.
Skobeltzyn, D.: 1927, *Zeit. Physik* **43**, 354.
Skobeltzyn, D.: 1929, *Zeit. Physik* **54**, 686.
1932, *Comptes Rendus* **195**, 315.
Structure et Propriétés des Noyeaux Atomiques, Rapports et Discussions du 7me Conseil de Physique Solvay: 1934, Gauthier-Villars, Paris, pp. 177–178.

ARCETRI, 1928 – 1932

Bruno ROSSI*

Early in 1928, a short time after receiving my doctorate at the University of Bologna, I was offered a position of assistant at the Physics Institute of the University of Florence. The Institute rose among olive trees, on the hill of Arcetri, a short distance from the villa where Galileo had spent the last years of his life as a political exile. The chair of physics was held by Professor Antonio Garbasso (Fig. 1) who, in earlier years, had done some creditable scientific work. But the first world war and the events of the post-war period had diverted his interests toward politics. He was now a senator and the mayor of Florence. However he still went to Arcetri three times a week to deliver his lectures and he still had a strong desire to see the Institute, which he had built, become an important center of research.

The group I found in Arcetri was quite small, but quality made up for size. Gilberto Bernardini (Fig. 2), a recent doctor in physics from the University of Pisa, had joined the group a short time before my arrival. Among the students, were Giuseppe Occhialini, Giulio Racah, Daria Bocciarelli, Guglielmo Righini, and Lorenzo Emo. And we had the benefit of regular visits of Professor Enrico Persico, who had undertaken to unravel for us the mysteries of wave mechanics.

Life at Arcetri was rather austere. My monthly salary was 600 lire (30 dollars at the then current rate of exchange). Not enough money was available in the laboratory budget to buy heating fuel — and winters in Florence are quite severe. To combat the cold, we would wear heavy woolen linings inside our laboratory smocks. The laboratory was always far behind with the payment of the electric bills, and the only reason why our electricity was not cut off was that the director of the laboratory was also mayor of the city.

Both Bernardini and I were very anxious to start some experimental program, and we spent our first year in Arcetri exploring, without much success, a few different lines of research.

For me, the turning point in the search came in the fall of 1929, with the appearance, in Zeitschrift für Physik, of the historical paper "Das Wesen der Höhenstrahlung" by W. Bothe and W. Kolhörster (Bothe and Kolhörster, 1929)

Until then, I had not been particularly interested in the phenomenon of the "Höhenstrahlung" or "cosmic radiation," using the suggestive expression introduced by Robert Millikan. I had not thought that it would offer, to me at least, a profitable field of research.

I had not been seduced by Millikan's well publicized theory, maintaining that cosmic rays were the "birth cry of atoms" in cosmic space, being born, in the form of γ-rays, when hydrogen atoms "fused" to form the heavier elements. To my skeptical mind, this was a romantic idea, lacking sound experimental support.

On the other hand, I had accepted, uncritically, the prevailing view that primary

* Massachusetts Institute of Technology (emeritus), Cambridge, U. S. A.

Y. Sekido and H. Elliot (eds.), Early History of Cosmic Ray Studies, 53–73.
© 1985 by D. Reidel Publishing Company.

Fig. 1. Professor Antonio Garbasso.

Fig. 2. A siesta in the sun, with Bernardini.

cosmic rays were high-energy γ-rays. Therefore I read with particularly keen interest the paper by Bothe and Kolhörster relating the first attempt to submit this assumption to a direct test.

The idea behind the experiment is well known. Gamma rays do not ionize directly. They do so through the intermediary of the secondary electrons which they generate in matter (at that time, Compton collisions were the only known interaction processes of gamma rays). The secondary electrons were thought to have a much smaller penetrating power than the parent γ-radiation. It followed that they would soon reach

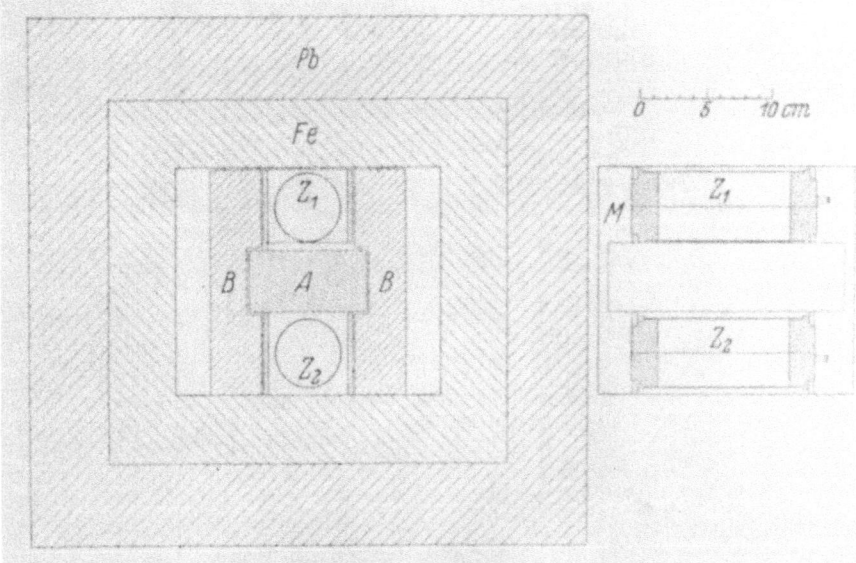

Fig. 3. The experiment of Bothe and Kolhörster (from an article by Bothe and Kolhörster, in *Zeitsch. für Physik,* **56**, 751, 1929).

equilibrium with this radiation, so that ordinary absorption experiments would measure the attenuation of the hypothetical γ-radiation, although, presumably, the ionizing agent recorded directly by the measuring instruments was the secondary corpuscular radiation.

Bothe and Kolhörster saw that a direct study of this corpuscular radiation offered the most crucial test of the current views about the nature of cosmic rays.

In their experiment (Fig. 3), they used "tube counters," of the kind that H. Geiger and W. Müller had invented the year before. Two counters were placed one above the other, a small distance apart. Simultaneous pulses (or "coincidences") were frequently observed and were interpreted as due to the passage through both counters of individual ionizing particles; according to current views, these were Compton electrons generated by the cosmic γ-rays in the matter above the counters or in the walls of the counters themselves.

If this had been the case, a very small thickness of absorber between the counters would have been sufficient to stop all coincidences, owing to the small panetrating power of Compton electrons. Instead, a 4.1 cm thick gold absorber produced only a moderate decrease in the counting rate. The authors concluded that the ionizing particles were not Compton electrons; going further, they argued that the primary cosmic radiation itself consisted of charged particles, and that the charged particles observed near sea level were those among the primary particles capable of traversing the atmosphere.

That this view proved to be an oversimplification does not detract from the pioneering character of the work of Bothe and Kolhörster.

For me, the paper by Bothe and Kolhörster opened a window upon a new, unknown territory, with unlimited opportunities for exploration. I quickly realized these

Fig. 4. On the steps of the laboratory; sitting, from left to right, Rossi, Occhialini, Bernardini and Bocciarelli.

opportunities, and started working, enlisting the help of my students, particularly Giuseppe Occhialini and Daria Bocciarelli. And thus began what I yet remember as one of the most meaningful and exhilarating periods of my life. Was it because of the excitement of venturing into a still virgin field of science? Was it because of the exceptional human and intellectual qualities of the young people I found myself associated with? Was it because of the subtle, poignant beauty of the tuscan countryside?

Our work proceeded rapidly. Within a few weeks we had our first G.M. counters in operation. To build a G.M. counter was, at that time, a kind of witchcraft. The tube was supposed to be made of zinc. Since no zinc tubing was available in Italy, we had to prepare it by bending a zinc sheet around a cylindrical surface and soldering it at the junction. The anode, according to the prescription, was to be a thin steel wire, slightly oxidized by immersion in nitric acid. The wire was held by two hard-rubber

Fig. 5. At lunch in the laboratory. The wife of the janitor had kindly agreed to feed us for nominal sum. From left to right; L. Emo, B. Crinó, G. Bernardini, A. Colacevich, D. Bocciarelli. Beatrice Crinó married later Guglielmo Righini. Colacevich was a young astronomer, working at the nearby astronomical observatory.

Fig. 6. Colacevich, Bernardini, Bocciarelli and Emo, sitting on the grass, behind the laboratory.

stoppers, closing the ends of the tube and made airtight with some sort of wax. The tube was evacuated through a thin glass pipe inserted in one of the stoppers and then filled to 1/10 of an atmosphere with dust-free dry air. Finally the glass pipe was sealed by melting its walls on a flame and the G.M. counter was taken off the filling system. To quench the discharge, we had to ground the wire through a very high resistor, over 10^9 ohms, which we prepared ourselves by filling a small glass tube with some appropriate mixture of organic fluids (remember that self-quenching counters were still in the future).

Bothe and Kolhörster had recorded the coincidences by connecting their counters to two separated fiber electrometers, which were imaged on a moving film. By some clever device, involving the use of a fast-oscillating screen, they had obtained a time

Fig. 7. Springtime in Arcetri.

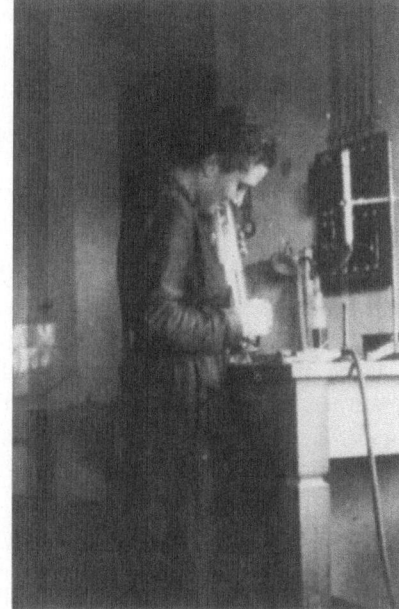

Fig. 8. Daria building a Geiger-Müller counter.

Fig. 9. In the laboratory, working at my early experiments. Several counters are visible on my right. I had not yet taken the time to learn how to build power supplies; therefore I used a number of dry cells in series to provide the high voltage for the counters, and storage batteries to provide the plate voltage for the vacuum tubes.

resolution of 1/100 of a second. I felt that the power of the coincidence method would be greatly enhanced if one could devise a method for recording coincidences which would be less cumbersome than that used by Bothe and Kolhörster, which would provide a better time resolution, and, especially, which could be extended to more than two counters. Thus the now classical coincidence circuit was born, which, for years to come,was to be by main research tool as well as that of many other experimenters (Rossi, 1930a). In its original version, it consisted of two or more triodes with the plates connected in parallel and the grids coupled electrostatically to the wires of the G. M. counters (Fig. 10). Only when the grids of all triodes were simultaneously driven to a negative potential by the coincident discharges of the G.M. counters would the current in all triodes stop and a large pulse appear at the plates of the triodes. In my earliest experiments, this event was detected by means of a gold-leaf electroscope, or by means of a headphone. In either case, either I or one of my collaborators had to sit by the equipment to count the coincidences. For all our enthusiasm, this turned out to be rather boring. So we switched to a photographic recording and, later still, we started using a mechanical counter.*

Having thus in my hands a satisfactory technique, I began some preliminary experiments. Before a year was over, I had measured the efficiency of G.M. counters and its

* For the history, I should note that, before I published my invention, Bothe reported the development of a coincidence circuit making use of a two-grid vacuum tube. This circuit, however, was not competitive with mine. It required a much more delicate adjustment and, more importantly, it could only be used to record twofold coincidences, while mine was capable of recording coincidences between any number of G.M. counters.

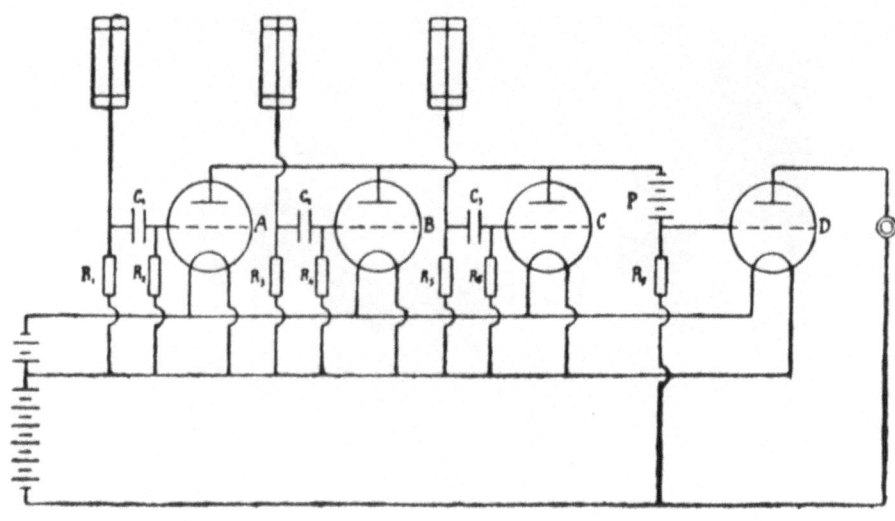

Fig. 10. The coincidence circuit (from an article by B. Rossi, in *Nature,* **125**, 636, 1930). $R_1, R_3, R_5 = 5 \cdot 10^9$ ohms. $R_2, R_4, R_6, R_7 = 8 \cdot 10^6$ ohms. $C_1, C_2, C_3 = 10^{-4} \mu F$.

Fig. 11. Professor Walther Bothe.

dependence on voltage (Rossi, 1930c) by comparing the rate of threefold coincidences between three counters placed vertically one above the other with the rate of twofold coincidences between the uppermost and lowermost counters. I had also checked that the penetrating particles came preferentially from the vertical direction (by comparing the coincidence rate between two counters placed vertically one above the other or horizontally one next to the other) (Rossi, 1930b).

Fig. 12. With Professor Hans Geiger in Tübingen.

Fig. 13. Asmara; setting up the shelter for the E-W experiment.

In another experiment I had attempted to detect a deflection of the penetrating particles in their passage through a bar of magnetized iron. (Rossi, 1930b). Not having obtained, from this experiment, any significant result, I later tried to observe the deflection of cosmic ray particles in magnetized iron, using a different, more sensitive arrangement suggested by Professor Puccianti of Pisa. This arrangement, which may be described as a magnetic lens, consisted of a closed-circuit magnet, formed by two oppositely magnetized iron bars, arranged next to one another. Two G.M. counters were placed horizontally, one above the other below the magnet. Depending on the direction of the magnetization and on the sign of the charge of the particles, it was expected that the particles, crossing the upper counter would be concentrated upon the lower counter, or deflected away from it (Rossi, 1930d, 1931a)

I observed a small effect, corresponding to positive particles, but I did not feel that this effect was statistically significant. Of course, I did not guess that my ambiguous result was largely due to the presence of both positive and negative particles in the cosmic radiation. In this connection, I may recall that several years later the "magnetic lens" was successfully used to separate positive from negative μ-mesons, in experiments designed to study the different behavior of the two kinds of particles.

In the meantime, I had communicated with Bothe (Fig. 11), describing what I had been doing, and expressing my strong desire to work in his laboratory for a while. Bothe's answer was exceedingly kind and encouraging. My boss, Professor Garbasso, who had been following my work with great interest, procured a fellowship which enabled me to spend the summer of 1930 in Bothe's laboratory at the Physikalische – Technische Reichsanstalt in Berlin – Charlottenburg. The memory of that summer is still vivid in my mind. Berlin was, at that time, the very heart of contemporary physics. I met there, Max Planck, Albert Einstein, Otto Hahn, Lise Meitner, Max von Laue, Walther Nernst, and Werner Heinsenberg to name just a few. Some of these scientists were at one or the other of the several Institutes in or around Berlin. Others would come from the neighboring cities for the weekly seminars. Patrick Blackett was also visiting there from England and my friendship with him began on that occasion. Bothe himself was most friendly and helpful; something which I learned to appreciate more and more as I realized that, by nature, he was not overly outgoing and trustful. In this connection, I still remember a curious episode. I had just begun to work in his laboratory and had realized that his G.M. counters were noticeably better than mine, in that they were more stable and had a longer plateau. I puzzled about this matter until one day Bothe took me aside with a mysterious air and began: "I will tell you a secret, but you must promise not to give it away to anyone." After I had promised, he continued: "My counters do not have a steel wire, as advertised; they have an aluminum wire!". I must confess to my shame that, when I returned to Italy, I felt that I could not keep the secret of the aluminum wire from my friends in Florence and in Rome; but I relieved my conscience by requiring of them the same oath of secrecy that Bothe had required of me.

During my comparatively short stay at the Reichsanstalt, I repeated, in an improved form, the experiment of Bothe and Kolhörster by comparing the coincidence rates of two G.M. counters with a given thickness of lead (9.7 cm) placed alternately above and between the counters. It turned out that the coincidence rate was not exactly the same in the two cases. With the lead above there was an excess of about 4% over the coincidence rate observed with the lead between. Thus things were not quite as simple as suggested by the assumption of Bothe and Kolhörster (Rossi, 1931b).

While I was at the Reichsanstalt, Bothe went, with Kolhörster, on an expedition to the North Sea and the northern Atlantic Ocean, in an attempt to discover a dependence of the cosmic-ray intensity on geomagnetic latitude (between 51° and 81°).

The prediction of such an effect was based simply on an argument of analogy with the phenomenon of the northern lights or aurorae. These, at that time, were believed to be due to high-energy electrons originating from the Sun and channeled toward the circumpolar regions of the Earth by the Earth's magnetic field; it was thought that if

Fig. 14. The equipment for the E-W experiment.

primary cosmic rays were indeed charged particles, the geomagnetic field should exert upon them a similar focusing action.

The experiment gave a negative result. But it aroused my interest in understanding more precisely what would actually happen when an initially isotropic stream of charged particles entered the magnetic field of the Earth.

Talking with Bothe, I learned that, in an attempt to explain the observed features of the aurorae, Störmer and his students had been working for many years on the mathematical problem of the motion of charged particles in a dipole field. But, as Bothe and Kolhörster remarked in the paper describing the results of their expedition, the theory developed by Störmer appeared to be so complex that it seemed hopeless to obtain from it quantitative conclusions pertinent to cosmic rays. However, studying Störmer's papers, I found that this was not at all the case and that, by just asking the proper question, it was quite easy to derive from Störmer's theory some simple and highly significant results.

Störmer and his collaborators, through years of painstaking numerical calculations, had computed the trajectories of hundreds of particles of different energies entering the geomagnetic field from the direction of the Sun, in order to determine where and from what directions they would hit the Earth (electronic computers of course, were undreamt of and Vannevar Bush, at MIT, was still in the process of developing his mechanical differential analyzer). But we, cosmic-ray physicists, were primarily interested in a simpler problem; we wanted to know, in the first place, whether particles of a given energy could or could not reach a given point of the Earth in a given direction. I found that at least a partial answer to this problem was contained in a simple formula derived by Störmer, which read

$$\sin \theta = \frac{300\,M}{R^2 V}\,\cos \lambda - \left(\frac{300\,M}{R^2 V}\right)^{1/2}\frac{2}{\cos \lambda}$$

On the right hand side, M is the magnetic moment of the Earth (in gauss $\cdot$ cm^3), R the Earth's radius (in cm), λ the geomagnetic latitude of the point of observation, and V the magnetic rigidity (in volt) of the particles under consideration. On the left hand side, θ is the angle between the trajectory of the incoming particle and the magnetic

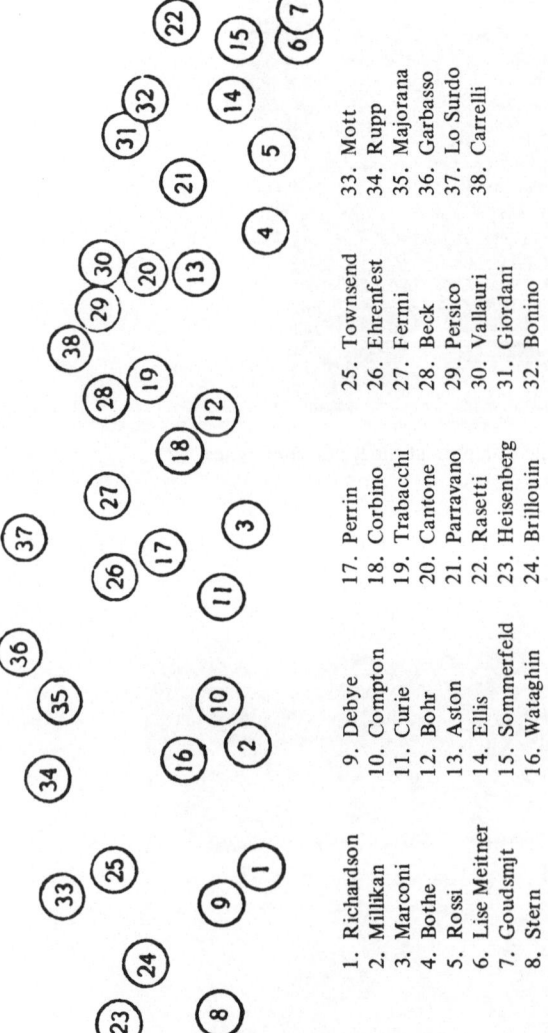

1. Richardson
2. Millikan
3. Marconi
4. Bothe
5. Rossi
6. Lise Meitner
7. Goudsmjt
8. Stern

9. Debye
10. Compton
11. Curie
12. Bohr
13. Aston
14. Ellis
15. Sommerfeld
16. Wataghin

17. Perrin
18. Corbino
19. Trabacchi
20. Cantone
21. Parravano
22. Rasetti
23. Heisenberg
24. Brillouin

25. Townsend
26. Ehrenfest
27. Fermi
28. Beck
29. Persico
30. Vallauri
31. Giordani
32. Bonino

33. Mott
34. Rupp
35. Majorana
36. Garbasso
37. Lo Surdo
38. Carrelli

Fig. 15. The participants in the 1931 Rome conference (from the Proceedings of the "Convegno di Fisica Nucleare," Reale Accademia d'Italia, 1932).

B. Rossi

Fig. 16. With Millikan and Compton at the Rome conference.

Fig. 17. With Walther Bothe and Lise Meitner, at the Lido (Venice) on their way back from the Rome conference.

Fig. 18.   Professor Eric Regener.

meridian plane, positive toward the east for negative particles, positive toward the west for positive particles. This angle is the semiaperture of a cone (known today as the Störmer cone) which separates the directions of incidence of the particles whose trajectories, followed backward, reach to infinity, from the directions of the particles whose trajectories always remain in the vicinity of the Earth. The latter trajectories cannot be possible trajectories of cosmic-ray particles, while the former are possible trajectories, unless they happen to cross the Earth. The "forbidden" directions are to the east or the west of the Störmer cone depending on whether the particles carry a positive or a negative charge.

This result (published in the summer of 1930 – see Rossi, 1930d) lead me to predict an E-W asymmetry in the angular distribution of cosmic ray particles and to estimate the energy of the particles for which a sizeable effect should be expected (depending on the latitude, it turned out to be of several times 10^9 or several times 10^{10} eV).

Upon my return to Arcetri, I attempted, unsuccessfully, to detect the predicted E-W asymmetry (Rossi, 1931c). Having realized that the effect would become more pronounced at low latitude and high altitude, I began to plan an expedition to Asmara (Eritrea; geomagnetic latitude $11°30'$; elevation 2370 m). Sergio de Benedetti joined me in this enterprise, which was made possible by the generous support of Professor Garbasso. Preparations took a fairly long time (and, in fact, the experiment was completed after I had left Arcetri). Finally, in the fall of 1933, Sergio and I reached East Africa and set up our experiment in a cabin on a hill near Asmara (Fig. 13). Soon we found that the cosmic ray intensity in the western directions was considerably greater than in the eastern directions, which proved unambiguously (1) that primary cosmic rays were, in part at least, charged particles and (2) that the charge of these particles was *positive* (a surprising result because most of us who had been supporting the corpuscular hypothesis, thought, more or less unconsciously, that primary cosmic rays would turn out to be electrons.) (Rossi, 1933b; 1934a).

Here I must admit that we were rather painfully disappointed when we found that,

Fig. 19.   Regener recovering a balloon payload from a farm house.

by just a few months, we had lost the priority of this important discovery. It so happened that, just as we were about to set out on our expedition, we read in Physical Review two articles one by Thomas Johnson, and another by Louis Alvarez and Arthur Compton, reporting the observation of an E-W effect in Mexico City. Moreover my 1930 prediction of this effect was ignored, and credit for this prediction was given to Lemaitre and Vallarta, whose paper had been published three years later. Though I am sure it was an oversight, it still added to my frustration.

One further result of our experiments at Asmara may be worth mentioning. I shall do so by quoting from my own paper "It would seem therefore (since doubts about possible disturbances were ruled out by appropriate control experiments) that once in a while there arrive on the instruments very extended showers of particles which produce coincidences between counters even rather far from each other. Unfortunately I lacked the time to study more closely this phenomenon in order to establish with certainty the existence of the supposed corpuscular showers and investigate their origin." (Rossi, 1934b). This, I believe, was the first observation of the air showers, which, a few years later were studied extensively by Pierre Auger.

After this digression, let me return to 1931.

In the fall of that year, the Italian Royal Academy sponsored a Conference on Nuclear Physics which brought to Rome, from all over the world, the most illustrious

Fig. 20. With Eric Regener and Lise Meither on the Bodensee, in a motorboat equipped for under water experiments. Regener had baptized the boat "Undula "to reaffirm his faith in the wave nature of cosmic rays. Here, following a discussion on this subject, he was telling me: "If it turned out that you are right, I would have to rename my boat "Korpuskel,"
which does not sound as nice as "Undula"!"

scientists interested in this and related fields of physics. At that time, under the direction of Fermi, the Rome group had just begun its research program in nuclear physics which was to produce such momentous results a few years later. At the invitation of Fermi, I gave an introductory speech on the problem of cosmic rays (Proceedings of the Nuclear Physics Conference, 1931). The main thrust of this talk was to present what, in my mind, were irrefutable arguments against Millikan's theory of the "birth cry" of atoms. Such a brash behavior on the part of a mere youngster (I was then 26 years old) clearly did not please Millikan who, for a number of years thereafter, chose to ignore my work altogether. On the other hand it awoke a strong interest on the part of Arthur Compton, who had never worked on cosmic rays before. He kindly told me, some time later, that my 1931 talk had provided the initial motivation for his research program in cosmic rays.

The Rome Conference provided the first occasion for the proponents of the new corpuscular hypothesis to present their case before the scientific community, still strongly attached to the old γ-ray hypothesis. So this conference marked the beginning of the great debate on the nature of cosmic rays, which was to continue for several years.

Fig. 21. Experimental arrangement for demonstrating the passage of cosmic-ray particles through one meter of lead.

In the United States the debate was at times bitter, involving the personal prestige of scientists committed to opposing views. For some reason, this did not happen on the other side of the ocean. In fact two of the strongest advocates of the γ-ray hypothesis – Lise Meitner and Eric Regener – were among my dearest and most respected friends.

From the previous discussion it is clear that the evidence concerning the penetrating power of the ionizing particles was a crucial argument in the controversy about the nature of cosmic rays. From direct experiments, it was known that most ionizing particles had ranges greater than 10 cm of lead. However, by indirect arguments, I had become convinced that many particles must have much greater ranges.

I felt that it was important to verify experimentally this conclusion; if the result of the experiment confirmed my expectations, it would kill once and for all the γ-ray hypothesis.

I planned to do this experiment by the usual method of counting coincidences between Geiger-Mueller counters arranged vertically one above the other and separated by a suitable absorber. The difficulty was that, with two counters far apart, the rate of "true" coincidences would have been smaller than the rate of chance coincidences. To overcome this difficulty, I decided to use threefold rather than twofold coincidences, thereby reducing the rate of chance coincidences to an almost negligible value.

The experimental arrangement is shown in Fig. 21. By varying the thickness of the lead absorber, I found that while the decrease in the rate of coincidences was fairly rapid in the first 10 cm of lead, it then became very slow, so that about 50% of the particles emerging from 10 cm of lead had ranges in excess of one meter of lead (Rossi, 1932a, 1933a).

It is difficult today to appreciate how hard it was for the majority of the scientific community to accept this result. After all, the most penetrating particles known at

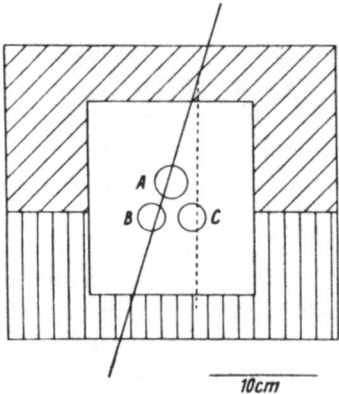

Fig. 22. The "triangular" arrangement of counters (from an article by B. Rossi, in *Phys. Zeitsch.*, 33, 304, 1932).

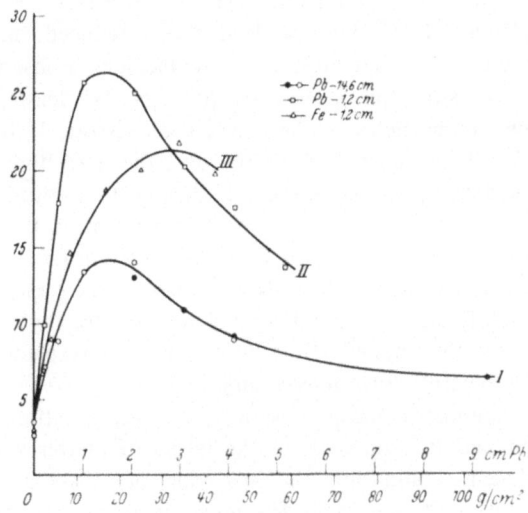

Fig. 23. Curves of the "secondary effects" of cosmic rays in lead and iron screens at 14.6 cm and 1.2 cm from the counters (from an article by B. Rossi, in
Zeitsch. für Phys., 82, 151, 1933).

that time (β-rays from radioactive substances) had ranges of a fraction of a millimeter of lead. Doubts were expressed as to the legitimacy of the coincidence method, and I had to perform further experiments to dispel these doubts.

Perhaps, at this point, I may be allowed to introduce a personal note. In retrospect, I realize today that an important asset in those early days was my completely open-minded approach to the problems I was working on. Unlike many other scientists, I did not feel constrained by concepts and results derived from the study of other phenomena, believed to be more or less akin to cosmic rays. At the Rome conference I began my discussion with the words:

"The most recent experiments have brought to light such strange facts that we are almost led to ask ourselves whether the penetrating radiation (cosmic radiation) may not be something fundamentally different from all other known radiations, or at least, whether, in the transition from the energies encountered in the ordinary radioactive processes to the energies encountered in the phenomena of the penetrating radiation, the behavior of particles and photons does not change much more deeply than one might have thought until now."

The discussions at the Rome conference provided the motivation for another experiment which I performed shortly after my return to Arcetri. At the Conference, both Bothe and I had emphasized the growing evidence for the production of a secondary radiation by cosmic-ray particles in matter. This evidence, however, was still of a rather indirect nature. So, upon my return to Arcetri from the Conference, I decided that I would try to detect the production process directly. For this purpose I arranged three G.M. counters in a triangular configuration (Fig. 22), so that a single particle traveling along a straight line could not traverse them all. With the counters enclosed in a lead shield a few centimeters thick, I did in fact observe a large number of threefold coincidences. Removing the lead shield reduced the coincidence rate considerably. It was thus clear that most of the coincidences observed with the shield in place were due to associated groups of particles (at least two) arising from interactions of cosmic-ray particles in the shield itself (Rossi, 1932b, 1933a). Qualitatively, this is the effect I had been looking for. But the magnitude of the effect, i.e., the high rate of coincidences, was astounding. It showed that cosmic-rays were capable of producing an enormously more abundant secondary radiation than any other known rays. So incredible were my results that a German magazine (if I remember correctly, it was Naturwissenshaften) refused to publish my paper. The paper was then accepted by Physikalische Zeitschrift after Heinsenberg had vouched for my credibility.

I was, of course, greatly excited. Here was a new, unexpected phenomenon, a surprising property of the still mysterious cosmic rays.

I continued my experiment using a variety of configurations of the counters, changing the position and the thickness of the layers of matter where the secondary particles were produced, comparing the behavior of different materials, placing absorbing shields in different positions. The most significant results of this work are summarized in Fig. 23 which shows the dependence of the coincidence rate of three counters in a triangular array upon the thickness (in mass per unit area) of the layer of matter (lead or iron) above them. One sees that the lead curves (I and III) reach a maximum between 10 and 20 g/cm^2 of lead, which shows that the secondary particles have a range in lead of this magnitude.

One then finds that, beyond the maximum, the curves drop much more rapidly than the absorption curve of cosmic rays at sea level; I interpreted this result as showing that the coincidences were, at least in part, produced by comparatively soft, secondary, rays generated by cosmic rays in the atmosphere.

Comparison of the curves for lead and iron, obtained with the shields placed at the same distance from the counters (curves II and III) showed that the rate of production of secondary interactions (per g/cm^2) was an increasing function of atomic number while the range of the secondary particles (again in g/cm^2) was a decreasing function of this quantity.

The next basic step in the understanding of the secondary interactions, which followed closely after my counter experiments, was the well known cloud chamber work by Blackett and Occhialini at the Cavendish Laboratory. It so happened that in 1931, at my suggestion, Occhialini had gone to the Cavendish Laboratory to work with Blackett. He was bringing with him the experience in the coincidence technique developed in Arcetri, and was planning to learn from Blackett the cloud chamber technique, in which, at that time, no one in Italy had any experience. The collaboration of Blackett and Occhialini had produced the counter-controlled cloud chamber. Among the first pictures obtained with this instrument, many showed groups of particles, which they named *showers*; these were clearly identical to the groups of particles produced in the secondary interactions which I had detected with my coincidence experiments. These cloud chamber pictures and the results of my counter work were to form the experimental basis of the theory of showers developed shortly thereafter.

In the late fall of 1932, having won a national competition for a chair in experimental physics in the Italian Universities, I was called to fill a vacancy at the University of Padua. I left Arcetri with a heavy heart. I was still young, and I know that there would be, in my life, other periods of interesting and rewarding work. I also knew, however, that none would have the very special flavor of my years in the Florentine hills.

REFERENCES

Bothe, W. and Kolhörster, W.: 1929, *Zeitsch. f. Physik* **56**, 751.
Rossi, B.: 1930a, *Nature* **125**, 636.
Rossi, B.: 1930b, *Rend. Lincei* **11**, 478.
Rossi, B.: 1930c, *Rend. Lincei* **11**, 831.
Rossi, B.: 1930d, *Phys. Rev.* **36**, 606.
Rossi, B.: 1931a, *Nature* **128**, 300.
Rossi, B.: 1931b, *Zeitsch. f. Physik* **68**, 64.
Rossi, B.: 1931c, *Nuovo Cim.* **8**, 85.
Rossi, B.: 1932a, *Naturwissenschaften* **20** 65.
Rossi, B.: 1932b, *Phys. Zeitsch.* **33**, 304.
Rossi, B.: 1933a, *Zeitsch. f. Physik* **82**, 151.
Rossi, B.: 1933b, *Ricerca Scient.* IV (2) 365; 1934a, *Phys. Rev.* **45**, 212.
Rossi, B.: 1934b, *Ricerca Scient.* V (1) 559.
Proceedings of the Nuclear Physics Conference, October 1931 (Royal Italian Academy, A. Volta Foundation).

ON ERICH REGENER'S COSMIC RAY WORK IN STUTTGART AND RELATED SUBJECTS

Georg PFOTZER*

1. Introduction

Scientists of numerous countries have contributed to our present understanding of the complex of cosmic rays. Evidently the weight of the contributions has varied at times and was effected besides of outstanding scientists also by prevailing favourable or obstructive circumstances. Keeping this in mind, the author does not intend to emphasize unduly the importance of a particular German effort but only to be reminiscent of some personal impressions on the time and place of when and where he had been an immediate witness. Nevertheless, he would like to take up a red thread which originates at a certain point in the past and will swing occasionally over the boundaries of Germany if it seems to be useful for recognizing motivations and stimulations. Timely this contribution will tie up to the situation immediately before the discovery of the cosmic rays in 1912 by the Austrian physicist Victor F. Hess and will be centered between 1928 and 1936. A further guideline may be that a plain recital of a straightforward positive development is not intended because, on the basis of our present knowledge, errors and sideways are often as fascinating as the definite solution of a problem.

2. Decisive Approach to the Discovery of Cosmic Rays

Of the history and prehistory of the discovery of cosmic rays, which can be read more or less exhaustively in most of the relevant monographs, the author would like to lay stress upon a step which played an important part in the first positive proof of the existence of an extremely penetrating radiation impinging from outside into the Earth's atmosphere. It was the introduction of a measuring instrument constructed specifically by the Jesuit priest, father Th. Wulf, of Aachen, Germany, member of the Ignatius College Valkenburg, Netherlands (Wulf, 1909) in order to solve the question how the intensity of the recognized strongly penetrating radiation changes versus the distance from the ground. This type of instrument came to be a reliable standard after its qualities had been carefully studied particularly by Bergwitz (1913) and Dorno (1913) and consequently improved by Wulf and Hess (Hess, 1912, 1912a) and finally particularly by Kolhörster (1913). A sketch of this device, called "Wulfscher Apparat" is seen in Fig. 1.** Its merits were that it was easily portable and that it could be air-tightly sealed and withstand over- and under-pressures that its reading was practically independent of pressure and temperature and hence that it was very suitable for measurements in crevasses, aboard of balloons and aircraft. In order to reduce the effective capacity, the two-fibre type electrometer was mounted inside of the sealed cylindrical vessel. It was visually read through a microscope and could be charged by a magnetically activated rod. Remarkable is the originally used narrow metallic cylinder i

* Max-Planck-Institut für Aeronomie (emeritus), Katlenburg-Lindau, W-Germany.
** <Ed.> See also page 34 (Old Photographs).

Y. Sekido and H. Elliot (eds.), Early History of Cosmic Ray Studies, 75–89.
© 1985 by D. Reidel Publishing Company.

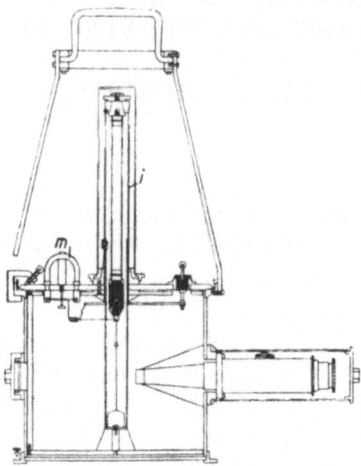

Fig. 1. Wulf apparatus (after Wigand, 1924). Guiding rail for the movable screening cylinder, which is shown here in the downward shifted position.

which could be shifted down to enclose the fibres of the electrometer suppressing thereby the ionization current from the main and greater part of the total volume. This permitted to check the quality of the isolation.

On the 7th August 1912, the day on which Victor Hess then succeeded to reach 5300 m altitude with a free-balloon and to discover the "Höhenstrahlung", he had three of these instruments on board (Hess, 1912). It is known that his far reaching conclusions encountered criticism concerning, among other arguments, the reliability of the Wulf-Apparatus. In that situation Kolhörster, 24 years old, devoted himself to this problem. He came from studies of radio-activity with Prof. E. Dorn at the University of Halle. Thereafter participating in a research project on aerophysics (Hallischer Aerophysikalischer Forschungsfonds (Wigand, 1917)), he improved the Wulf-apparatus further (Kolhörster, 1913) and undertook five balloon flights in the years 1913 and 1914 by which he could brilliantly confirm the results of Hess and extend the ionization curve up to 9300 m. There he measured an ionization strength of fifty times the ground value. The readings of the two Wulf-Kolhörster apparatuses deviated from each other only by 5% at most (Kolhörster, 1913a, 1914, 1914a). The quality and extension of this measurement in the open gondola of a free balloon by Kolhörster remained unsurpassed for almost two decades.

3. Emotions on Absorption Coefficients and High Altitude Measurements

During and some years after World War I little progress in the exploration of the new phenomenon was made. Moreover, a tendency prevailed to interpret the relevant measurements in terms of radio-active constituents accumulated in certain strata, either around the tropopause or higher up in the stratosphere. A particular controversy culminating between 1926 and 1928 arose due to the pitfalls in the measurements of the absorption of the "ultra-gamma radiation."

Kolhörster subsequently to his balloon flights had derived an absorption coefficient first from his measurements between altitudes of 1.5 to 6.2 km of $\mu = 7.1 \cdot 10^{-6}$ cm^{-1} standard air (Kolhörster *et al.,* 1914). His flight data, up to 9.2 km (Kolhörster, 1914, 1914a), pointed to a little higher value but "confirmed the order of magnitude $\mu \approx 1 \cdot 10^{-5}$ cm^{-1}." A refined computation of v. Schweidler (1915) after Kolhörster's data which eliminated the influence of the radiation from the soil lead to $\mu = 7.46 \cdot 10^{-6}$ cm^{-1} standard air corresponding to a (fictitious) mass absorption coefficient $\mu/\rho = 5.77 \cdot 10^{-3}$ cm^2 g^{-1}. Fictitious because Kolhörster as also v. Schweidler assumed only vertical incidence and so could fit the balloon data from 1.5 to 9.2 km fairly well by a pure exponential versus absorber thickness in cm of standard air or g/cm^2.

Whatever the significance of this rather formal coefficient might have been, the conclusion was certainly justified that the radiation in question was at least 6 times more penetrating than the RaC-γ-radiation with $\mu = 4.5 \cdot 10^{-5}$ cm^{-1} air after Hess (1911). Moreover, it was shown by Seeliger (1918) that the assumption of a mere exponential by which the Hess-Kolhörster curve was fitted yields only an upper limit of the true absorption coefficient. He computed a physically more meaningful value under the assumption of an omnidirectional incidence of $\mu = 4.9 \cdot 10^{-6}$ cm^{-1} air $\hat{=} \mu/\rho = 3.8 \cdot 10^{-3}$ cm^2 g^{-1}. In January 1923 Kolhörster and G. v. Salis, animated by Nernst, who had hypothetisized an origin of the "ultra-gamma radiation" in cosmical processes of matter formation in red giant starts (Nernst, 1921), carried out absorption measurements in alpine ice holes under 1.5 to 9.7 m absorber thicknesses and at sea level under the surface of water between depths of 1.5 − 5 m. There they found for vertical incidence $\mu/\rho = 2.7 \cdot 10^{-3}$ cm^2 g^{-1} and $= 2 \cdot 10^{-3}$ cm^2 g^{-1} respectively (Kolhörster, 1923) and this meant about a seventeenth of the absorption coefficient of RaC and ThC gamma rays. (< Ed.> See photos in "Early Propagation of Cosmic Rays Studies" in this book.)

This was the situation when Millikan and Bowen (1923) published a short notice about a new approach to measurements of the ionization strength in altitudes higher than could be reached by manned free balloons. They had constructed an automatically recording ionization chamber of the Wulf type but with a total weight of only 190 g, which in spring 1922, carried aloft by pilot balloons, ascended to an altitude of 15.5 km. This flight and its result were not much noticed after the first publication, because details for a critical judgement had not yet been published. A later paper (Millikan and Bowen, 1926), obviously written in the light of the famous absorption measurements in and between high mountain lakes performed in 1925 by Millikan and Cameron (1926), and subsequent papers of Millikan and co-workers (e.g. Millikan, 1926; Millikan and Cameron, 1928) caused however much irritation and concern among the German and Austrian colleagues.

What had happened? Due to a "marked temperature effect of the electroscope" (Millikan and Bowen, 1926) during the above-mentioned flight, an extension of the Hess-Kolhörster curve was not obtained but only an average value of the ionization strength between 5 and 15 km, taken from the difference of the two electrometer readings at 5 km on the ascent and descent. This value of 46.2 J amounted to only 25% of what had been expected from an extrapolation of the Hess-Kolhörster curve according to the fictitious absorption coefficient of $5.8 \cdot 10^{-3}$ cm^2 g^{-1} after Kolhörster-Schweidler. The apparent small increase of ionization strength seemed now to point to a much lower fictitious absorption coefficient than assumed, as a matter of fact, if taken seriously, to about $2.2 \cdot 10^{-3}$ cm^2 g^{-1} (the author). This motivated Millikan and co-workers to further measurements on board of airplanes and with free balloons up to 5300 m (Otis, 1923) and on mountains up to 4500 m (Otis, 1923a). But still in 1924 they interpreted the outcome of these investigations "as meaning that the whole of the penetrating radiation is of local origin" (Otis and Millikan, 1924).

Only one year afterwards Millikan and Cameron came to a radically different view on account of their above-mentioned measurements. These yielded absorption coefficients between $3 \cdot 10^{-3}$ and $1.8 \cdot 10^{-3}$ cm^2 g^{-1} in fair agreement with the corresponding measurements of Kolhörster and v. Salis in the Alpes and at sea-level in 1923. The lowest absorption coefficient corresponded now to the twentyfold of the penetrability of RaC and ThC-γ-radiation. This did not only convince Millikan and Cameron of the extraterrestrial origin of such a penetrating radiation, moreover, they considered the result of their systematic and very dependable measurements as its first indubitable proof and proposed the name "Cosmic Rays." Millikan and Bowen found now that the lower absorption coefficients are nicely compatible with their flight result, much better than the value adequate to the Hess-Kolhörster curve. Hence they felt justified to trust their own measurements, although admittedly obtained under questionable conditions. Hence they wrote (Millikan and Bowen, 1926): "The results then of the whole Kelly field work (launching place near San Antonio, Texas) constitute definite proof that there exists no radiation of cosmic origin having such characteristics as we had assumed." This conclusion in connection with the remark that the flight was undertaken "in order to obtain a crucial test as to whether there is such a cosmic radiation as the Hess-Kolhörster data seemed to require" provoked of course sharp reactions of Hess and Kolhörster (Hess, 1926; Kolhörster, 1926a, b). The dispute revived again subsequently to a new paper of Millikan and Cameron (1928a). They presented further and very elaborate absorption measurements down to water depths of 50 m and an extension of the absorption coefficient from $2.5 \cdot 10^{-3}$ cm^2 g^{-1} down to now $1 \cdot 10^{-3}$ cm^2 g^{-1}. They emphasized the inhomogeneity of the rays and attributed their origin to induced or spontaneous nuclear changes in the cosmos. On the other hand they emphasized again the significance of the suspected high altitude flight, discussed and claimed priorities concerning the recognitions of the cosmical origin and of the inhomogeneity of the radiation beam. Thereafter, some passages which questioned priorities of the involved Austrian and German colleagues lead to a detailed reply with rectifications by Bergwitz, Hess, Kolhörster and v. Schweidler (Bergwitz et al., 1928). Today the whole matter deserves only to be taken out because, on the one hand, it throws some light on the competition and emotions in the early history of cosmic rays and it stimulated, on the other hand, further sounding of the limits of penetrability of the radiation and the extension of the measurements to still higher altitudes. These were particularly the aims of Prof. Regener who started his work on cosmic radiation in autumn 1928. As the author, onward from 1929, spent two periods of his scientific life in Regener's domain, as student and as a senior scientist, before and after World War II, respectively, it is quite natural that he will be particularly reminiscent of the work and personality of his venerate "doctor father."

4. Regener and His Institute in Stuttgart

Regener was offered the chair on Experimental Physics of the Technical High School Stuttgart and appointed to the director of the Physical Institute in 1920. He had completed his studies at the University of Berlin in 1905 with a thesis on the production and destruction of ozone by certain ultraviolet spectral bands; thereafter he had worked on radio-activity and on the problem of the electron charge, respectively the possible existence of "subelectrons". During the first World War he completed his military service as a "field-X-ray mechanician". As elsewhere is appreciated (Paetzold *et al.*, 1974), his original ideas and his merits on all these subjects were remarkable; lately he became however particularly renowned through his work on cosmic rays, but also owing to the first measurement of the atmospheric ozone profile which he had carried out together with his son Victor by means of a balloon-borne spectrograph (Regener, E. and V.H., 1934).

His scientific staff in the Physical Institute of Stuttgart consisted of four assistents and about ten doctorands and candidates for the first graduation (Diplom-Ingenieur). Although he guided his students with light hand, his authority was always undoubted. After the beginning of cosmic ray research, most of the students had to work on the same subject for their theses either directly or by investigation of the various influences which had to be considered in regard to the measurements with ionization chambers or Geiger counters. Normally, Regener showed up rather hastily in the rooms of the advanced students. But as he had his official residence in the institute and, as always some students used to return after dinner and were busy far into the night or sometimes also merry-making, he now and then dropped in for a chat or to give advices if problems had arisen; merry-making was of course immediately damped by his presence. In the whole the atmosphere in the institute was very pleasant until in the late thirties the political situation brought about severe troubles for Regener. This is, however, a different story. In the context of this narrative on early Cosmic Ray work, it remains only to state, that Regener, after having been suspended from his chair and directorship, at the end of 1938 was appointed director of a research laboratory in the "Kaiser-Wilhelm-Gesellschaft". After World War II he had the satisfaction of an institutio in integrum also in Stuttgart. When the "Kaiser-Wilhelm-Gesellschaft" was changed into the "Max-Planck-Gesellschaft" in 1948, Regener was nominated Vice-president and his research laboratory was upgraded to the "Max-Planck-Institut für Physik der Stratosphäre" in 1953. He deceased in 1955.

5. Regener's Absorption Measurements in the Lake Constance

Returning to the year 1928, we note that the speculations on cosmical transmutations of energy being the sources of the ultra-gamma radiation by Nernst (1921), Antropoff (1924), Millikan and Cameron (1928) had raised the question as to the lower limits of the wave length spectrum respectively the corresponding high quantum energies. As according to the state of knowledge at that time the penetrability of gamma-rays was to be a function of the lowest wavelength, Millikan and Cameron (1928a) in the United States and Steinke (1928, 1929) in Germany had systematically extended their measurements below the surfaces of lakes until no further decrease of the ionization strength could be observed. This occurred between 60 m and 40 m respectively. Regener, impressed by the far reaching importance of such measurements in the light of the above-mentioned hypotheses, felt confident to achieve further progress. As he was an outstanding experimentalist he used to approach a new problem always remarkably unprejudiced. He constructed an automatically recording ionization chamber consisting of a cylindrical steel bomb, filled with carbon dioxide at a pressure of 30 bar. The central electrode was connected to a one-fibre electrometer of Regener's special construction. A dark-field illumination of the fibre was hourly switched on by a clock for a few seconds and caused thus a picture to be taken of the position of the fibre on a fixed photographic plate (Fig. 2c). Hence the only moving elements of the automatic recording device were the pointer of the clock and the fibre of the electroscope, the distances between the fibre images being proportional to the ionization current. The zero effect of the ionization chamber could be suppressed by a special treatment of the walls so efficiently, that a clear decrease of the ionization strength down to 235 m below the surface of the Lake Constance could be recorded.

At the beginning of the experiments the 130 kg weighing "bomb" hanging on a wire rope was let down by a hand-winch to the desired depth through an opening of a rowing boat (Fig. 2a), which was anchored over the deepest point of the Lake Constance. This method proved however not to be satisfactory, because the rocking motion of the boat on a moving sea propagated down to the bomb and caused the blurring of the fibre images. The next step, as shown in Fig. 2b, was then to suspend the apparatus below an iron tank. This had a buoyancy sufficiently high for raising the apparatus to the surface if the heavy anchor-weight which kept it down was lifted. The distance from the ground was predetermined by the length of the wire rope between the anchor and the apparatus. Another wire rope connected the anchor weight with a buoy on the surface by which the place of the "bomb" was marked and from which the rope could be taken up for weighing anchor. This worked in principle excellently, as shown in Fig. 2b, the water waves did not penetrate deeply below the surface and hence the apparatus floated completely undisturbedly above the anchor-weight. The buoy caused however a lot of trouble, because floating fishing nets became sometimes entangled and damaged and the fishermen understandably excited. It was therefore necessary to develop another mode for the refinding of the "bomb" and weighing anchor.

This was achieved by navigational means. The point of the apparatus was fixed by the crossing of two lines of sighting each defined by two suitable landmarks. Instead of fastening the anchor hoisting rope to the buoy, it was laid out on the ground about rectangular to one of the lines of sighting. The end of the rope was connected to a second light anchor.

For re-surfacing of the apparatus in order to exchange the photographic plate, and to recharge the electroscope, a search anchor was let down and towed over the ground behind the boat until it grasped the laid out wire rope. This was then lifted and connected with the windlass. Then the heavy anchor was hoisted causing the apparatus to emerge.

The low intensity of the hardest components of cosmic rays demanded extreme measures to exclude external influences which could fake the decrease of ionization rate with depth. Such an influence could be that a small radio-active contamination of

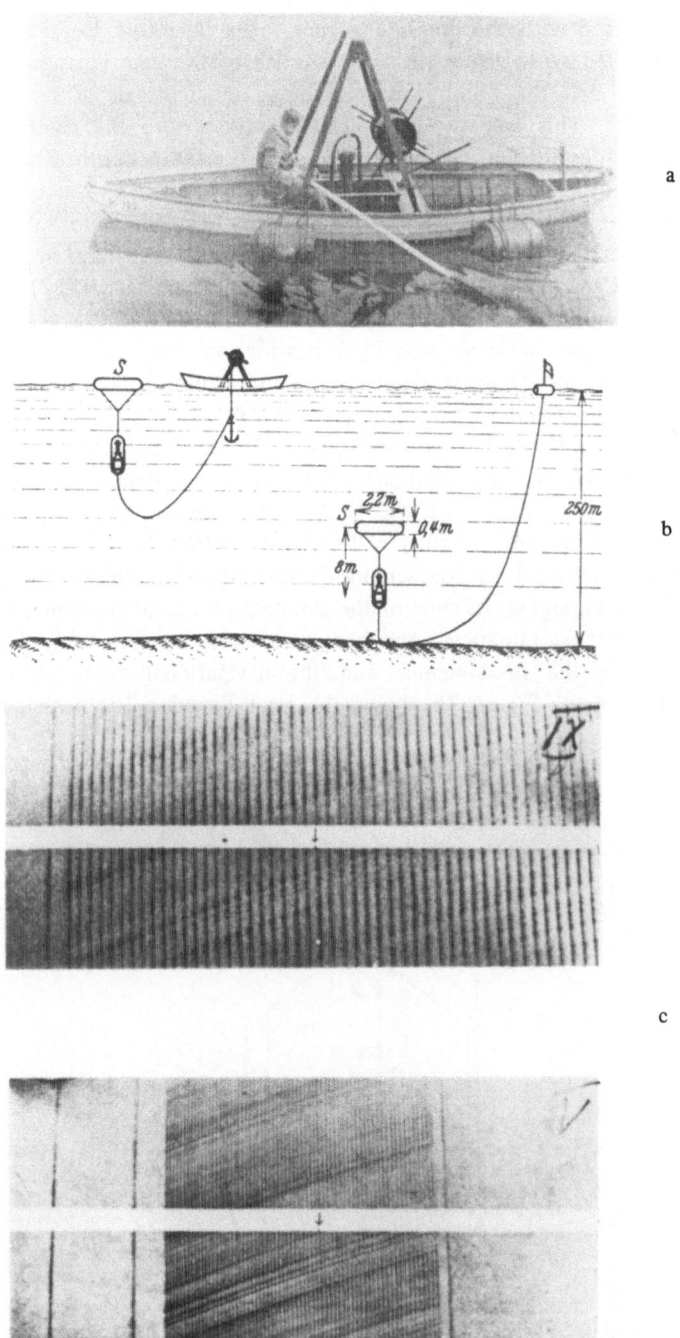

Fig. 2. a) Rowing boat with the ionization chamber (called "Bodensee-Bombe") ready for diving. b) Scheme of the floating suspension. c) Examples of recording plates. Recordings in 47.8 m and 167.8 m depths, upper and lower plates, respectively. The distances of the fibre images taken hourly are proportional to the ionization strength.

the water changed with the depth. In order to exclude this, Regener put the "bomb" into a tank filled with water from the surface of the lake (Fig. 3a). The size of the tank was such that the bomb was surrounded by a screen of surface water of one meter thickness. This was enough to sufficiently absorb the gamma radiation from outside for eliminating an influence of its variation with depth. Six cylindrical floats were attached to the tank providing the required buoyancy.

The first series of measurements in late autumn 1928 were successful in proving that a component of the radiation was far more penetrating than, in the limits of error, Millikan and Cameron had been able to detect (Regener, 1932). It was however desirable to confirm this result and to improve the accuracy of the absorption coefficients in view of their suspected bearing on the spectrum and origin of the radiation, at least of the most penetrating components. This required a planning on a long term basis. Thanks to the recognition of Regener's success, the plan was supported by the "Deutsche Notgemeinschaft der Wissenschaft" and by the friends of the "Technische Hochschule Stuttgart". This permitted Regener to replace the rowing boat by a handsome motor boat, which contained a small laboratory with a dark-room, power supplies, a vacuum pump, etc. It was also fitted out with a motor-winch and a framework of steel at the stern, serving the sinking and lifting of the apparatus and its fixing after the emerging. In view of the assumed nature of the penetrating component of the "ultra-gamma radiation" the boat was given the name "Undula" what means little wave. For the maintenance and the navigational manoeuvres a very skilled boatsman was hired. Figure 3b shows the boat dragging the vessel, Fig. 3c Regener approaching the emerged vessel.

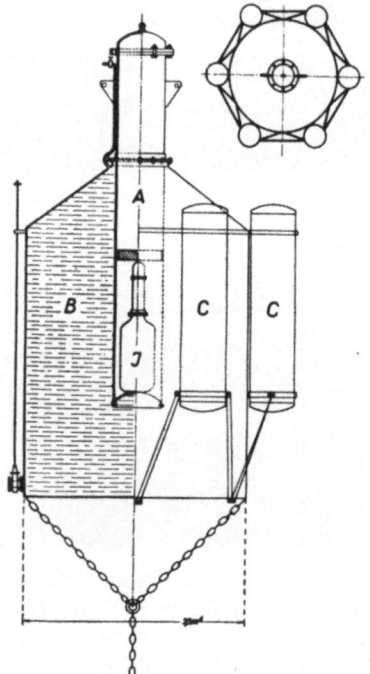

Fig. 3a

Fig. 3b

Fig. 3c

Fig. 3. a) Sketch of the big vessel, filled with water from the surface B containing the ionization chamber J in the cylindric tube A. C floats providing the buoyancy. b) The "Undula" having the big vessel in tow. c) The big vessel emerged and fixed at the stern of the "Undula". Regener posing, before opening the lid.

This investment proved to be very useful, for the measurements with the ionization chamber under continuous improvement of their accuracy (Weischedel, 1936) and later on with a GM-telescope (Ehmert, 1937) dragged on nearly a decade.

6. Hypotheses of Element Formation or Annihilation as Sources of Cosmic Rays

The interpretation of the ionization measurements in terms of a wave radiation was quite natural at that time. It was also very tempting to attribute the origin of the different components to specific cosmical energy transmutations. Millikan and Cameron (1928a) took the view that the quantum energies of the spectral components that they had filtered out corresponded to the mass defects of He, O, Si and Fe at their build-up in the cosmos. The contemporary colleagues know how poetically they had expressed this view: "The observed cosmic rays are the signals broadcasted throughout the heavens of the births of the common elements out of positive and negative electrons" (Millikan, 1928). Jeans, contrary to this hypothesis, had calculated that the high energies to be attributed to Regener's two hardest components, could best be explained by the annihilation of helium and hydrogen (Jeans, 1931). Regener thought that it was hard to believe that the surprisingly good correspondence between the calculated quantum energies of these two components and the rest energies of helium and hydrogen respectively should be accidental (Regener, 1933). Although at that time the particulate nature at least of a fraction of the primary beam was evidenced by the discovery of the latitude effect (Clay, 1927, 1928, 1930) and corroborated by the coincidence experiments of Bothe and Kolhörster (1929), the hypothesis of the wave nature of the hardest components could not easily be overcome. It is in retrospect no more interesting as such but in view of the seemingly logical arguments by which it was supported not without appeal. The general belief was that very hard gamma rays were essentially absorbed by the Compton process and that a relation between the wavelength, respectively the energy of the quanta and the absorption coefficient exists. Finally, the Klein-Nishina formula (Klein and Nishina, 1929) was generally accepted as the best relation. It was, however, already experienced that the nuclei played also their part in the absorption of very energetic radiation "of whatever type" (Rutherford, see Discussion, 1931). Jeans (1931) made therefore allowance for the contribution of the nucleus by including also "the nuclear electrons" in his calculations of the quantum energies of Regener's radiation components (Regener, 1932).

7. Speculations on the Nature of the Cosmic Rays at a Turning Point

The situation in 1931 is characterized extremely well in a "Discussion on Ultra-Penetrating-Rays" which took place in London on May 14, 1931 (Discussion, 1931). It was opened by Geiger with a review. Other participants were such eminent scientists as C.T.R. Wilson, Soddy, Rutherford, O.W. Richardson, Lindemann, Eddington, Dobson, Chapman, Chadwick, Bragg. It can only be recommended to those interested in the history of cosmic rays, occasionally to take a look at this paper. Geiger in his review stressed the then available three sources of knowledge on cosmic rays: "the absorption measurements, the coincidence measurements, and the effects taking place at the boundary of two adjacing media" and he emphasized their consequences with respect to a wave or corpuscular nature. Rutherford, after having reminded to his remark at a former occasion "what we wanted in reference to cosmic radiation was more work and less talk", continued then, stating that "we have now arrived at a stage, when we have reliable data. Thanks to the fine experiments of Professor Millikan and the even more far-reaching experiments of Professor Regener, we have now got for the first time, a curve of absorption of these radiations in water which we may safely rely upon." And a little bit later he said: "It seems to me that the penetrating radiation is of the gamma-ray type until we have definite evidence of the contrary." In the context of the annihilation hypothesis it is interesting to note that Chadwick emphasized first that "by assuming that the radiation is absorbed by the nuclear electrons as well as by the outer electrons, Jeans obtains a surprising agreement with the experimental absorption coefficient," but then he critisized that in an one-quantum emission by annihilation, as assumed by Jeans, only the energy and not the momentum is conserved and therefore "we must suppose that two quanta of radiation, not one, are emitted in the process. The agreement between the calculation and the experiment is then no longer so good, but not unsatisfactory." As a matter of fact, it can be shown (the author) that the agreement with the measurement remains as good as it could be, if one assumes a two gamma decay (per quantum half of the annihilation energy) under the assumption that only the atomic shell electrons are taken into account for Compton scattering. According to our present knowledge, this would have been more adequate in considering Compton scattering at all and only. However, in the light of the then already recognized influence of the nucleus (Chao, 1930; Meitner and Hupfeld, 1931) the inclusion of the nuclear electrons corresponded indeed to a consequent logical deduction, but was based on premises still lacking the real background which remained to be explored for the future.

Concluding the reminiscences of the "annihilation episode". it should yet be mentioned that Rutherford made an experimental test in the laboratory whether perhaps also the helium in the atmosphere might contribute to the penetrating radiation (Discussion, 1931). For that purpose he let place all the helium he could obtain at a pressure of 100 atmospheres round a suitable electroscope. This corresponded to a layer equivalent to 13 m thickness as to be compared with 2 to 10 cm total helium content in the atmosphere at atmospheric pressure. As we can imagine today no measurable effect was observed.

8. The Onset of Successful Flights with Pilot-Balloons in Cosmic Ray Research

Regener had realized from the beginning of his cosmic ray work that a reliable extension of the Hess-Kolhörster curve to high altitudes was also of great importance. He had clearly recognized that, as Millikan and Bowen had already attempted, the use of automatic instruments, carried aloft by pilot-balloons, would be the most promising way not only with respect to the attainable altitude, but also with respect to the expenses. He constructed then his famous "Ballonelektrometer," the ingenious simplicity of which is admirable yet in our days. The electrometer and the photographic recording of the fibre position on a fixed photographic plate were of the same type as used in the great "ionization bomb" for the measurements under water (Fig. 4a).

The air pressure and the temperature of the instrument were recorded by lineals, which, controlled by an aneroid barometer and a bimetallic thermometer, moved rectangular to the image of the fibre. They obscured it in such a manner that the length of the fibre image between the shadows of the lineals and a fiducial bar indicated the pressure and temperature. The instrument was tested for the first time in spring 1928. It was carried by a captive balloon of a meteorological station and worked as expected until the clock which controlled the illumination of the electrometer fibre stopped short due to the influence of the coldness. As further trials brought to light that by the same reason the soldering turned brittle and untight, the instrument was put into a light gondola which consisted of an elastic wooden framework covered with cellophane. The gondola served as greenhouse against the coldness and as buffer spring on landing. Another problem was to overcome an undue

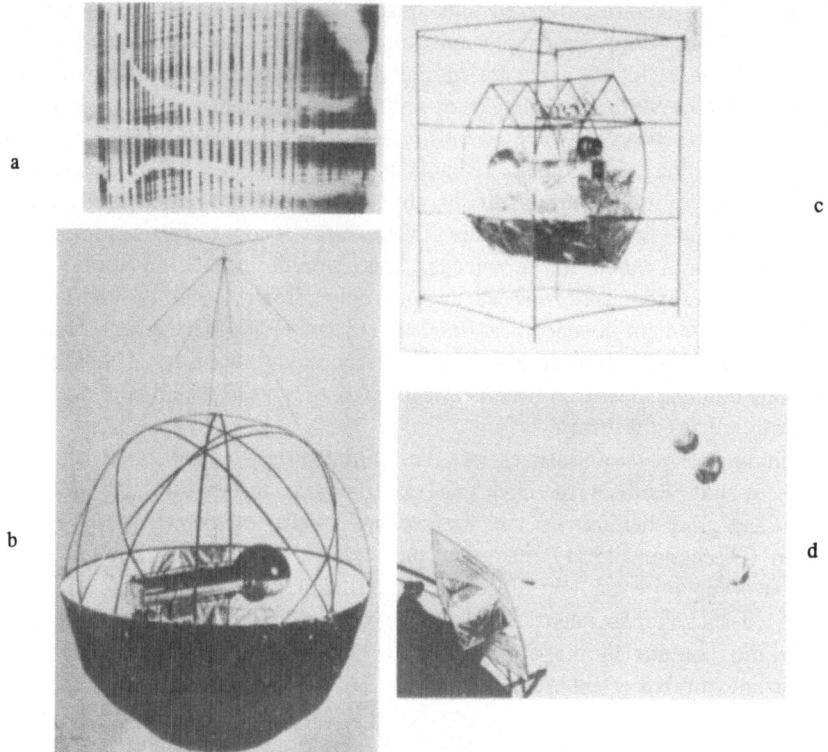

Fig. 4. a) Recording plate of the first successful flight with the "Balloon-Elektrometer". The upper shadow trace indicates the pressure, the lower one the temperature of the apparatus. The step of the pressure curve shows the beginning of the descent, which occurred rather turbulently as indicated by the complete blurring of the fibre images. b), c) and d) Modifications of the housing to improve the stream lining.

blurring of the fibre image by preventing or damping vibrations caused by turbulences at the ascent and descent. As Regener was very ambitious to achieve perfection, the shape of the gondola underwent many changes in the course of time (Fig. 4b, c, d).

At the beginning of the balloon work the flights with the ionization chamber, which, together with the clock and battery had a weight of 1.5 kg, were carried out with a tandem of rubber balloons. The attainable altitude was limited by the bursting of the balloon with the weakest envelope, whereafter the apparatus began to sink, in most cases, with approximately the same speed as on the ascent. At later flights with heavier payloads, teams up to four balloons were used. After the delivery of the balloons from the factory, they had to be carefully scrutinized with a magnifying glass for tiny pinholes, which could not yet be avoided according to the state of the art at that time. These pinholes widened during the ascent and caused the leakage of hydrogen and, even worse, gave rise to a too early bursting of the balloons. Patching of the pinholes with adhesive rubber plaster proved to be useful. In order to detect the pinholes the balloons were inflated with compressed air in the dark auditorium and illuminated with a search light. This made the pinholes stand out as clearly recognizable light points. Some of the balloons with surfaces of 20 to 30 m² at inspection had finally hundreds of patches. This tiresome work was shared by the janitor and the students. Much of the success of Regener's and his students' balloon flights with respect to the attained altitude, was due to this anything else but popular measure.

The balloons may be characterized by their weight and the bursting thickness of the envelope. The weights of the balloons at the beginning of the balloon work amounted to 4 kg, later on, for heavier payloads, up to 12 kg. The bursting thickness of the rubber plate varied between 15 and 20 μm, the bursting diameter between 8 and 15 m. The price of a 4 kg and a 12 kg balloon amounted to about 100 RM and 500 RM respectively (1 \$ ≙ 4 RM at that time).

The launching of a balloon track usually performed from the narrow courtyard of the Physical Institute was always an exciting procedure (Fig. 5). The weather had to be fine, wind speed and wind direction as function of the altitude had to be determined in the early morning of the x-day with small pilot balloons. After launch the balloon track was surveyed with theodolites. At the beginning of the descent the vertical speed was determined and, using the data of the ascent, the expected launching area anticipated. Thereafter, the landing of a scientific payload was broadcasted and sometimes a crew started with an autocar to the supposed landing place. Usually after 4 to 5 hours the landing was announced by a telephone call from the finder. After this scheme, in the whole about 30 flights took place between 1932 and 1936. The distances between the institute and the landing area ranged between normally 30 to 50 km, the largest distance was 220 km.

Fig. 5. Launching of a balloon train from the courtyard of the institute.

During the preparations of Regener's sounding flights Piccard and Kipfer made the first sensational flight inside of a hermetically closed spherical cabin up to 16 km altitude (Piccard *et al.*, 1932). This was certainly an admirable achievement in the whole. But as the crew had been engaged in closing a dangerous leak most of the time, they could measure only one value of the ionization strength in 16 km altitude, anyway the highest point obtained up to that time. This flight stimulated of course Regener to enter the competition between manned and pilot balloons and to speed up the efforts of getting ahead with his instrument. His first successful flight up to 24 km altitude (29 mb) took then place on August 12, 1932 (Regener, 1932a), just one week before Piccard and Cosyns (1932) started for the second, now in every respect also successful flight, again up to 16 km (103 mb). Thus the Hess-Kolhörster curve could be extended by Piccard, Cosyns and Regener up to 16 km and by Regener up to 25 km (Regener, 1933). The essential new result of Regener was, apart from the extension of the measurements to this high altitude, that the ionization strength approached a nearly constant value towards the top of the atmosphere. Bowen and Millikan (1933) in continuation of their early attempt in 1922 had also carried out two flights in summer 1932 with an improved instrument and obtained "dependable electrometer readings down to pressures of 88 mmHg (15 km) and 61 mmHg (18 km) respectively." Their results confirmed essentially also the findings of the European colleagues. Herewith all former misunderstandings had been overcome.

9. Epilog

Until the outbreak of World War II the continuation of the stratospheric flights by Millikan's group led to the excellent high precision measurements in different latitudes, which showed the increasing importance of the latitude effect in high altitudes (Millikan *et al.*, 1936; Bowen *et al.*, 1937 and 1938).

Regener's group took up measurements with GM-counters (Regener and Pfotzer, 1934) and GM-telescopes (Regener and Pfotzer, 1935; Pfotzer, 1935, 1936). These succeeded in the discovery of the maximum of secondary particles in the vertical beam at about 100 mb. Regener and Ehmert measured the altitude dependency of the "soft" and "hard" components with counter arrangements for shower detection and lead filters respectively (Regener and Ehmert, 1939; Ehmert, 1940). An early flight with paraffin covered nuclear emulsions up to 18 km altitude was performed by Schopper in order to investigate the neutron component via the tracks of the recoil protons expelled from the paraffin layer (Schopper, 1937). These early successful flights with pilot balloons and automatic instruments have opened a wide field of innumerable applications (see e.g. Pfotzer, 1972). The increasing complexity of the payloads and the corresponding size of the balloons has ensured that the launching of balloons with up to date payloads has remained to be an exciting event.

REFERENCES

Antropoff von, A.: 1924, *Z. Angew. Chemie* 37, 827.
Bergwitz, K.: 1913, *Physik. Z.* 14, 953.
Bergwitz, K., Hess, V. F., Kolhörster, W., and Schweidler von, E.: 1928, *Physik. Z.* 29, 705.

Bothe, W. and Kolhörster, W.: 1929, *Z. Physik.* **56**, 571.

Bowen, I. S. and Millikan, R. A.: 1933, *Phys. Rev.* **43**, 695.

Bowen, I. S., Millikan, R. A. and Neher, H. V.: 1937, *Phys. Rev.* **52**, 80; 1938, *Phys. Rev.* **53**, 855.

Chao, C. Y.: 1930, *Proc. Natl. Acad. Am.* **16**, 431.

Clay, J.: 1927, *Naturwiss.* **15**, 356; 1928, *Proc. Amsterdam* **31**, 1091; 1930, *Proc. Amsterdam* **33**, 711.

Discussion on Ultra-Penetrating Rays: 1931, *Proc. Roy. Soc., London, A* **132**, 331.

Dorno, C.: 1913, *Physik. Z.* **14**, 956.

Ehmert, A.: 1937, *Z. Physik.* **106**, 751; 1940, *Physik. Z.* **115**, 326.

Hess, V. F.: 1911, *Physik. Z.* **12**, 998; 1912, *Wiener Ber. IIa* **121**, 2001; 1912a, *Physik. Z.* **13**, 1084; 1926, *Physik. Z.* **27**, 159.

Jeans, J. H.: 1931, *Nature* **128**, 103.

Klein, O. and Nishina, Y.: 1929, *Z. Physik.* **52**, 853.

Kolhörster, W.: 1913, *Physik. Z.* **14**, 1066; 1913a, *Physik. Z.* **14**, 1153; 1914, *Verhandl. d. Deutsch. Phys. Ges.* **16**, 719; 1914a, *Beitr. z. Phys. d. fr. Atm.* 7,87; 1926a, *Naturwiss.* **14**, 371; 1926b, Ann. d. *Physik* **80**, 621.

Kolhörster, W., Wigand, A. and Stoye, K.: 1914, *Abh. d. Naturf. Ges. Halle*, N.F.Nr.4.

Kolhörster, W. and Salis, G. v.: 1923, *Berlin. Ber. phys.-math. Kl.*, p. 366.

Meitner, L. and Hupfeld, H.: 1931, *Z. Physik.* **67**, 147.

Millikan, R. A. and Bowen, I. S.: 1923, *Phys. Rev.* (2), **22**, 198; 1926, *Phys. Rev.* (2), **27**, 353.

Millikan, R. A.: 1926, *Ann. d. Physik.* **79**, 572; 1928, *Science* **68**, 279.

Millikan, R. A. and Cameron, G. H.: 1926, *Phys. Rev.* **28**, 851; 1928, *Nature* **121**, 19, Suppl.; 1928a, *Phys. Rev.* **32**, 533.

Millikan, R. A., Neher, H. V. and Haynes, S. K.: 1936, *Phys. Rev.* **50**, 992.

Nernst, W., *Das Weltgebäude im Lichte der neueren Forschung*, J. Springer, Berlin 1921.

Otis, R. M.: 1923, *Phys. Rev.* **22**, 198; 1923a, *Phys. Rev.* **22**, 199.

Otis, R. M. and Millikan, R. A.: 1924, *Phys. Rev.* **23**, 778.

Paetzold, H. K., Pfotzer, G. and Schopper, E.: 1974, *Zur Geschichte der Geophysik*, ed. by H. Birett, K. Helbig, W. Kertz and U. Schmucker, Springer-Verlag Berlin, Heidelberg, New York, p. 167.

Pfotzer, G.: 1935, *Z. techn. Phys.* **16**, 400; 1936, *Z. Physik.* **102**, 23 and 1936, **102**, 41; 1972, *Space Sci. Rev.* **13**, 199.

Piccard, A., Stahel, E. and Kipfer, P.: 1932, *Naturwiss.* **20**, 592.

Piccard, A. and Cosyns, M.: 1932, *Compt. Rend. Acad. Sci.* **195**, 604.

Regener, E.: 1932, *Z. Physik.* **74**, 433; 1932a, *Naturwiss.* **20**, 695; 1933, *Physik.Z.* **34**, 306.

Regener, E. and Regener, V. H.: 1934, *Physik. Z.* **35**, 788.

Regener, E. and Pfotzer, G.: 1934, *Physik. Z.* **35**, 779; 1935, *Nature* **136**, 718.

Regener, E. and Ehmert, A.: 1939, *Z. Physik.* **111**, 501.

Schopper, E.: 1937, *Naturwiss.* **25**, 557.

Schweidler von, E.: 1915, *Elster u. Geitel Festschr.*, Braunschweig, p. 411.

Seeliger, R.: 1918, *Münch. Ber. Math.-Phys. Kl.*, p. 1.

Steinke, E.: 1928, *Z. Physik.* **48**, 672; 1929, *Z. Physik.* **58**, 183.

Weischedel, F.: 1936, *Z. Physik.* **101**, 732.

Wigand, A.: 1917, *Beitr. z. Phys. d. fr. Atm.* 7, 189; 1924, *Physik. Z.* **25**, 445.

Wulf, Th.: 1909, *Physik. Z.* **10**, 152.

SOME OF THE PROBLEMS AND DIFFICULTIES ENCOUNTERED IN THE EARLY YEARS OF COSMIC-RAY RESEARCH

H. Victor NEHER*

The history of cosmic rays is a history of failures and difficulties as well as successes. This is true of all fields of research but seems to be especially true in the case of cosmic rays. This arises partly from the difficulty of detecting the radiation in the first place but also from the wholly unexpected nature of the many phenomena encountered. Almost everyone in the game made mistakes. These involved errors not only in the taking of the data because of unknown influences at work but also in the interpretation of even the good data.

One of the early sources of confusion was due to the radioactivity in the soils and rocks. It wasn't until the balloon flights of Gockel, Hess and Kolhörster in the period 1910 to 1914 that the effects of the earth's radioactivity were essentially eliminated by using the earth's atmosphere below the balloons as an absorber. Other workers, particularly after World War I, tried to use high mountains for places of observation but in many cases were not aware of the full effects of local radiation. In the U.S.A. some early workers tried to use Pike's Peak (elevation 14100 ft) to make observations not realizing that the radioactivity of the rocks was several times as large as the soils in the valley below. The gamma rays from Thorium C" are particularly penetrating.

The instruments used to detect cosmic rays up to 1929 were all ionization chambers. In 1929 Bothe and Kolhörster introduced the Geiger counter. Their experiment will be discussed later. As for ionization chambers, they continued to be used for many years and have yielded some very valuable information about cosmic rays. When an ionization chamber is used one must measure electric currents at sea level of the order of 10^{-14} amperes. If measurements are to be accurate to 1 percent, then any extraneous effects must correspond to currents of less than 10^{-16} amperes. Of the two choices one has of placing the detecting device exterior to the ion chamber or placing it inside the ion chamber, it is obvious that the latter arrangement has decided advantages. In the former case the ion current must be taken through the wall of the chamber and this involves insulators. Not only can humidity cause troubles but polarization of the dielectric can also be serious. Humidity can be particularly troublesome in the tropics. Millikan developed the internal type of system which was particularly useful for his early underwater work. The early instruments suffered from a temperature effect when air was used under high pressure. This was due to a change of recombination of the ions with temperature. This difficulty was remedied when argon became available in about 1933.

As soon as it became evident that cosmic rays in our atmosphere were due to radiation coming to the earth from outer space the questions immediately asked were: (1) "What is the nature of the radiation?" and (2) "From where do they come?" Millikan, as well as others in the 1920's thought that the primaries were photonic in

*California Institute of Technology (emeritus), Pasadena, U.S.A.

Y. Sekido and H. Elliot (eds.), Early History of Cosmic Ray Studies, 91–97.
© *1985 by D. Reidel Publishing Company.*

nature, largely because of the way they were absorbed in the atmosphere. Millikan had speculated that there might be a latitude effect due to the action of the earth's magnetic field on the charged particles that would be formed by Compton collisions as the photons travelled through space. He and Cameron went by ship from Los Angeles to Peru in 1926 in an attempt to measure such an effect. Due to several difficulties they found no effect with an experimental uncertainty which Millikan estimated at 6 percent. Later it was found that there is a latitude effect of about 7 percent in the ionization at sea level between Los Angeles and Peru; so that he was not far wrong.

There were many wrong guesses as to the nature of cosmic rays and some right guesses. Some of these right guesses were for the wrong reasons. The Geiger counter, coincidence experiment of Bothe and Kolhörster in 1929, where they found charged particles penetrating 4.1 cm of gold was a good illustration of the right answer for the wrong reasons. They concluded that the primary radiation must consist of charged particles because photons would be highly absorbed by that much gold. We now know, of course, that the charged particles they were detecting were not primaries but mu-mesons which are secondaries.

The nature of the radiation incident on the earth was not really determined until the late 1940's. There were indications during the 1930's that protons were a constituent of the primaries, but it remained for the experiments with photographic emulsions, sent to high altitudes with balloons, to determine that the primaries consisted of protons, α-particles and heavier nuclei.

Fig. 1. Millikan (right) and Neher with early cosmic ray ionization chambers. Instruments similar to the one in the partially assembled lead shield were used in early airplane flights and in sea level measurements. Instrument on right of the table had 30 atm of air pressure. Constants had been accurately determined. Instrument still in good condition.
Acted as standard for many years. Photo: Oct. 1932.

Fig. 2. Manned balloon sent up from the state of South Dakota in U.S.A. in 1934. Sponsored by National Geographic Society and U.S. Air Corps. Gondola carried scientific equipment including three ionization chambers. Reached 62000 ft elevation.

The place of origin of cosmic rays is still a matter of conjecture. In the late 1920's and early 1930's, Millikan, as stated above, thought that cosmic rays were photonic in nature. He also thought that they originated in interstellar space. The chief reasons for these assumptions were as follows: (1) Using the then available theoretical expressions, the manner in which the radiation was absorbed in the atmosphere indicated that the primaries were photons with more or less specific energies and (2) the absence of a measured latitude effect indicated that these photons had not passed through an appreciable amount of matter and hence must have originated in interstellar space. The specific energies seemed to coincide with those to be expected from the fusion of protons to form the heavier nuclei. When the latitude effect was found and especially when the very large latitude effect at balloon altitudes was found, it became necessary to assign a major part of the incident radiation to charged particles. Millikan then suggested that the radiation was the result of a spontaneous annihilation of various nuclei in space. To account for the east-west effect, the large latitude effect at high altitudes and the nature of the atmospheric absorption of the radiation, Millikan, in his later years thought that the primaries consisted mainly of electrons (+ and −) with positrons predominating. The remaining radiation, not affected by the earth's magnetic field, could be photons.

These ideas became untenable when it was definitely shown that the primaries were protons and heavier nuclei. Some of the mechanisms involved, when these particles interact with the nuclei of the atmosphere, were beginning to be unraveled in the late

1930's. No one knew, in those years, where the study of the myriad of interactions would lead. Those events seen in cloud chambers that were fairly frequent such as showers, some of the more common mesons as well as some of the shorter lived particles, could be studied in cosmic rays. To study the rarer events it was necessary to control the energy and intensity of the bombarding particles. To this end huge particle accelerating machines were developed. But one should be aware that these developments had their origins in trying to understand what was taking place in Nature's big machine which still outperforms man-made machines as far as individual particle energies are concerned.

Many of the difficulties of the mid-1930's had to do with the unexpected behavior of the radiation as found in the atmosphere. To interpret the results of an experiment one could only use the ideas and concepts then known. It turned out that these were completely inadequate. One of the first breakthroughs came in the theoretical works of Bethe and Heitler. They worked out the quantum mechanical, relativistic mechanisms by which electrons are absorbed in matter. While this led to an understanding of the birth and death of much of the radiation of the atmosphere, there remained the very puzzling behavior of the penetrating part of the radiation. Working with a cloud chamber near sea level, Anderson and Neddermeyer, in 1936, showed that there are two distinct groups of singly charged particles. Using those particles whose curvature in their magnetic field could be measured, one group of particles was rapidly absorbed in a 1 cm thick piece of platinum. The other group, with the same curvatures, penetrated the piece of platinum. The first group, they identified as electrons. They decided that the second group must consist of a new kind of particle, more massive than the

Fig. 3. Making balloon flights from Madras, India in 1936. Instrument had to be recovered to obtain record.

electron, but not as heavy as a proton. This new particle became known as the mu-meson. Thus it became possible to understand the penetrating part of the radiation found in the atmosphere.

But, there were still difficulties. A study of the numbers of charged particles at various altitudes compared with the numbers found in passing through dense media indicated that an extended medium like the atmosphere had an apparently higher absorption than dense materials with approximately the same atomic weight. It was suggested that this could be understood by assuming that the mu-meson decayed in flight. In 1939 this idea was shown to be definitely correct by several experimental groups.

The decay of the mu-meson also led to the understanding of other data, in particular the different latitude effect in the ionization at sea level found in summer compared with that found in winter. Another way of describing the difficulty is that there is very little, if any, annual variation of cosmic rays at the equator, but the difference between winter and summer may amount to several percent at high latitudes, being larger in winter. The reason for this behavior, of course, is that a colder atmosphere means a less extended atmosphere and hence less chance for meson decay. This explanation was first suggested by Blackett. Due to these seasonal changes in ionization, most of the monitoring of cosmic rays at sea level and mountain altitudes, since the early 1950's, has been done with neutron detectors which are far less sensitive to atmospheric temperature changes.

Fig. 4. Millikan was one of the early workers in cosmic rays. When it was experimentally established that penetrating rays were coming to the earth from outer space, he suggested they be called cosmic rays. Photo by Talbot Waterman, 1940

Fig. 5. Carl Anderson (right) and Victor Neher with ionization chambers used in cosmic-ray research. Photo taken about 1953.

One of the properties assigned to cosmic rays in the early days was that they do not change with time, or very little. Experiments seemed to bear this out. In fact when giving the properties of cosmic rays, many authors listed their constancy in time as one of their important characteristics. By cosmic rays, most authors were thinking of the radiation found in the lower part of the atmosphere. As we learned later, this radiation is almost entirely secondary in origin. In thinking of cosmic rays, one should be concerned about the radiation in space in the vicinity of the earth. Then one considers what happens when this radiation interacts with the earth's magnetic field and its atmosphere. Although there are secular changes at and near sea level as demonstrated by Forbush, these show up in a much more pronounced way at balloon altitudes and high latitudes. Forbush showed that there is a definite correlation of these changes with solar activity, in particular with the 11 year solar cycle. This modulation is such that when the sun is most active cosmic rays are at a minimum and when the sun is least active, cosmic rays are at a maximum. This is not quite true for there is a lag of the change of cosmic rays of roughly one year behind solar activity. While the changes in ionization at mountain altitudes may be something like 10 percent during a solar cycle, the actual change in cosmic-ray particles in space near the earth may be a factor of 4 or 5.

The above is by no means a complete list of the difficulties encountered by cosmic-ray research workers during those eventful years of the 1930's. The list is meant to include some of the problems that arose in which the author had a personal interest, both from the gathering of data and its meaningful interpretation. I would guess that this period will be considered as one of the most eventful and challenging in the history of the development of physics.

EDINBURGH, CAMBRIDGE, AND BAFFIN BAY

Hugh CARMICHAEL[*]

1. Quartz Fiber Electroscope

My postgraduate work started in 1930 in Edinburgh with Professor C.G. Barkla on the scattering of X-rays. A primary goal was somehow to win acceptance by Lord Rutherford for work in the Cavendish Laboratory, Cambridge. Being interested in instruments, I spent much of my time perfecting the gold-leaf tilted electroscope of C.T.R. Wilson. Eventually I produced a very quick-acting, critically damped quartz-fiber[**]electroscope with sensitivity limited only by the Brownian motion of the fiber (Carmichael, 1934). Ionization pulses from individual alpha particles could be observed at response times of a fraction of a second and the fiber could not be damaged by being driven off-scale towards the plate. As regards the latter feature, Wilson told me later, at Cambridge, that the tilted electroscope had provided him with a portable instrument for his outdoor work. By charging the plate of the electroscope to a high voltage, the instrument could be carried safely because "the leaf remained steadily pointing towards the plate which it was just too short to touch."

E.G. Dymond, who was then working in the same Edinburgh laboratory, suggested that I should use the new electroscope to investigate the cosmic-ray ionization bursts that had been discovered by G. Hoffman and co-workers. A paper by Steinke (1932) had just appeared. We had all been fascinated by cosmic rays, of course, since 1926 when R.A. Millikan had first acknowledged the existence of the penetrating radiation and had begun to dominate the field with his series of enthusiastic papers.

Early in 1933 Professor Barkla told me that he had obtained my admission to the Cavendish Laboratory, "as a research student from October 1933 under the supervision of Lord Rutherford." I wrote to Rutherford in February 1933, suggesting the problem of ionization bursts. He replied in March indicating, characteristically, that they were already working on the ionization bursts. "During the last five months, two of our men have been trying experiments of an analogous character using a very large flat ionization chamber more than a meter across and trying to record the bursts of ionization by means of the counting method developed for α-particles. These experiments are still in progress, the main difficulty being to get rid of the sound disturbances. It is quite possible that your electrical method might have certain advantages for such a systematic investigation, particularly as it would largely avoid the sound disturbances which showed markedly in the valve method."

[*] 9 Beach Avenue, Deep River, Canada.
[**] Fused quartz or silica glass.

Y. Sekido and H. Elliot (eds.), Early History of Cosmic Ray Studies, 99–113.
© 1985 by D. Reidel Publishing Company.

2. Cavendish Laboratory

I arrived at the Cavendish Laboratory in September 1933. Before anything could be decided about a research project, Rutherford wanted a demonstration of the electroscope that had come from Edinburgh. In recognition of my preference to work under a thin roof, space was allotted beneath a skylight in the middle of a very large room called the Drawing Office* adjacent to where Chadwick about two years previously had discovered the neutron. I found a suitable vacuum pump, connected the electrometer to a small ionization chamber having a thin mica window, filled the chamber with hydrogen to produce sharper alpha particle ionization pulses and pumped the electrometer down to the critical damping pressure. W.B. Lewis provided a weak radium active deposit as a source of alpha particles. From a letter to my parents: "Rutherford has seen the alpha kicks. He was quite impressed. Rutherford examined the electroscope, first with the fiber earthed, then with the ionization chamber in action. He spotted one natural alpha particle and I was going on but he had to wait to see another. I then put up the source of α-particles near the little window and he saw lots of kicks. He made me draw the source further back to get only the long range particles and he counted them out loud as they came in. Then he said, 'What are you going to do with it now that you have it working?' I reminded him of the cosmic ray experiment and showed him a drawing of an ionization chamber. He told me to consult Cockcroft [J.D.] about it."

Very soon after my arrival, a small, thin, elderly man approached me and, in a soft hesitant voice asked, politely, what I had been doing. I said I had been developing an improved Wilson tilted electroscope. "I, uh, I am Professor Wilson," he said and offered his hand. The electroscope was on the bench and he examined it with interest. At that time C.T.R. Wilson was Jacksonian Professor at the Cavendish and he worked with research students on atmospheric electricity and the cloud chamber. One of his students (J.G.W.) had the same surname and spent much of his time dropping a cloud chamber and camera onto an air-filled rubber bag in the hope of obtaining pictures unaffected by gravity. This was actually a clever idea but it became a source of much hilarity during our skit at the annual Cavendish Dinner when, after a loud crash, the actor representing the younger Wilson said, "Professor, I forgot to inflate the rubber bag."

Very little money, less than £40 each per year, was available to research students at the Cavendish Laboratory for the purchase of equipment and supplies. Students were expected to assemble what they could from other parts of the laboratory and to construct the rest themselves with the help of the workshop. Progress was painfully slow. It was clear that the existing ionization chamber was not rigid enough because of its drumlike shape. When this was mentioned to Cockcroft, he procured a large hot-water cistern with a diameter of 60 cm and with convex ends. This allowed the flat chamber to be quietly forgotton. A reliable camera, designed to move 16 mm film at a uniform slow speed, had to be fabricated. Graphite-on-quartz resistors of more than

---

* Working in this room were J. Chadwick, W. B. Lewis, B. V. Bowden, C. E. Wynn-Williams, A. N. May, C. B. O. Mohr, W. E. Bennett, L. H. Gray, G. T. P. Tarrant, W. E. Duncanson, and later O. Klemperer, G. H. Briggs (briefly), M. Goldhaber and A. E. Leipunsky.

10^{11} ohms had to be made. My struggle with the ancient bank of small wet cells used in the laboratory to provide about 2000 volts lasted for several weeks, but most of the cells no longer retained their charge. Chadwick said that the laboratory could not afford to purchase a new bank of cells, but he did tell me where a small (1.75 liters) ionization chamber filled with argon at 80 atm had become available and while I was testing this chamber with 500 volts from selected cells, a huge ionization burst fortuitously occurred and was recorded (Fig. 1). Immediately Chadwick authorized the purchase of 12 dry batteries to provide 1500 volts.

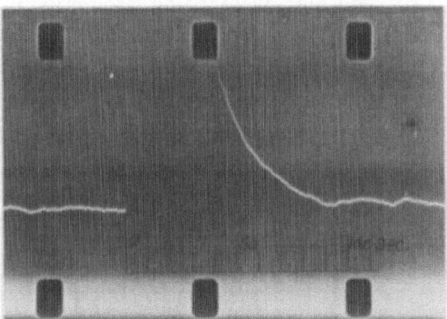

Fig. 1. Cosmic-ray ionization-burst of 2×10^8 ion pairs recorded on 16 mm film before the time-marking circuit was working.

3. Investigation of Ionization Bursts

The large ionization chamber operated satisfactorily and it was shown along with the photograph of the large burst (Fig. 1) to most of the many visitors continually being taken round the laboratory. The size of off-scale bursts was found by extrapolation of the exponential recovery of the fiber; in this case, the size of the burst was about 25 times the visible part. Actually, the fiber moved offscale only towards Wilson's safe location, "pointing towards the plate which it was just too short to touch." From a letter to my parents, dated March 12, 1935: "The apparatus continues to run well. I have almost finished with nitrogen. I have done three films of 100 feet each on it and Dr. Chadwick has suggested that I do a fourth. I have been making enlargements of some of the best bursts and pasting them into a book. Professor Blackett [P.M.S.], from London, was here last weekend and saw some of the photos. He asked me to let him have some slides. Professor Wilson is becoming a positive trouble he comes to see me so often! On Friday, probably the most famous American Professor of Physics, Professor Compton [A.H.] of Chicago was in Cambridge staying with Professor Fowler [R.H.]. Fowler came to see me on Friday morning and said he was bringing Compton in on Saturday morning. I was at a very good lecture that Compton gave on Friday night and then he came to see me on Saturday at about 10 a.m. He sat down in front of the apparatus on my tool chest and we started off. I began by describing the electroscope which I was surprised to find he had not heard about, as he himself has invented one of the most sensitive electrometers, the Compton electrometer. 'I am very glad to have my attention directed to this,' he said. Fowler then came round and Compton told him, 'We are just getting busy.' I had him for an hour and he approved of my work and gave me a number of ideas besides getting a few surprises himself. 'Say, that *is* a swell idea,' he said at one point. He is, himself, working on Cosmic Rays and knows far more about them than any of the Cavendish men so that I found him very easy to talk to and very appreciative."

4. Paper and Thesis

While some 1500 hours of recording were being accumulated under different thicknesses of lead and with argon, nitrogen or hydrogen in the ionization chamber, Rutherford became impatient and increasingly so when I had to spend time away from the laboratory and the visitors to write a paper and also a doctoral thesis. At Cambridge, the Ph.D. has to be completed within three years. However, as soon as Rutherford had read the paper, he came to the Drawing Office and said, "I think you have got that up rather well, Carmichael." Rutherford communicated the paper to the Proceedings of the Royal Society in November 1935 (Carmichael, 1936) and I received an invitation to read the paper to the Society, a highly coveted distinction. At the reception before the meeting, I mentioned to Blackett that a slide projector had not been provided. His face froze, and he hastened to find a projector which he then asked one of his students to operate.

Rutherford and C.D. Ellis were the examiners for my doctorate. In my thesis, I had mentioned the cascade process of C.G. and D.D. Montgomery (1935) along with several ideas now best forgotten. Rutherford called me into his office and said he thought it was time for me to have discussions with the theoretical people. He suggested H. Bhabha whom I knew very well. "Don't worry," he said, "if anything comes of it, we will know where it began." On May 2, 1936, I took the degree of Doctor of Philosophy and, two days later, I was elected to a Research Fellowship at St. John's College. Evidently Rutherford now wished me to continue in the Cavendish. Not long before this he had said to me, "Of course, Carmichael, you are only a bird of passage."

5. Cascade Theory

Bhabha had a strong urge to do experimental physics and had often suggested that I should help him get started. He was a remarkable man, erudite, energetic, artistic, fun-loving and a highly lucid lecturer. After Sir Lawrence Bragg became Cavendish Professor, Bhabha had me help him paint a light-hearted picture of Bragg (Fig. 2) for the Cavendish dinner, taking the likeness from a portrait on the cover of Time Magazine.

In 1935–36, it was not yet possible for anyone to understand the occurrence of ionization bursts because the nature and composition of the incident cosmic radiation and of the radiation within the atmosphere had not been established. Mesons had not been discovered. Excitement was intense. W. Heitler came to work in Cambridge. The measurements showed that the bursts developed and were absorbed in remarkably few centimeters of lead, and that they were complex examples of the shower phenomenon

Carmichael   Bhabha

Fig. 2. A light-hearted picture of Cavendish Professor W.L. Bragg, F.R.S., painted for the 1938
Cavendish Dinner by Homi Bhabha assisted by Hugh Carmichael.

involving positive and negative electrons and, presumably, gamma rays. The predictions
of quantum electrodynamics were being questioned. In his Halley Lecture in June
1936, Blackett said, "I would suggest that a partial solution may be obtained by the
assumption that the incident particles are electronic and that the probability of shower
formation varies with their energy in an anomalous manner for energies of the order 2
or 3×10^9 eV."

Some months later Heitler placed a lead block on my bench and asked me
pointedly if I thought that classical electromagnetic theory could yield a burst in the
lead starting from a single incident electron or photon. He obviously then knew that it
could. Unfortunately I replied, "No." All I can claim on the theoretical side is that I
delighted Bhabha by passing on the word 'cascade.' In 1937, Bhabha and Heitler
published their well-known cascade theory paper.

6. Arctic Expedition

In the summer of 1936, J.M. Wordie, a Fellow of St. John's College and a well-known
explorer, announced an Arctic Expedition for the summer of 1937. A similar
expedition in 1934 had been stopped by unbroken pack-ice in Baffin Bay and had
failed to reach Thule and Ellesmere Island. A small 172-ton wooden Norwegian ship,
used in winter for hunting seals in pack-ice, was to be chartered for three months
complete with its crew of twelve. Several young geologists, archaeologists, or other
scientists with projects, were needed to provide a complement of ten. Each participant
was expected to contribute £300.

In 1936 measurements of cosmic radiation in the atmosphere at high altitudes had
not been made north of geomagnetic latitude 49° in Europe or 52° in North America.
E. Regener had made numerous measurements at Stuttgart with a self-registering
ionization chamber taken aloft by free balloons sometimes reaching an altitude of

30 km. Millikan and co-workers, in 1933 had had a self-registering ionization chamber taken above 20 km in the gondola of the Explorer 1 Fordney-Settle manned balloon flight. Millikan had also, in 1936, sent an ionization chamber aloft near geomagnetic latitude 40°N. From comparison of these measurements at different latitudes, it seemed that farther north, where primary cosmic rays (then believed to be electrons) of lower energies could penetrate the earth's magnetic field, very much higher intensities might be found at altitudes in excess of 20 km. Half-way through Baffin Bay, the ship would be sufficiently close to the North Geomagnetic Pole in N.W. Greenland to make full use of an altitude of 30 km, the best then obtainable with free balloons. I therefore proposed that high altitude measurements of cosmic radiation with free balloons be attempted provided that a dependable full-time collaborator, preferably from another laboratory, could be found.

Dymond, at Edinburgh, readily agreed to collaborate and came to Cambridge for a meeting with E.V. Appleton, Wordie and myself. The proposal was accepted by Wordie and, later, by both Rutherford and Barkla. Development work began in both laboratories in October 1936. The ship was due to sail from Edinburgh in June 1937.

7. Regener

The eight months available for development and fabrication of lightweight instruments provided no time to experiment with the launching of free balloons. Dymond, who had studied in Germany, and also Blackett, wrote to Professor Regener and Regener kindly invited Dymond and myself to visit his laboratory at Stuttgart. In December 1936, Dymond and I spent 7 days with Regener and met Ehmert, Hoerlin, Rathgeber, Victor Regener and others, but Pfotzer was away. Regener spent most of his time with us during the visit, seized a good-weather opportunity within hours of our arrival to release his 51st free balloon flight, and arranged our daily programme with precision, including a visit to Freidrichshafen and a trip on his boat on Lake Constance to inspect the underwater ionization chamber. It must be emphasized that without the detailed results of years of experience so patiently communicated to us by Regener during this visit, we could not have carried out our proposed measurements.

Regener was, of course, well known to Rutherford. He was the first, in 1908, to devise methods of counting alpha particle scintillations and, in 1909, he had found 4.79×10^{-10} esu for the charge on the electron which Rutherford and Geiger in 1908 had given as 4.65×10^{-10} esu. From my manuscript of a Cavendish colloquium in February 1937, presided over by Rutherford, we have: "You remember that the wavelength of X-rays has recently been determined very exactly by Compton and others by diffraction from gratings ruled on glass. Thus it is possible, using the known angles of diffraction of the X-rays from crystals, to evaluate 'e'. The value obtained is about 4.800×10^{-10} esu which is not in agreement with the hitherto accepted value of 'e' obtained by Millikan, 4.774×10^{-10} esu. The Millikan experiment has been repeated by Kelletrop (Uppsala) and it seems that 4.800×10^{-10} esu is the better value—Regener is presently having three of his men repeat his original measurement with all possible modern refinements." I outlined the refinements, but Rutherford in the discussion said that he remained doubtful about the alpha particle method.

8. Rubber Balloons

From the same manuscript: "We were fortunate in that the weather on the first day we were at Stuttgart was suitable for a flight—the hydrogen cylinders for filling the balloons were already in the courtyard of the Institute when we arrived. This slide shows the scene a little later. A single balloon is never used—always two, three, or even four. One of the balloons bursts at the top and the unburst balloon or balloons control the speed of the descent. The balloon in the foreground is being inflated until it can just lift the weights in the scale pan. The other balloon is having the eight cords that connect the balloons attached to the eight equatorial lugs. The hydrogen cylinder can be seen, the paper spread on the ground to protect the rubber, and Professor Regener, back view. Inside the Institute a student is nervously winding up the contact clock and putting the apparatus into a cellophane case which acts like a green-house to keep the apparatus warm. The next slide shows the launch. You can see the upper balloon and the eight cords to the lower balloon which is beging controlled by a launching cord. The launching cord is slipped by releasing one end as the balloons leave. Then there are eight cords to the 'bremse' or brake, a small parachute which serves to steady the violent jerking and swinging of the apparatus during the ascent. Finally, far below the bremse, comes the apparatus in its cellophane case."

The overall height of the equipment we saw was 85 m and the weight about 25 kg. Clearly such an outfit could not be launched from a small ship; launching would have to be from the shore or from pack-ice.

The balloons were followed by telescope to a distance of 120 km. They suddenly stopped ascending when 12½ km above the ground owing to the development of a small hole in one of the balloons.

9. Plastic Balloons

Again from the same manuscript: "Even with the care which he takes in carefully examining the balloons for pin-holes and weak places before a flight, all Regener's flights do not reach great heights. Most flights end at 20 to 25 km and a flight which gets quickly past this height usually goes on to 30 km. Regener thinks that this is the effect of the ozone layer in the atmosphere. Tests in the laboratory (which we saw) show that the rubber used in balloons quickly bursts in the presence of ozone. Even 31 km is 'not the limit set by the stretching of the rubber but it seems to be the practical limit of such flights. This slide shows the results of measurements by Regener of the amount of ozone per kilometer in the atmosphere. There is a maximum density of ozone at 25 km.

"There is another possibility of reaching great heights. We must use a substance not affected by ozone or strong sunlight. For about three years Regener has been experimenting with cellophane for balloons. These cellophane balloons must, of course, be full size at ground level and be filled with less than 1/100th of their volume of hydrogen.

"Here is a cellophane balloon at ground level (Fig. 3). The cellophane sheets are cemented together and the joins strengthened with tape. Regener had no success with

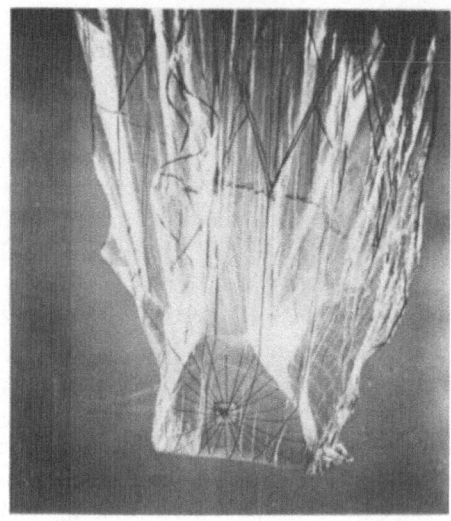

$120 \, m^3 \; 15. \, \overline{VIII}. 36$ Regener

Fig. 3. Photo of a large plastic balloon, dated and signed by Professor Regener who first made and used such balloons.

this work until he employed two women dressmakers to make the balloons. The partially filled cellophane balloon takes all kinds of queer shapes and, one day, some girls who were passing along the street, as the balloon was being filled, exclaimed, 'Oh! Look at the white ghost!' Regener hopes to reach 40 or even 50 km with cellophane balloons. The first task is to measure the temperature there.

"Cellophane balloons have one peculiarity; they do not like to come down. One of Regener's flights went to Italy and when, after much trouble, the apparatus was recovered he found that the temperature and pressure record had been censored! To avoid trouble of this kind, he now uses a device which operates at a preset time. A razor blade splits a stretched rubber membrane and opens a hole at the top of the balloon. A balloon is masculine in German and when this happens 'he' is said to commit 'Hara-kiri'."

10. Counter Telescope

Also from the same manuscript: "Last year Pfotzer succeeded in measuring, with triple coincidence counters, the vertical rays up to 29 km as shown on the next slide. This experiment is probably the finest cosmic ray experiment of the year. He obtained the remarkable result that there is a pronounced maximum in the number of vertical rays at a pressure of about 8 cm Hg. This maximum must be due to the production of large numbers of secondary cosmic rays in the atmosphere. A slight hump is also observable at 30 cm Hg.

"In this experiment, the coincidences were amplified by a 3-stage amplifier and recorded by a mechanical counter, the dial of which was photographed at regular intervals. The pressure and temperature were photographed at the same time. Very light dry batteries weighing only 1½ gm per volt were used to supply the counter potential of over 1000 volts. The whole apparatus weighed 6 kg.

11. Progress to Date

Finally the same manuscript describes the state of our preparations on 24 February 1937: "We propose to carry out experiments both with coincidence counters and with ionization chambers. The counter experiments are being conducted by Dymond and here you see some counters which have been made by him. They are counters of the new alcohol vapour type developed in Regener's laboratory by Trost. They contain argon at a pressure of about 90 mm Hg. This mixture makes a self-acting counter, that is to say, one which stops its own discharge and so can be used with a low external resistance giving kicks of short duration. Dymond does not propose to send up an amplifier and mechanical counter like Regener and Pfotzer did but, instead, he hopes to be able to transmit the coincidences and also the atmospheric pressure by radio and pick up the signals with a receiver on the ship. This will allow a lighter balloon apparatus and, also, the balloon apparatus does not need to be recovered after a flight. Here is a model of the radio transmitter which has been developed in the Cavendish Laboratory by J.A. Ratcliffe.

"The weight of the coincidence counter radio transmitting apparatus is chiefly that of the batteries required for the counter potential and for the radio valves. Dymond and Ratcliffe are going to use the new accumulators of Väisälä which you see here. They are very small and weigh only about 1½ gm per volt after the acid has been put in.

"I, myself, am making ionization-chamber apparatus of the type which registers photographically and so must be recovered after a flight. We hope that it will be possible by working in a wide expanse of open sea to have the apparatus come down on to the sea where it can be picked up with the ship.

"I am imitating the apparatus of Millikan and Neher more than that of Regener. Here is a steel sphere which has been made by a metal spinner in London. The wall is less than 0.5 mm thick and yet this sphere has withstood an internal hydraulic pressure of more than 30 atmospheres. It will thus be quite safe to use it at 5 atmospheres pressure. Inside the sphere there will be a collecting electrode connected through an amber insulator in the wall of the sphere to an electrometer of the Neher type.

"The next slide shows the Neher electroscope. It is made of quartz and all the joints are fused so that what you see is essentially one piece of fused quartz. You will observe, from the centimeter scale shown below, it is very small. It is rather difficult to make but I have been able to construct a crude first model and I hope to be able to make a large number before the Expedition leaves in June. I am planning to make the electrometer recharge itself automatically after the manner of the ticking electroscope of Zeleny. Thus, no matter how intense the cosmic radiation near the geomagnetic pole, the apparatus will be able to record it. The record will be made on a strip of film or paper moved by clockwork and the pressure, and the temperature inside the cellophane case, will be recorded at the same time."

When I had finished speaking, Rutherford rose to thank me. He was pleased to have heard about his friend Regener and had enjoyed the stories from the Stuttgart Laboratory. He had been rather worried because, for two months, he had seldom seen me in the laboratory. Then he had found me upstairs in a small room with a three-dimensional array of mechanical manipulators (based upon the precision optical bench developed in the Cavendish by G.F.C. Searle) trying to make the Neher electroscope. He commented upon the possible importance of the self-quenching counter which they were hearing about for the first time. He then moved over to examine the exhibits. He delicately picked up the small accumulator but hastily put it down again and mopped his fingers with his handkerchief. "You must always be careful if you handle an infant!" he said. (Loud laughter).

Shortly after this, having heard that Barkla had made additional laboratory funds available to Dymond, Rutherford paid £100 into my personal bank account to be used for the purchase of the numerous smaller items of equipment that were needed, with any balance to be refunded.

12. Millikan

Late in May, Millikan visited the Cavendish and gave a lecture illustrated by Anderson and Neddermeyer's cloud-chamber photographs. There was also a striking picture of the launch of H.V. Neher's ionization chamber using a long string of rather small balloons. We had already tested some of those seamless American balloons and intended to use the 1 m (unstretched) size as a pilot balloon to probe the atmosphere before releasing apparatus. Millikan, a kind man but also an opportunist, after he had seen the state of my preparations, decided that I must take one of his survey electroscopes so that, as Fowler said, I would "get something out of the expedition." The heavy metal castings used to shield the ionization chamber from ambient gamma radiation appeared to present a problem. After discussion, it was decided that in a fixed position in a wooden ship a gamma ray shield would not be needed. Millikan wrote to Neher, and he, in a letter dated 3 June 1937, told me that one of their cosmic ray instruments had been shipped to Edinburgh from Los Angeles. On 14th June, Millikan telephoned from London and on 15th June he sent a hand-written letter, "I am putting into writing all that I told you yesterday,—(5) Since no shield accompanies the instrument see that it is not placed on the deck close to radiolite dials,—(7) If convenient it would be useful to send it back running via Panama." The instrument was transhipped by passenger train from Southampton to Edinburgh on 24—25 June in time for our departure on 26 June. Unfortunately we were unable to control the large number of gamma ray emitters on that ship, wrist watches, prismatic compasses and the sources used for testing our equipment, and so all we got from Millikan's apparatus was a good continuous recording of the barometric pressure throughout the voyage!

Nearly four months after our return, a very smoothly worded letter from Millikan dated 17 January 1938 came, wondering if there was any likelihood of our shipping the electroscope back in the near future, running, on some boat that was going to Los Angeles, "so that we could get another line on just where the dip begins to set in on that side of the Atlantic." Fortunately, I had written on 10 January to say that the electroscope was already on a ship to Los Angeles "in working order as you wished" with "all charges forward in your name." On 15 February, Millikan wrote that the film spool had jammed and they got no record: "It was wholly our fault that the spools were not identical."

13. Final Preparations

Happily during the last two months before departure, the complex nature of the project became apparent, and a marvelous spirit of excitement and willingness to help pervaded both the Edinburgh and Cambridge laboratories. All the good laboratory technicians were made available and often worked overtime. Eager students offered help and were asked to construct gondolas, cover them with cellophane, inflate the apparatus-carrying balloons with air and examine them for pin holes, make the parachutes, prepare sets of eight cords for each flight with 'droppers,' 6 inches long, at measured places, and so on. Everything, then, had to be packed in a waterproof manner in wooden boxes with contents listed on the outside.

At Cambridge, 10 sets of the ionization-chamber apparatus were completed and, in Edinburgh, 5 sets of the counter apparatus. The technicians finally began to cheer every time I brought them another ticking Neher electroscope from my manipulator bench. With only a month to go, a satisfactory lightweight source of about 200 volts for recharging the electroscope had not been found. I did not want to add a box of accumulators weighing 400 g to an instrument weighing only 1100 g. We were trying to find out about a condenser used by Neher and Hayes. Then C.T.R. Wilson solved the problem by saying that, in 1906, he had blown very thin bulbs of fused quartz and silvered them inside and outside getting often 10000 cm capacity, excellent retention of charge, and a weight of less than 20 g. A chemist friend finished silvering 14 quartz-bulb condensers twelve days before we sailed.

From a letter to my parents: "I was invited to a dance—somebody's engagement dance—and I duly accepted and noted the fact in my diary. I was also invited to a house to dinner a few evenings later. This I also accepted and, do you know, I forgot all about both!

14. Baffin Bay

The 'Isbjörn' was small and uncomfortable. The ten of us could squeeze round the table in the aft cabin for meals when the weather was fine but there were table frames for only eight when the fiddle was in place for rough weather. The ninth square in the center of the table usually held a huge brown teapot which could be flipped completely upside down by a big wave. When this happened, there was no escape for those at the corners but fortunately our clothing was heavy. The diesel engine was used only when the ship could not make 6 knots in the proper direction under sail. The ship was hove-to when headwinds became too strong. We took 18 days to reach Godhavn from Edinburgh. Only one of the Norwegians could speak English. One of our party was continually sea-sick.

The Wordie Expedition of 1937 (Wordie, 1938) and the results of the High Altitude Cosmic Radiation Measurements (Carmichael and Dymond, 1939) have been fully reported. We attempted eight flights and two of them were successful, one with the counter apparatus and one with the ionization chamber. Two weeks after the successful ionization chamber flight at 85°N, Bowen, Millikan and Neher, unknown to us, made a similar flight at Saskatoon at geomagnetic latitude 60°N. This provided for comparison which added to the significance of our work.

It may be of interest to quote Wordie's description of the successful ionization chamber flight: "We were abreast of Ryder Island [off the coast of Greenland, near 74° N geographic] in roughish weather late on August 2—The next three days are an interesting record of rapid weather changes and the ups and downs of fortune. We had anchored early on August 3 to take shelter from a cold southerly wind but by the forenoon the weather was mild and encouraging, the sky cleared and the sea began to go down. Meantime various members had been left ashore, some on South Ryder, others on the North-West Island, at places where archaeological investigations had been made in 1934. By the afternoon, however, it became evident that the weather conditions had steadied quite unexpectedly and were, by then, exceptionally good; the scattered parties were hurriedly recalled and preparations made for a flight with Carmichael's apparatus. It was already late in the day but Carmichael's apparatus did not require the same warm conditions as Dymond's. A pilot balloon was flown about 3 p.m. and in its one-and-a-half hour ascent to a height of about twenty miles it drifted very slightly to the NNW. A rocky ledge only a little above high tide level on the west side of North-West Ryder Island was chosen as the launching ground and hydrogen piped to the balloons from a rowing boat, the swell being too heavy to handle the cylinders ashore. We worked as quickly as we could, for everyone knew his particular tasks by now, but it was not until about 8:30 p.m. that the balloons were finally released. The flight that night was most impressive as the big yellow balloons rose into the clear sky; the air was calm but a restless sea was still breaking around the launching ledge and against some great stranded icebergs. The balloons reached their maximum height after about two hours, when the upper one burst, and then, in a somewhat shorter space of time, the remaining balloon and the attached apparatus descended with a slight eddying motion and were recovered shortly after midnight less than two miles from the place of launching. This was a completely successful flight to a height of nearly 18 miles (28 km) and gave Carmichael a record of practically all he wanted to find—It seemed that our luck had turned and we planned that a Dymond flight should take place next day but, on the 4th, the pilot balloon was almost immediately carried north-eastward towards the ice cap only 15 miles away. The flight was cancelled but we had at least the satisfaction of knowing that we had not failed to take quick advantage of the odd chance of still weather on the previous day."

After the ionization chamber had been picked up from the sea and I had noted that the clock mechanism had turned the film pulley the required 3½ revolutions to the stop position, I wanted to develop the film but Wordie persuaded me to go to my bunk. I could only hope that the midnight sun had provided enough light to expose the film. Later that day, in the cook's pantry, with Dymond standing guard at the door, I developed the film. The delay between pouring on the developer and the first appearance of the image under the safelight was agonizing. Then, slowly, I could see that the electroscope had discharged many times (14) and that we had a good clear recording.

On 8 August a successful flight with the counter apparatus was made from Wolstenholme Island, near Thule, at 88°N (Fig. 4). The balloons were under simultaneous observations by theodolites at either end of a base line, so that positions and heights were determined during their flight. The height reached was 20 km, or 5.2 cm Hg. They were observed to fall about 10 miles inland and we recovered this apparatus. The variation of intensity with pressure was similar to that found by Pfotzer at 49°N geomagnetic latitude including the maximum near 8 cm Hg.

Fig. 4. Preparation of Dymond's successful August 8 flight with counter telescope, showing the two balloons, the rowboat with hydrogen cylinders and, in the back-ground, the Isbjörn at anchor.

15. More Millikan

Four months after a preliminary report of our results appeared in the 21 May, 1938, issue of *Nature*, Millikan wrote, not realising that we had not yet heard about his nearly simultaneous work at Saskatoon published in June 1938. I replied on 16 October giving him our numerical values. He wrote again on 24 November, "When you wrote your *Nature* article of course you had nothing but our Fordney-Settle flight data and Regener's curve to compare with—if you raise all your actually observed values by 16½%, as you say you have reason for doing, then your directly observed values at 85° hit our mean results at Saskatoon—right on the nose—your results are actually closer to our mean curve than the fluctuations which we find." I replied on 23 January 1939, "What I meant to say was, merely, that if our results were *arbitrarily* increased by 16% then good agreement with your flights would be found." On 9 February 1939 Millikan wrote, "The absolute values—all go back to my observations in which I obtained the ionization at sea level in an electroscope at 30 atmospheres—in comparison with 1 atmosphere—page 244, *Physical Review*, **37**, 1931. If I knew just the steps you take in your reduction—I might see possibilities of reconciliation." On 25th February I described how our absolute calibration had been made and then wrote, "It would be good if our absolute values could be brought a little nearer together especially as our curves are so similar in shape. I think that it is now up to you to make a direct recalibration of your present apparatus!" There was no reply to this letter.

At the Echo Lake Symposium in Colorado in 1949, I met Millikan again. I was invited by J.A. Wheeler at short notice to present an impromptu paper. Having nothing new, I reported the 1937 Arctic results and then said, "We have not been able to resolve this discrepancy—in absolute value—I would like to suggest therefore, with due respect and deference to Dr. Millikan, who is here in front, that perhaps the original calibration of the Millikan-Neher apparatus might be checked to see if this discrepancy, which affects the value assigned to the total energy of cosmic radiation, can be resolved." When I returned to my seat beside Millikan, he said, "Be sure you get all that written up for publication."

It was exciting for me to meet Dr. Neher for the first time. He spoke very nicely about our Arctic exploit and how surprised he had been that it had succeeded. He agreed that a recalibration on their part was indicated. However, at this time, Neher was still working under the shadow of Millikan. In their joint papers, the half-column usually allowed to Neher for reporting his measurements contains in fact all the information in these papers that is now of value.

16. Twenty Years of Waiting

Our 1937 curve of ionization as a function of depth in the atmosphere, obtained at 85°N geomagnetic latitude, had a well-defined maximum between the pressures 20 and 40 g cm^{-2}. The presence of such a maximum indicates the absence near the orbit of earth of a high flux of cosmic rays of low energy. A similar maximum near the geomagnetic pole was not seen again for 20 years—a long time to wait for confirmation of a measurement. It is known now that the maximum is present for only three to five years, around a time of sunspot maximum as marked on Fig. 5. The 1947 maximum must have lasted almost to Neher's first measurement at Thule, Greenland, in 1951. Then, in 1957, during his remarkable unbroken sequence of July-August measurements at Thule from 1954 to 1969, the ionization maximum re-appeared.

17. Last Word

Figure 5, developed during the writing of this article, provides an estimate of the ionization at the 20 g cm^{-2} depth at high latitudes for the years 1930 to 1953. The estimated curve is seen to pass close to our value measured in 1937. However, in constructing Fig. 5, both Neher's and our own published rates were divided by the correction factors given in MOD-67, at the 12th International Conference on Cosmic Rays, Hobart, 1971. These factors are 1.17 for Neher's data and 1.11 for ours; the latter due to an improper conversion from ionization in argon to ionization in air.

Figure 5 appears to confirm the essential correctness of our 1937 measurements. As discussed in MOD-67, Millikan's Saskatoon result was higher than our Baffin Bay result by a factor 1.19 (rather than the 16½% quoted earlier), ours being a factor of 1.11 too high as just noted. Therefore it seems probable that the original absolute determination of Millikan and Cameron (1931), briefly mentioned in Section 15, was

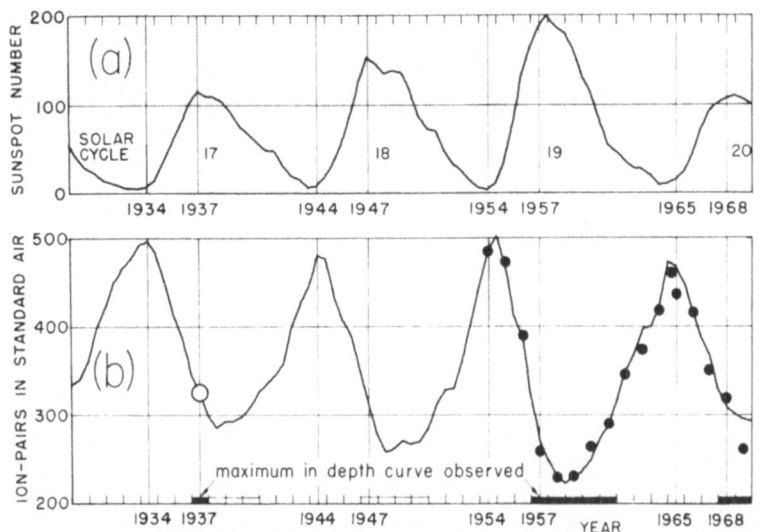

Fig. 5. Curve (a) was drawn through the Zurich smoothed sunspot numbers for June and December each year from 1930 till 1969. The solid circles are the measurements of Neher (*J. Geophys. Res.*, **76**, 1637 (1971), Table 4) of cosmic ray ionization at atmospheric pressure 20 g cm^{-2}, converted to standard air and divided by the correction factor mentioned in the text. A regression line, not reproduced here, was established between 16 of these values and appropriate smoothed sunspot numbers. The 1969 value and also the 1951 observation referred to above (section 16) could not be reconciled and were omitted from the regression on the ground that even and odd numbered solar cycles may behave differently. For sunspot numbers larger than 30 it was found best to use those of June of the previous year: for smaller, those of the previous December. The regression was a curved line because the ionization rate changed more rapidly as the sun became quieter. The expected cosmic ray ionization was read off the regression line from 1930 to 1969 to form curve (b). The open circle in 1937 is the ionization published by Carmichael and Dymond (1939) after division by the correction factor mentioned in the text.

too high by the factor $1.19 \times 1.11 = 1.32$. This factor seems now to have been confirmed (Kyker and Liboff, 1978).

I am very grateful to the friends and relatives who helped me with this article and to all those who have enabled me to live such a satisfying life. As regards Millikan, I am anxious only to get the record scientifically correct. It is regretable that such a large error appears to have occurred. It is certainly not an error of measurement; there must be an undisclosed factor in his method.

REFERENCES

Carmichael, H.: 1934, *Proc. Phys. Soc.* **46**, 169; 1936, *Proc. Roy. Soc. A* **154**, 223.
Carmichael, H. and Dymond, E. G.: 1939, *Proc. Roy. Soc. A* **171**, 321.
Kyker, G.C. and Liboff, A.R.: 1978, *J. Geophys. Res.* **83**, 5539.
Millikan, R. A. and Cameron, G. H.: 1931, *Phys. Rev.* **37**, 235.
Montgomery, C. G. and D. D.: 1935, *Phys. Rev.* **47**, 429.
Steinke, E. G.: 1932, *Z. Physik,* **75**, 115.
Wordie, J. M.: 1938, *Geogr. J.* **92**, 385.

PART 3   POSITRON AND MESOTRON

EARLY DISCOVERIES OF NEW PARTICLES

UNRAVELING THE PARTICLE CONTENT OF THE COSMIC RAYS

Carl D. ANDERSON*

1. Introduction

I understand my task is to report on the part played by cosmic ray research in providing new information on the particles of physics, until such work was gradually driven into oblivion by the advent of the new accelerators. Because of my health, I have not had access to a library except under very trying and difficult circumstances, and hence this paper was prepared at my home, based wholly on a limited amount of material and my own recollections. This has inevitably tended to make for omissions in the paper, to make it less objective and more personal than it otherwise might or should have been. I regret the omission of the contributions provided by the extensive programs of research by Geiger-counters, ionization chambers, and the extremely important results from the photographic emulsion techniques originally devised by Cecil Powell, and later continued by many other investigators. In particular, I regret the omission of the innovative and important series of experiments carried out by Bruno Rossi beginning about 1930 based mainly on the adaptation of Geiger counter techniques to a study of the complex character of the cosmic radiation. His results as well as those of other investigators have had a strong influence on my thinking, and were very helpful in interpreting some of our own results.

The name of Dr. Robert A. Millikan appears early in this paper. Later on, I describe some differences of opinion with respect to the interpretation of certain cosmic-ray data which occurred between Dr. Millikan and me, and later on between Dr. Millikan and Seth Neddermeyer, who was my first graduate student and afterwards my close collaborator for several years. I wish to emphasize that those differences with Dr. Millikan were very minor, and in no way have affected the respect and admiration that I have always had for this great man to whom I owe so much.

2. Background

At about the end of 1929, when it became clear to me that I was likely to receive my Ph. D. degree at Caltech in June 1930, I made an appointment to see Dr. Millikan. The purpose of my visit was to see if it were at all possible for me to spend one more year at Caltech as a postdoctoral research fellow. My reasons for doing so were two-fold: to

* California Institute of Technology (emeritus), Pasadena, USA.

<Ed.> This article was originally prepared for the International Symposium on the History of Elementary Particles held at Fermilab, Batavia, Illinois, USA in May 1980. Proceedings of the Symposium on the History of Elementary Particles, edited by L. Brown and L. Hoddeson, will be published by Cambridge University Press. We are grateful to the editors of the Proceedings and to Professor Anderson for their kind permission to include this article also as a contribution to the present book.

Beside this article, three photographs were contributed by Professor Anderson to the present book. We are grateful also to Professor R. S. White who helped to locate the photographs.

Y. Sekido and H. Elliot (eds.), Early History of Cosmic Ray Studies, 117–132.
© *1985 by D. Reidel Publishing Company.*

carry out an experiment I had in mind, and to learn something about quantum mechanics.

After a brief discussion with Dr. Millikan, in which I described the experiment and my desire to study quantum mechanics, he informed me that this would not be possible. The gist of his remarks was that, having had both my undergraduate and graduate training at Caltech, I was very provincial, and should plan to continue my work at some other institution under a National Research Council Fellowship, about the only fellowship available at this time for postdoctoral studies. Thus, I had no choice but to apply for the Fellowship, and I wrote to Arthur H. Compton at the University of Chicago. I received a very cordial reply and began planning for my sojourn at Chicago, an idea that appealed to me more and more as time went on.

My thesis work as a graduate student consisted of studying, by means of a Wilson cloud chamber, the space-distribution in various gases of photoelectrons produced by X-rays. At the time (1926 to 1929) that I was doing this work, Dr. Chung-yao Chao, working in a room close to mine, was using an electroscope to measure the absorption and scattering of γ-rays from ThC". His findings interested me greatly. At that time, it was generally believed that the absorption of "high energy" γ-rays (2.6 MeV from ThC") was almost wholly by Compton collisions as governed by the Klein-Nishina formula. Dr. Chao's results showed clearly that both the absorption and scattering were substantially greater than that calculated by the Klein-Nishina formula. A detailed explanation of these anomalous effects was not possible from his experiments because of the lack of detailed information that one could obtain from an electroscope. My proposed experiment was to study the interaction of ThC" γ-rays with matter using a cloud chamber operated in a magnetic field to study the secondary electrons produced in a thin lead plate inserted in the cloud chamber, to measure their energy distribution, and to see what further light could be thrown on Chao's results. While still a graduate student, I obtained the ThC" source used by Chao and attempted to photograph the secondary electron tracks produced by the ThC" γ-rays in the same cloud chamber I had previously used for the X-ray studies. I found the tracks produced by the ThC" difficult to photograph; the tracks were very thin on account of the much higher speed of the ThC" electrons as compared with those from the X-rays (80 keV), which ionized heavily and therefore produced a heavy, easily-photographable track.

After trying various things, I happened one day to pick up a bottle containing ethyl alcohol and poured some of it into the water normally used as a condensing liquid in the cloud chamber. This immediately produced tracks of much higher visibility which were very easy to photograph. To my knowledge, it was the first time a mixture of alcohol and water, rather than pure water, was used in a cloud chamber. With this problem solved, I was ready to design and build the equipment for the ThC" studies, consisting of a cloud chamber operated in a magnetic field.

It is my firm conviction that had this experiement been carried out, the positive electron would have been discovered at an earlier date than its actual discovery, for about ten per cent of the electrons emerging from the lead plate would have had a positive charge. It is, of course now well-known that Chao's excess absorption was caused by pair-production, and the excess "scattering" by the γ-rays produced from positive-negative electron annihilation.

3. Cosmic Rays

One day I received a call from Dr. Millikan asking me to see him in his office. The gist of his comments on this occasion was that he wanted me to spend one more year at Caltech and to build an instrument to measure the energies of the electrons present in the cosmic radiation (more about this instrument later). By this time, Chicago was clearly my first choice, and I used all the arguments that he had previously made as to why I should not stay at Caltech. He replied that all these arguments were very valid and cogent, but that my chances of receiving an NRC Fellowship would be much greater after one more year at Caltech. He was a member of the NRC Fellowship selection committee at the time.

Again, I seemed to have no choice in the matter, so without further ado, I began work on the design of the instrument he had proposed for the cosmic ray studies. It was to consist of a cloud chamber operated in a magnetic field. This equipment, however, would require a very powerful magnetic field, for the cosmic ray electrons were expected to have energies in the range of at least several hundred million eV, rather than about one million eV for the ThC'' experiment.

4. The Magnet Cloud Chamber

The existence at that time of the Guggenheim Aeronautical Laboratory on the Caltech campus helped to dictate the design of the new magnet cloud chamber instrument, for the laboratory was equipped with a 450 kW DC generator used for supplying power to a wind tunnel. Under overload conditions the generator was capable of safely delivering 600 kW for extended periods of time.

The magnet, as actually designed and built, was essentially a pair of aircore solenoids capable of operating at 600 kW. Cold-rolled steel was used to form a frame to support the solenoid coils rigidly enough to withstand the rather large forces expected. Only the pole pieces were made of high quality permeable iron, and one of these contained a large square hole in order to permit the cloud chamber at the center to be photographed. When operated at full power the magnetic field through the whole of the cloud chamber was slightly over 25000 gauss. The solenoid coils were wound with copper tubing and tap water was used for cooling, making possible continuous operation at full power.

Funding was in short supply and was the underlying factor that determined how one built scientific equipment in those days. For example, to build a magnet of conventional design would have been completely out of the question. < see p. 132 >

5. First Results

The first results from the magnet cloud chamber were dramatic and completely unexpected. There was an approximately equal number of particles of positive and negative charge, in sharp contrast to the Compton electrons expected from simply the absorption of high energy photons. Dr. Millikan was on a visit to England at the time the first results were obtained, and I sent him a group of eleven photographs. The accompanying letter describing the photographs revealed my own excitement − the concluding sentences of the letter were:

"A hundred questions concerning the details of the effects immediately come to mind. Such questions as the loss of energy by high energy electrons, loss of energy by high energy protons, presence or absence of heavier nuclei of high energy, energy distribution among the particles in the case of double or triple ejections, momentum relations, etc. It promises to be a fruitful field, and no doubt much information of a very fundamental character will come out of it."

It was, of course, important to provide unambiguous identification of these unexpected particles of positive charge, and this could best be done by gathering whatever information was possible on the mass of the particles, inasmuch as the photographs clearly showed that in all cases these particles carried a single unit of electric charge. Experimental conditions were such that no information as to the particle's mass could be ascertained except in those cases in which the particle's velocity was appreciably smaller than the velocity of light, and this was true for only a small fraction of the events. Only a few of the low-velocity particles were clearly identified as protons.

One of the first tasks undertaken with the first photographs, in fact the original purpose of the exeriment, was to determine an energy distribution of the particles by means of the curvature they showed in traversing the powerful magnetic field. My original measurements showed an energy distribution extending from very low energies (~100 Mev) up to above 1 Bev with the great majority of particles having energies in the range of several hundred Mev.

At about this time, Neddermeyer joined me as my first graduate student, and I assigned him the task of continuing the curvature measurements, paying particular attention to obtaining as precise measurements as possible for those of highest energy, i.e., in the range above 1 Bev. As we will see later, the results of these energy measurements were to lead to some very interesting discussions with Dr. Millikan.

6. Positive Electron

As more data were accumulated, however, a situation began to develop which had its awkward aspects in that practically all of the low-velocity cases were particles whose mass seemed to be too small to permit their interpretation as protons. The alternative interpretations in these cases were that these particles were either electrons (of negative charge) moving upward or some unknown lightweight particles of positive charge moving downward. In the spirit of scientific conservatism we tended at first toward the former interpretation, i.e. that these particles were upward-moving negative electrons. This led to frequent, and at times somewhat heated, discussions between Professor Millikan and myself, in which he repeatedly pointed out that everyone knows that cosmic ray particles travel downward, and not upward, except in extremely rare instances, and that, therefore, these particles must be downward-moving protons. This point of view was very difficult to accept, however, since in nearly all cases the specific ionization of these particles was too low for particles of proton mass.

To resolve this apparent paradox a lead plate was inserted across the center of the chamber in order to ascertain the direction in which these low-velocity particles were traveling, and to distinguish between upward-moving negatives and downward-moving positives. It was not long after the insertion of the plate that a fine example was obtained in which a low-energy lightweight particle of positive charge was observed to traverse the plate, entering the chamber from below and moving upward through the lead plate. Ionization and curvature measurements clearly showed this particle to have a mass much smaller than that of a proton and, indeed, a mass entirely consistent with an electron mass. Curiously enough, despite the strong admonitions of Dr. Millikan that upward-moving cosmic ray particles were rare, this indeed was an example of one of those very rare upward-moving cosmic ray particles.

Soon additional cases of light-weight positive particles traversing the plate were observed, and in addition, events in which several particles were simultaneously emitted from a common source were observed. Clearly in both types of cases the direction of motion was known, and it was therefore possible to identify the presence of several more light-weight positive particles whose mass was consistent with that of an electron but not with that of a proton.

After the existence of positrons was clearly indicated, the question naturally arose as to how they came into being. Just what was the mechanism responsible for their production?

It has often been stated in the literature that the discovery of the positron was a consequence of its theoretical prediction by Paul A.M. Dirac, but this is not true. The discovery of the positron was wholly accidental. Despite the fact that Dirac's relativistic theory of the electron was an excellent theory of the positron, and despite the fact that the existence of this theory was well known to nearly all physicists, including myself, it played no part whatsoever in the discovery of the positron.

It was not immediately obvious to me, however, as to just what the detailed mechanism was in the production of positrons. Did they somehow acquire their positive charge from the nucleus? Could they be ejected from the nucleus when there were presumably no positrons present in the nucleus? The idea that they were created out of the radiation itself did not occur to me at that time, and it was not until several months later when Patrick M.S. Blackett and Giuseppe P.S. Occhialini suggested the pair-creation hypothesis that this seemed the obvious answer to the production of positrons in the cosmic radiation. Blackett and Occhialini suggested the pair-production hypothesis in their paper published in the spring of 1933, in which they reported their beautiful experiments on cosmic rays using the first cloud chamber which was controlled by Geiger counters.

Soon after that, experiments in which gamma rays were used showed that a pair of electrons, one positive and one negative, could be created in the coulomb field of a nucleus in such a way that the energy required to create the mass of the pair, $2\ mc^2$, and their kinetic energies as well, was supplied by the incident radiation, thus giving quantitative support to the pair-creation hypothesis.

The positron thus represents the first example of a particle consisting of antimatter. It is now generally believed that all particles have their corresponding antiparticles, and, in fact, several have been identified. < see p. 131 >

7. Comment

If one goes back a few years, say to just after the Dirac theory was announced, it is interesting then to speculate on what a sagacious person working in this field might have done. Had he been working in any well equipped laboratory, and had he taken the Dirac theory at face value, he could have discovered the positron in a single afternoon. The reason for this is that the Dirac theory could have provided an excellent guide as to just how to proceed to form positron-electron pairs out of a beam of gamma-ray photons. History did not proceed in such a direct and efficient manner, probably because the Dirac theory, in spite of its successes, carried with it so many novel and seemingly unphysical ideas, such as negative mass, negative energy, infinite charge density, etc. Its highly esoteric character was apparently not in turne with most of the scientific thinking of that day. Furthermore, positive electrons apparently were not needed to explain any other observations. Clearly the proton was the fundamental unit of positive charge, and the electron the corresponding unit of negative charge. This kind of thinking prevented most experimenters from accepting the Dirac theory whole-heartedly and relating it to the real physical world until after the existence of the positron was established on an experimental basis, although the Dirac theory has since proved to be a great milestone in early twentieth century physics.

The discovery of the positron is also an example of a situation which is so often present in physics, in which the same discovery is made, or could easily have been made, in experiments simultaneously underway but carried out for quite different purposes. One such example is the famous experiment of Walter Bothe and H. Becker in which a light nucleus such as Be was bombarded by α-particles from a radioactive source. This experiment was first performed in 1930 by Bothe and Becker and later repeated by a number of investigators. As was shown later, this single simple experiment produced neutrons, positrons, and induced radioactivity.

8. Energy Measurements

Let us return to the matter of the energy measurements of the cosmic ray particles. I think it should be said at this point that Millikan, in his previous studies of cosmic rays in which he used electroscopes, had found what he interpreted as a "banded structure" in the absorption coefficients of the cosmic rays as they passed through the atmosphere. This led him to his atom building theory of the origin of the cosmic radiation, and to the conclusion that the primary cosmic ray beam consisted of photons in the energy range of several hundreds of millions of eV. ((Ed.) See Daniel J. Kevles, *The Physicists* (New York; Alfred A. Knopf, 1978), pp. 179–180.) To explain the origin of cosmic rays he postulated a process by which electrons and protons in outer space would somehow combine and coalesce to form nuclei of atoms, with the "packing fraction" energy released as photons forming the primary cosmic ray beam which impinged on the earth's atmosphere. He thus expected the magnet cloud chamber experiments to reveal secondary electrons produced in Compton collisions by the primary photons constituting the incoming cosmic-ray beam. According to his hypothesis, the energies of the electrons observed should be in the general energy range of about a hundred million eV, but not to exceed some 400 to 500 million eV.

On many occasions Neddermeyer and I would meet with Millikan to discuss energy measurements and their interpretation. Millikan was a very busy man and although not officially president of Caltech, he performed that complex function and many more. Not the least of the demands on him was raising money. Thus, because of Millikan's many duties, our meetings often occurred late at night in the laboratory after the conclusion of one of his many evening social engagements. He would remove his tuxedo-necktie, open his collar, rest and relax during these discussions.

Although the "atom building" hypothesis did not appeal to Neddermeyer and me, it seemed to be very firmly fixed in Millikan's mind. Millikan seemed steadfastly to think in terms of the atom building hypothesis which did not permit energies above 400 million eV. I remember on one occasion Neddermeyer was relating energy measurements he had made on a series of tracks and he came to one over a billion eV. Millikan virtually hit the ceiling and gave Neddermeyer a rather tough, third-degree type questioning. Both Neddermeyer and I tried to argue with Millikan but it seemed impossible to change the direction of his thinking — his mind's momentum seemed close to infinite. It was only after many of these meetings that Millikan readily accepted energies in the range of several billion eV.

9. Paradoxes

During the months that followed, Neddermeyer and I accumulated much more data and at least for a while believed the bulk of the high energy particles to be electrons about equally divided between positive and negative charge. But doubts soon began to develop, and it was only through the discovery of the meson that these doubts were finally resolved.

The discovery of the meson, unlike that of the positron, was not sudden and unexpected. Its discovery resulted from a two-year series of careful, systematic investigations all arranged to follow certain clues and to resolve some prominent paradoxes which were present in the cosmic rays.

The gist of the matter was as follows. Neddermeyer and I were continuing the study of cosmic-ray particles using the same magnet cloud chamber in which the positron was discovered. In these experiments it was found that most of the cosmic-ray particles at sea level were highly penetrating in the sense that they could traverse large thicknesses of heavy materials like lead and lose energy only by the directly produced ionization which amounted to something like 20 million eV per cm of lead. A principle aim of the experiments was to identify these penetrating cosmic-ray particles. They had unit electric charge and were therefore presumably either positive or negative electrons or protons, the only singly charged particles known at that time.

There were difficulties, however, with any interpretation in terms of known particles, as was pointed out as early as 1934 in a paper presented to the International Conference on Physics held in London of that year (Anderson and Neddermeyer, 1934).

The most important objection to their interpretation as protons lay in the fact that the energy of the electron secondaries produced by the direct impact of these particles as observed in a cloud chamber contained too many "knock-on" secondaries of high energy to correspond with the known energy spectrum if particles as massive as protons were producing these secondaries. ((Ed.) Particles of similar mass tend to share their energy equally in "knock-on" collisions.) On the other hand the spectrum was just that to be expected if the particles producing the secondaries were much lighter than protons. Furthermore, to interpret these particles as protons would mean assuming the existence of protons of negative charge since these sea-level particles occurred equally divided between negative and positive charges, and at that time there was no evidence for the existence of protons of negative charge.

There were difficulties also in interpreting these sea-level penetrating cosmic-ray particles as positive and negative electrons. The most important objections to their being electrons arose from three considerations. Firstly, theoretical calculations by Hans Bethe, Walter Heitler, and Fritz Sauter on the energy loss of electrons led to the conclusion that high-energy electrons should lose large amounts of energy through the production of radiation, which the penetrating particles in question were observed not to do. Secondly, we had found individual cases of electrons which did, in fact, show large energy losses through radiation, in some cases 100 million eV or more per cm of lead. Clearly in these cases the electrons showed a behavior quite different from that of the penetrating particles. And thirdly, the so-called highly absorbable component of the cosmic rays and the existence of electron showers could find an appealing explanation in terms of electrons if electrons did, in fact, suffer large radiative losses at high energies as demanded by the above-mentioned theory.

This then was the situation in 1934 in which the sea-level penetrating particles had this paradoxical behavior. They seemed to be neither electrons nor protons. We tended, however, to lean toward their interpretation as electrons and "resolved" the paradox in our informal discussions by speaking of "green" electrons and "red" electrons – the green electrons being the penetrating type, and the red the absorbable type which lost large amounts of energy through the production of radiation. < see p. 131 >

Evidence of an entirely new type was soon obtained. In experiments carried out on the summit of Pikes Peak in 1935 a number of cases of cosmic-ray-produced nuclear disintegrations were observed from which many protons were ejected, but showing also in a few cases particles which, from ionization and curvature measurements, were lighter than protons and heavier than electrons. These observations were not conclusive evidence in themselves for the existence of a new type of particle, but they did tend to lend support to this assumption in view of the other difficulties involved in interpreting the date in terms of known particles.

The next year or so brought further evidence on all the above points and only tended to strengthen the paradox further. The hypothesis that the penetrating particles were protons was further weakened by the observation of many cases of particles which did not suffer appreciable radiative collisions and still which could not be as massive as protons, as evidenced by the ionization-curvature relations of their cloud-chamber tracks. These cases could, however, be interpreted as electrons, but only if electrons ceased to radiate appreciably above a certain energy, such as say a hundred million eV.

The crux of the matter then was whether electrons above a certain energy did or did not experience a large energy loss through radiative impacts. In other words the paradoxical character of our data could be removed if one assumed that the Bethe-Heitler theory, although correct for electrons of energies below a few hundred million eV, in some way became invalid for electrons of high energy, thus permitting high-energy electrons to have a much greater penetrating power and thus perhaps to permit interpreting the highly penetrating sea-level cosmic-ray particles as positive and negative electrons.

This was the view held by Blackett but it did not seem to fit most of the facts as known at that time.

10. New Particles

In the summer of 1936, Neddermeyer and I were quite firmly convinced that all the data on cosmic-rays as known at that time nearly forced upon us the conclusion that the penetrating sea-level particles could be neither electrons nor protons and must therefore consist of particles of a new type. Let me give two quotes from our previous papers which strongly support this view. That the interpretation of the penetrating particles as protons seemed untenable:

"It was shown above that the number and distribution in energy of the negatrons ejected as secondaries by particles traversing plates of lead and carbon agree with those to be expected theoretically for extranuclear encounters if the incoming particles possess electronic mass. On the other hand, the assumption that the incoming particles possess protonic mass and have the curvature distribution in the magnetic field given above, would lead to an electron-secondary energy distribution noticeably different from that actually found" (Anderson and Neddermeyer, 1934, p. 182).

That the penetrating particles could not readily be interpreted as electrons:

"This large absorbability of electrons and photons is difficult to reconcile with the highly penetrating character of a large fraction of the sea-level particles on the view that the latter are electrons" (Anderson and Neddermeyer, 1936).

The first formal publication relating to the new particles appeared in the spring of 1937 (Neddermeyer and Anderson, 1937). However, before publishing this paper, we wished to obtain as convincing proof as possible that the Bethe-Heitler theory was indeed valid at energies sufficiently high to require that the penetrating particles could not be electrons but must be particles of a new type. To do this we took an additional 6000 photographs in which we measured separately the loss in energy of "single" and "associated" particles in their traversing a platinum plate of 1 cm thickness inserted in the cloud chamber. In this paper our conclusion was "that there exist particles of unit charge, but with a mass (which may not have a unique value) larger than that of a free electron and much smaller than that of a proton."

Evidence for the existence of new particles of intermediate mass was first presented in a colloquium at Caltech on November 12, 1936; a brief summary of this colloquium was sent out by Watson Davis of *Science Service* on November 13, 1936. A brief report also appeared in *Science,* November 20, 1936, page 9 of the supplement. Perhaps the first reference in the "literature" to the new particles was the last sentence in my Nobel Lecture on the positron delivered in Stockholm on December 12, 1936. In the forty-plus years since the delivery of that address, I have received no reaction at all from it, so I will quote that sentence here:

"These highly penetrating particles, although not free positive and negative electrons, will provide interesting material for future study."

At about this time, Jabez Street and E.C. Stevenson reported experimental results from which they concluded:

"From the data in the table it is evident that the penetrating particles cannot be described as electrons obeying the Heitler theory nor can an appreciable fraction be protons" (Street and Stevenson, 1937).

That the validity of the theory was not generally accepted is clear from a quotation from a paper by Blackett and J.G. Wilson at about the same time.

"Since it has long been clear that the radiation formula must break down at high energies, though the exact value of the energy at which the breakdown started was not known, attempts have been made by Williams (1934), by Oppenheimer (1935), and by Nordheim (1936), to modify the formula so as to reduce the calculated energy loss" (Blackett and Wilson, 1937).

11. Nomenclature

For awhile the new particles were known by various names such as, baryon, Yukawa particle, X-particle, heavy electron, etc. One day, Neddermeyer and I sent off a note to *Nature* suggesting the name *mesoton* (meso for intermediate). At the time, Millikan was away, and after his return we showed him a copy of our note to Nature. He immediately reacted unfavorably and said the name should be *mesotron.* He said, consider *electron* and *neutron.* I said, consider *proton.* Neddermeyer and I sent off the "r" in a cable to Nature. Fortunately or not, the "r" arrived in time and the article appeared containing the word *mesotron.* Neither Neddermeyer nor I liked the word, nor did anyone else that I know of. Further discoveries, such as the pion and the passage of time have greatly improved the matter of nomenclature as related to particles.

There is an interesting story (which I won't relate here) concerning the origin of the name *positron.* I don't like it particularly well, and although I have never discussed it with Prof. Dirac, my feeling is that he would have found it wholly inelegant.

12. Comment

In discussing the discovery of the meson I have not so far mentioned anything about the theoretical aspects of the situation. We saw previously how the Dirac theory predicted the existence of positrons, although it played no role in their discovery. The discovery of mesons, similarly, was based on experimental measurements and procedures, with no guide from any theoretical predictions.

As with the positorn, this need not have been the case. For before the discovery of the meson had been finally achieved a novel idea was published in a Japanese journal by the Japanese physicist Hideki Yukawa. Reasoning by analogy with quantum electrodynamics, he made the suggestion that perhaps nuclear forces, which are not electromagnetic in character, could be described in terms of a particle carrier of these nuclear forces, analogous to the photon being the carrier of electromagnetic forces. Nuclear forces, however, differ from electromagnetic forces in that they possess only a short range of action. This means that if nuclear forces are described in terms of a particle carrier, this particle carrier must have a finite rest mass unlike the photon of zero rest mass which is appropriate to the long range electromagnetic forces. Yukawa estimated from the known range of nuclear forces that this carrier should have a rest mass about 200 times that of an electron.

This novel suggestion of Yukawa's was unknown to the workers engaged in the experiments on the meson until after the meson's existence was established. Although Yukawa's suggestion preceeded the experimental discovery of the meson, he published it in a Japanese journal which did not have a general circulation in this country. It is interesting to speculate on just how much Yukawa's suggestion, had it been known, would have influenced the progress of the experimental work on the meson. My own opinion is that this influence would have been considerable even though Dirac's theory, which was much more specific than Yukawa's, did not have any effect on the positron's discovery. My reason for believing this is that for a period of almost two years there was strong and accumulating evidence for the meson's existence, and it was only the caution of the experimental workers that prevented an earlier announcement of its existence. I believe that a theoretical idea like Yukawa's would have appealed to the people carrying out the experiments, and would have provided them with a belief that maybe after all there is some need for a particle as strange as a meson, expecially if it could help explain something as interesting as the enigmatic nuclear forces.

It was clear almost from the beginning, however, that the Yukawa particle and the cosmic ray meson cannot possibly be the same particle at all. The Yukawa particle was invented to explain the strong nuclear forces while even in the very early experiments the cosmic ray mesons seemed to ignore nuclear forces completely and to interact with matter only through electromagnetic forces.

For example (London Conference, 1934), a total of 2437 transversals of carbon and lead plates in the cloud chamber showed no evidence of a nuclear reaction. In this instance a total of 1215 cm of carbon and 663 cm of lead was traversed. Thus, these data provided very strong support for the conclusion that the experimental mesons could not be quanta of the nuclear force field. Other experiments carried out in later years provided proof for this conclusion, for example, the important experiments of Marcello Conversi, Ettore Pancini and Oreste Piccioni in which they observed the capture and decay of negative and positive mesons stopped in an iron absorber.

It was not until the discovery of the pion in 1947 by Powell and his group that a particle existed that could be identified with the Yukawa particle, although the subsequent discovery of a host of other particles have shown the extremely complex character of nuclear forces, a problem still unsolved.

13. B-29 Airplane

I will now jump to the period following World War II. During the war years, I worked on the Artillery Rocket Research and Development Project based at Caltech. This program, the brain-child of Charles C. Lauritsen, under his leadership and with the able assistance of William A. Fowler and Thomas Lauritsen, grew into a large and very successful operation, and made a substantial contribution to the whole war effort. My own part in the program was the responsibility of adapting the firing of various types of Caltech rockets from suitable Army and Navy aircraft. It was through this work that I made close contacts with the proper military authorities to obtain the use of a B-29 airplane in 1946 for cosmic-ray experiments.

The plan was to continue the cosmic-ray studies in the same magnet-cloud-chamber formerly used in Pasadena and on Pike's Peak, but in the airplane at an altitude as high as possible. The B-29 was stripped of all armaments at the facory and modified for operation at extreme altitudes. It was equipped with optional electric generators on each engine capable of providing 750 amperes at 28 volts giving a magnetic field strength in the cloud chamber of 7500 gauss.

My collaborators in the B-29 experiments were Raymond V. Adams, R. Ronald Rau, Ram C. Saxena (graduate students) and Paul E. Lloyd (a post-doctoral research fellow.) I do not remember the date of our maiden flight, but I do remember that as I was climbing aboard the B-29, the pilot asked me what our route and destination were to be. This took me aback because in no previous airplane flight had this question come up. I quickly answered, however, by saying let's fly first to Mt. Shasta, circle it once, then to San Diego, circle it once, then to Pike's Peak, circle it once, and then back home to the Naval Ordnance Test Station, China Lake, California, where the B-29 was stationed. The cloud chamber operated perfectly throughout the flight.

In all, 35 flights of about 5 hours duration each were made at altitudes of between 30000 and 40000 ft. (Anderson *et al.*, 1947; Adams *et al.*, 1948), while a few flights were made at higher altitudes when the B-29 was capable of flying that high. The highest altitude we ever reached was 41000 ft.

14. Comment

At this point I cannot resist saying that although the B-29 flights were successful and gave considerable new information on the cosmic radiation, they could have accomplished much more. Now, of course, I am speaking of hindsight. Many of the engineering problems associated with the B-29 flights were formidable and, unfortunately, received all our attention. Had we forgotten the B-29, and spent a week in the nearby High Sierra arguing only about cosmic-rays and physics we might have done better. All the clues were present and published, one of the most important being the experiments by Lajos Janossy in which he used counter arrays separated by various thickness of lead chosen to select nuclear collisions of high energy.

Such a week, solely devoted to physics, might have given us a proper goal, i.e., to study the nucleonic component of cosmic rays. The modification required in the B-29 equipment would have been very minor, and would have taken no more than an hour or so to do. We need only have added a small block of lead about 20 cms thick, an additional counter, and required triple rather than double coincidences. This would have selected nuclear events, and undoubtedly would have given us hundreds of examples of the new unstable particles, heavy mesons and hyperons, subsequently discovered by George D. Rochester and Clifford C. Butler. The relative intensity of high energy nuclear events in the cosmic rays at 30000 ft compared to sea level must be in the vicinity of several hundred to one. In any case, this is an example of a superb piece of experimental equipment not used to its maximum advantage.

15. Strange Particles

While the B-29 experiments were in progress, Robert B. Leighton and Eugene W. Cowan joined me in a program of building a variety of cloud chambers for further studies of cosmic rays in Pasadena and at White Mt., Calif. This program was active for several years, until the advent of the accelerator era made the use of cosmic rays for particle studies more and more difficult. I will not attempt to cover the work carried out during this period of time except to refer to one early experiment.

In 1947, Rochester and Butler (1947) reported observing in a cloud chamber two cases of "forked tracks," one charged and one neutral, which they interpreted as examples of the spontaneous decay of particles of a new type. No further cloud chamber photographs of events such as these were observed for about two years until, in a series of 11000 photographs made at Pasadena and White Mt., Calif., we observed thirty-four additional cases of "forked tracks" and were led to the same conclusion as that drawn by Rochester and Butler, i.e. namely the existence of unstable particles of a new type (Leighton *et al.*, 1949; Seriff *et al.*, 1950). From these thirty-four cases we obtained some information as to the nature of the decay products, and found the lift-time of the neutral particles to be about 3×10^{-10} sec. A variety of further investigations were carried out in an attempt to elucidate the properties of the hyperons and heavy mesons, their modes of decay, life times, associated production, relative abundance, etc. Robert W. Thompson of the University of Indians was also active during this period and was especially successful in his precision work on the neutral theta meson.

However, the ever encroaching larger and larger accelerators clearly indicated the imminence of the death knell of cosmic rays as a useful tool in studies of particle physics. To exemplify this threat, I should like to give two quotations from Robert Marshak's article on the Rochester Conferences (Marshak, 1970). Commenting on the 1952 conference Marshak said:

"The machine results were beginning to overtake the cosmic ray results, certainly in quantity, albeit to a lesser extent in importance (Anderson, Leighton and Thompson were coming through strong and clear on the two unstable particles), and the theorists were contributing their wisdom with regard to selection rules in particle reactions, with particular reference to isospin invariance."

And on the 1956 conference, he wrote:

"It was the year when machine results were coming in so fast that R.B. Leighton was led to remark that "next year those people still studying strange particles using cosmic rays had better hold a rump session of the Rochester Conference somewhere else."

However, undaunted by the irresistable encroachment of the accelerators, Cowan built a complex arrangement of eight flat ionization chambers and twelve flat cloud chambers of a total height about twenty feet, designed for investigations at energies above those obtainable in any accelerator, and continued his studies of cosmic ray particle events until 1971 (Cowan and Moe, 1967; Cowan and Matthews, 1971).

In reading this paper, as I warned in the Introduction, I find that it is devoted almost wholly to cloud-chamber investigations, and to a too great extent to those of our own laboratory. For this, again, I apologize.*

* This paper was based in part on an earlier one, C.D. Anderson, "Early Work on the Positron and Muon", *Am. J. Phys.* **29** (1961), 825–830. (Ed.) The editors wish to thank John S. Rigden, Editor, American Journal of Physics, for permission to include a portion of Dr. Anderson's 1961 article.

REFERENCES

Adams, R. V., Anderson, C. D. Lloyd, P. E., Rau, R. and Saxena, R. C.: 1948, *Rev. Mod. Phys.* **20**, 334

Anderson, C. D. and Neddermeyer, S. H.: 1934, *Papers and Discussions of the International Conference on Physics* (*London*) **1**, 171; *ibid.*, 182.

Anderson, C. D. and Neddermeyer, S. H.: 1936, *Phys. Rev.* **50**, 263.

Anderson, C. D., Adams, R. V., Lloyd, P. E. and Rau, R.: 1947, *Phys. Rev.* **72**, 724.

Blackett, P.M.S. and Occhialini, G: 1933, *Proc. Roy. Soc.* **A139**, 699.

Blackett, P. M. S. and Wilson, J. G.: 1937, *Proc. Roy. Soc.* **160**, 304.

Cowan, E. W. and Moe, M. K.: 1967, *Rev. Sci. Inst.* **38**, 874.

Cowan, E. W. and Matthews, K.: 1971, *Phys. Rev.* **D4**, 37.

Leighton, R. B., Anderson, C. D. and Seriff, A. J.: 1949, *Phys. Rev.* **75**, 1432.

Marshak, R. E. *Bull. Atomic Sci., June 1970*, 92.

Neddermeyer, S. H. and Anderson, C. D.: 1937, *Phys. Rev.* **51**, 884.

Rochester, G. D. and Butler, C. C.: 1947, *Nature* **160**, 855.

Seriff, A. J., Leighton, R. B., Hsiao, C., Cowan, E. W. and Anderson, C. D.: 1950, *Phys. Rev.* **78**, 290.

Street, J. C. and Stevenson, E. C.: 1937, *Phys. Rev.* **51**, 1005.

Three photographs follow next pages.

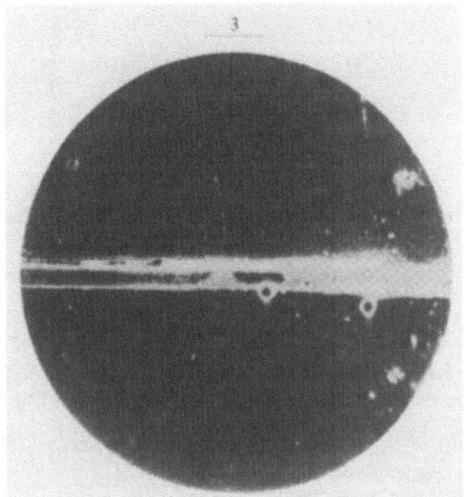

Courtesy of the Archives, California Institute of Technology, for caption also. Positron from Carl Anderson's paper, 1932, which won the Nobel Prize in 1936. A 63 million electron-volt positron passing through a 6 mm lead plate and emerging with an energy of 23 million electron-volts. The length of this latter path is at least ten times greater than the length of a proton track of this curvature (15000 gauss magnetic field).

Courtesy of the Archives, California Institute of Technology, for caption also. Carl Anderson with Robert A. Millikan on Pike's Peak 1935.

Courtesy of the Archives, California Institute of Technology, for caption also. Dr. Carl Anderson beside his apparatus with which he discovered the positron in 1932. This apparatus measures the energies of the individual cosmic-ray particles — positive and negative electrons. In the foreground is shown a spotlight which shines through the central black box into the cloud-chamber inside the magnet. Wholly automatic, the cloud chamber is actuated by the incoming cosmic ray itself, and at the same time a camera, which sits on the bench at the right end of the magnet, takes a photograph of the particle's path through the cloud chamber. Experiments have been made using this equipment on the summit of Pike's Peak at an elevation of 14000 ft, at Coco Solo, Panama Canal Zone in the equatorial regions, and also in a B-29 airplane at elevations up to 40,000 ft. photo 1934.

COSMIC RAYS AND THE BEGINNING OF THE MESON THEORY

Hideki YUKAWA*

I feel unwell probably due to my severe illness of four years ago. It is difficult to refresh my memory on events in the beginning period of the meson theory. However, one impression still remains clearly. It was the discovery of the cosmic ray mesotron, which is now known as the mu-meson. This experimental discovery of the mu-meson encouraged the meson theory, and the theory developed actively and rapidly by this stimulation. I remember clearly how really effective it was.

However, during a recent conversation with Y. Sekido some other events were recalled to my mind. Was I conscious of cosmic rays since long before the mesotron theory? Yes, naturally, it is written in "Tabibito."[1] (Soon after I graduated from Kyoto University in 1929, I saw Heisenberg and Pauli's paper on quantum electro-

* Kyoto University (emeritus), Kyoto, Japan.

\<Ed.\> Narrated by H. Yukawa on Oct. 5, 1979 at his home. The hearer, Y. Sekido, arranged the memo with supplements written here in parentheses, added footnotes and translated into English. (Postscript: This manuscript was seen by H. Yukawa on Nov. 4, 1979.)

[1] The source of all supplements in this article, written in parentheses, is *"Tabibito"* which is a book written by H. Yukawa in Japanese and published in 1958 by The Asahi Shinbun Company and Tabibito means a traveller, and the book contains his reminiscences on his life during the period from his birth (1907) to 1934. The English version of this book was published by World Scientific Publishing Co. Pte. Ltd., Singapore, in 1982.

Y. Sekido and H. Elliot (eds.), Early History of Cosmic Ray Studies, 133–135.
© *1985 by D. Reidel Publishing Company.*

dynamics,[2] and tried for a time to overcome the infinity in the paper.[3] It was so difficult that I then looked for an easier problem. At that time, the application of quantum mechanics was developing in many new fields, but my eyes were still pointed in the direction of undeveloped fields such as nuclear physics and cosmic rays,[4] though I took a rest for I did not know where to begin.)

To what extent, or from what angle was I conscious of cosmic rays? Now, I cannot remember clearly. Have I any remembrance of study from a motive to explain a phenomenon characteristic to cosmic rays? No. The theory of meson came from the study of nuclear force. (Neutron and positron were discovered in 1932, and I devoted myself to the problem of nuclear force during the two years since that autumn.[5])

(At first, I assumed the particle virtually emitted and absorbed by proton and neutron to be an electron.[6] I did not write a paper on this theory for I was aware of its defect in spin and statistics, but talked of the idea in April 1933 at the Sendai Meeting, where Dr. Nishina suggested to me to consider an electron which would satisfy Bose's statistics.)

In 1934, (one day after the beginning of the new term[7]) I found a paper by Fermi on beta-decay[8] (in newly-arrived periodicals. It was based on the neutrino assumption which was proposed by Pauli at a meeting[9] in 1931, though I did not know of his assumption at that time.) I thought (immediately) that his assumption might solve the problem of nuclear force also (for a pair of electron and neutrino might be considered as the particles vertually emitted and absorbed.[10]) The same idea was thought by foreign researchers also, and papers by Iwanenko[11] and Tamm[12] came out. Their result was negative, and this negative result inspired me.

(I began to consider not to seek for the medium of neclear force among the known particles, but to pursue the character of the nuclear field[13] which might determine the properties of a new particle suited for the field. But it meant a long and hard time of groping just around the goal. One night in early October, an idea occurred to my mind unexpectedly. The range of nuclear force, which was known to be very short, should

---

2) "Text" p.129, Heisenberg u. Pauli: 1929, *Zeits. Phys.* **56**, 1; 1929, *ibid.*, **59**, 168. A text book *"Theory of Atomic Nuclei and Cosmic Rays"* was written by H. Yukawa and S. Sakata in Japanese and published in 1942 by Iwanami Shoten. This book was useful for preparing the footnotes here, and abbreviated as "Text." The book is a revised edition of their book with the same title written in 1940 as a part of series *"Text Books of Physics"* compiled by Iwanami Shoten in 1941.

3) *Tabibito,* p.236.

4) *Tabibito,* p.238.

5) *Tabibito,* p.261.

6) *Tabibito,* p.270.

7) *Tabibito,* p.277. New school term starts in April.

8) Text, p.100, Fermi: 1934, *Zeits. Phys.* **88**, 161.

9) Text, p.98, Lecture at Pasadena in June 1931.

10) *Tabibito,* p.278.

11) Text, p.125, Iwanenko: 1934, *Nature* **133**, 981.

12) Text p.125, Tamm: 1934, *Nature* **133**, 981; 1936, *Sow. Phys.* **10**, 567.

13) Text, p.128, The only one wave equation, which is relativistically invariant and linear without derivatives higher than the second, is $(\Delta - \frac{1}{c^2} \frac{\partial^2}{\partial t^2} - \kappa^2) \phi = 0.$

be inversely proportional to the mass of the new particle.[14] I wondered why I had no thought of this before. Next morning, I calculated the mass of the particle. It should be a charged particle with about 200 electron mass.[15])

(Why had this particle remained unknown? I asked in return to myself. An accelerator of 100 MeV, which is necessary to produce this particle, was not available at that time. I became self-confident, and[16]) talked of the new theory based on this idea at the colloquium of Kikuchi Laboratory of Osaka University. Prof. Kikuchi said, "If it is a charged particle, it should be caught by cloud chamber." I answered, "Yes, it might be found in the case of cosmic rays." I remember this conversation even now. Was my answer prepared before his question? Now, I cannot recall since when such a thought had been in my mind. However, cosmic ray particles had been in my mind, irrespective of the method of direct detection such as the cloud chamber.

(After the colloquium, I talked at the meeting of the Osaka branch of the Physico-Mathematical Society of Japan, and) in November at the Tokyo Monthly Meeting of the Society. I believe I stated there also that the particle might be found in cosmic rays. Dr. Nishina was immediately[17] interested in this theory and encouraged me. (I wrote a paper[18] in English during November and sent it to the Society.)

After that, many people discovered the mesons experimentally in cosmic rays. The first information I received was Anderson's discovery,[19] of which I was informed by a letter from him before I saw his paper. I became intimate with him, and kept up an intimate friendship with him for a long time. His manners are natural and honest. He has shown great kindness even to the likes of me. He was a man of a favourable impression. Anderson's discovery encouraged me remarkably, and gave me the opportunity to promote the meson theory of cosmic rays.

Because the meson has a lifetime, cosmic rays must be effected by the altitude variation of the meson producing layer, as also proposed by me. I had an intercourse with Rossi on this matter. In the beginning of the meson theory, I thought it corresponded to the cosmic ray mesotron which is now known as the mu-meson. But discrepancies in its lifetime and so on became evident some years later. It seems to be about 1939–40, as I recall. It is my desire that Anderson and Rossi would also write some reminiscences of this occasion.

14) Text, p.129, $m_u = \frac{\hbar \kappa}{c}$.
15) *Tabibito,* p.280.
16) *Tabibito,* p.281.
17) Y. Sekido, who attended this meeting by chance, remembers the rare event that the chairman, Dr. Nishina, stood up and praised Yukawa's work.
18) Text, p.128, Yukawa: 1935, *Proc. Phys.-Math. Soc. Jpn.* 17, 48.
19) Neddermeyer and Anderson: 1937, *Phys. Rev.* 51, 884.

COSMIC RAY STUDY IN NISHINA LABORATORY

Masa TAKEUCHI*

1.

The cosmic ray research in Japan was initiated by Dr. Yoshio Nishina in 1931. In the summer of that year, he was nominated chief of a new laboratory (Nishina Laboratory) of The Institute of Physical and Chemical Research (I.P.C.R.). There, R. Sagane soon started to make Geiger-Müller counters. The first counter was made of a copper tube 5 cm in length and 2 cm in diameter. Both ends of the tube were shielded by ebonite plugs, and a steel wire was stretched between the plugs as a central electrode. This counter was of the shape that two "Spitzen Zähler"'s were connected facing each other. Fortunately the discharge characteristics of this counter showed a plateau as large as 50 volts, and it was primarily used to measure the intensity of radiation in the laboratory, When the center of a typhoon passsed through Tokyo area in the fall of this year, it was found that the counting rate varied with decreasing atmospheric pressure, and later returned to the normal rate when the typhoon had gone. Apparently there was a significant correlation between the counting rate and the atmospheric pressure.

While this kind of preliminary experiment was continued, we concentrated our effort to make larger G.M. counters of different type. One of these counters had such a structure that a brass tube and a steel wire were sealed in a glass tube, which was then filled with a low pressure air. In those days the good counter with a 50 − 100 v plateau could be obtained only by trial and error, as the discharge mechanism of G.M. counter was not clear.

In 1932, we started a cloud chamber study of cosmic rays; we planned to measure the energies of cosmic ray particles in a magnetic field. In the beginning, however, nobody in the laboratory was convinced whether cosmic ray tracks could be obtained in the cloud chamber.

The first cloud chamber we made was expanded by a horizontally moving piston, and was mounted in a magnetic field of 2000 Gauss, which was generated by a Helmholtz coil. At the end of 1932, we succeeded in taking photographs of cosmic ray tracks. The counter controlled method for cloud chamber operation was invented by R. Sagane, and was applied to this experiment as mentioned in November 1932 at the meeting of I.P.C.R. The next year, we found that the Blackett group had also invented similar method and applied it to cosmic ray experiment, when their publication of March 1933 arrived at our laboratory.

One of the photographs thus taken showed a pair of circular tracks emitted from a point where a straight track crossed the inner surface of the glass wall. Another interesting photograph demonstrated the simultaneous incidence of multiple tracks with different raddi of curvature.

* Yokohama National University (emeritus), Yokohama, Japan.

Y. Sekido and H. Elliot (eds.), Early History of Cosmic Ray Studies, 137–143.
© *1985 by D. Reidel Publishing Company.*

In March of 1933, Blackett published the result obtained by the counter controlled cloud chamber in a magnetic field (Blackett and Occhialini, 1933). Their experiment was similar to ours, but with a stronger magnetic field. Having read this paper, we found that the pair of circular tracks in our photograph of 1932 was nothing but a pair creation and that the picture with multiple tracks showed a shower phenomenon. However, we could not but conclude that our apparatus was not good enough to continue further experiments on energy spectrum, and that a larger cloud chamber in a stronger magnetic field was necessary for the future study of high energy cosmic ray particles. Thus we decided to suspend the cloud chamber experiment for a while.

2.

At the end of 1933, The Japan Society For The Promotion of Science* started "The 10th subcommittee" to promote cosmic ray study in Japan. The subcommittee soon discussed what would be fruitful research projects, and the first project they chose was an observation of cosmic ray intensities by ionization chambers. Later, the G.M. counter experiment was added.

As Dr. Nishina was a member of this subcommittee, he decided to start this project in the Nishina Laboratory. At first, C. Ishii and other members constructed the ionization chambers of the Compton type. In 1935, for a measurement of altitude variation of cosmic ray intensities, the Compton type ionization chamber was taken up to Mt. Fuji to observe the intensity at mountain altitude, which was then compared with the sea level intensity measured at Tokyo by F. Yamasaki. In June 1936, on the occasion of the solar eclipse, Y. Nishina, C. Ishii, and Y. Sekido observed the cosmic ray intensity at Mt. Syari in Hokkaido. In the fall of that year, the observation of underground burst was carried out in Shimizu tunnel by the Neher type ionization chamber (Nishina and Ishii, 1936).

According to the progress of research activity in those days, two cosmic ray groups started in the Nishina Laboratory. One was to observe cosmic ray intensities by ionization chambers and by G.M. counters. The other group (Takeuchi and others) was to carry out the cloud chamber experiment. As the history of the former group will be

---

* Japan Society for the Promotion of Science: This society was established in 1931 by the Japanese government primarily in response to strong demands by famous Japanese scientists, and later the discussion made at the National Diet.

The aim of this society was to promote academic research activities in Japan: aiming at the development of culture and industry, the repletion of national defence, and the contribution to nations prosperity and benefit of the human being. The programms that the society carried out in the beginning were to offer financial support to scientists, to organize the committee for investigation of research projects, to issue academic publications, and to recommend its support to important scientific projects.

The society was first organized as a foundation supported by the Japanese government, the royal family, and a rich company. The first president was Prince Chichibu, the first chairman the Prime Minister, and the first director was Prof. Joji Sakurai, who was later replaced by Prof. Hantaro Nagaoka (Fujioka, 1973).

Since its establishment, the society had continued its activity for 36 years. In 1967, the society was reorganized according to a law passed by the Diet, and placed under the supervision of the Minister of Education.

described by Dr. Y. Sekido in this book, I shall give a brief outline of the history of the cloud chamber group.

3.

In 1935, Dr. Nishina planned to make a large cloud chamber and a strong magnet; according to his plan, at first I made a model magnet with the size 1/10 of the proposed magnet, and studied the shape of pole pieces that could generate a magnetic field as uniform as possible over the whole effective volume of the cloud chamber. Finally I was able to find the best shape for the pole pieces.

In the meantime, I constructed a cloud chamber of disc shape of 40 cm diameter in collaboration with the engineering staff of I.P.C.R.. This chamber was expanded by moving an aluminum plate, which is supported by a rubber sheet connected to the outer structure of the chamber. The plate was moved by taking in and out the pressurized air. The light source was composed of an arc light and a parabolic mirror of 50 cm diameter with a focal length of 10 cm.

The magnet was manufactured by Tokyo Shibaura Electric Co. The maximum diameter of the pole pieces was 70 cm, and one of the pole pieces had a cone shaped hole for photographing (Fig.1). Each unit of the magnetic coil was made of dual layers of swirling copper tube, which was cooled by letting water flow inside the tube. All units were connected in series electrically and the tubes in parallel for cooling. When a current of 1000 A was passed through the coil, a magnetic field of 16500 Gauss was obtained with a 3 percent uniformity over the whole volume of the cloud chamber (Fig.1).

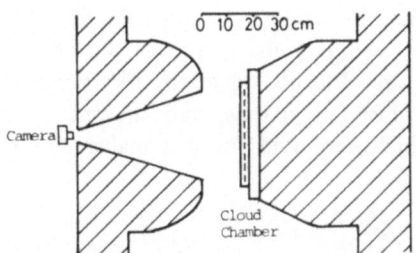

Fig. 1 (a). The cloud chamber apparatus in Nishina Laboratory, I.P.C.R.

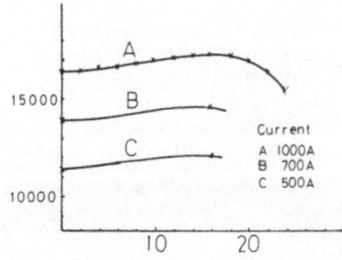

Fig. 1 (b). The magnetic field along the broken line shown in Fig. 1 (a) (average position of tracks).

Next a large D.C. generator that could generate a D.C. power as large as 300 — 400 kw was needed for the operation of this magnet. Because we could not find such a powerful generator in our institute, we asked the Japanese Navy to help us. Eventually the Navy permitted us to use their generator which had been used to charge submarine batteries. They agreed to let us use the generator at Yokosuka Naval Arsenal only when they were not using it.

Then Takeuchi and Ichimiya moved to Yokosuka with the cloud chamber in the spring of 1936, and for about one year we carried out our experiment. The purpose of this experiment was to measure the energy spectrum of cosmic ray particles. In 1937 Blackett published his result on the energy spectrum obtained by his cloud chamber again prior to our publication (Blackett, 1937). We found that our result agreed completely with his result, and presented our result only orally at the meeting of I.P.C.R., but did not published it.

4.

After that we started an experiment to measure the energy loss of cosmic ray particles putting a lead plate of 3.5 cm or 5 cm thickness inside the cloud chamber.

In 1937, Neddermeyer and Anderson published their observation on the energy loss of cosmic ray particles through an 1 cm platinum plate, and suggested the existence of a particle with a mass heavier than that of an electron (Neddermeyer and Anderson, 1937). Other papers also indicated the evidences of such particles that were apparently not protons but did not create electron pairs when they passed through the absorbing material (Street and Stevenson, 1937; Crussard and Leprince-Ringuet, 1937a, b). These papers were published in May, however, Nishina had heard some news of them from N. Bohr who visited Japan in April via United States. As we were doing a similar experiment, we tried to estimate the masses of the observed particles. M. Kobayashi then calculated, using the Bloch's equation, the energy losses in lead of charged particles with several different masses, and showed us a graph of energy loss — momentum relation which was used for our analysis. In August we found a particle

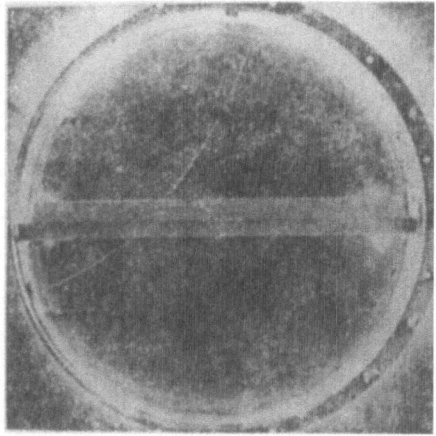

Fig. 2

Fig. 3. The author and the large magnet in Nishina Laboratory. (November 22, 1938)

(Fig. 2) with a mass $1/7 - 1/10$ of that of a proton (Nishina *et al.*, 1937). This paper was received by the Physical Review on August 28 and was published on December 1 after shortening the description, while Street and Stevenson's estimation of mass was reported in their letter to the Editor on October 6 which was published on November 1.

In 1938 we returned to the campus of I.P.C.R., and continued our cloud chamber experiment (Fig.3). At that time a large generator was introduced in our campus for the cyclotron project, and could also be used for our experiment. In those days, our interest turned towards the measurement of the mass of the mesotron. At the end of this year, we obtained an another example of a particle with an intermediate mass (Nishina *et al.*, 1938).

5.

Since then we went forward to measure the ionization density of particles in the cloud chamber, and planned to determine the mass of the particles by the simultaneous measurement of momentum $(H\rho)$ and ionization density (I). In the meantime we obtained an interesting event in which $H\rho = 1.2 \times 10^6$ Gauss.cm and $I = 3$ times minimum ionization which corresponds to twice the ionization density of an electron with the same momentum (Fig. 4). We estimated the mass of this particle to be a half of that of a proton, and reported this result at the meeting of I.P.C.R. However the existence of such a particle was not predicted at that time, and furthermore some people had doubts about the accuracy of our measurement of ionization density. As we observed only one such event, we did not publish it.

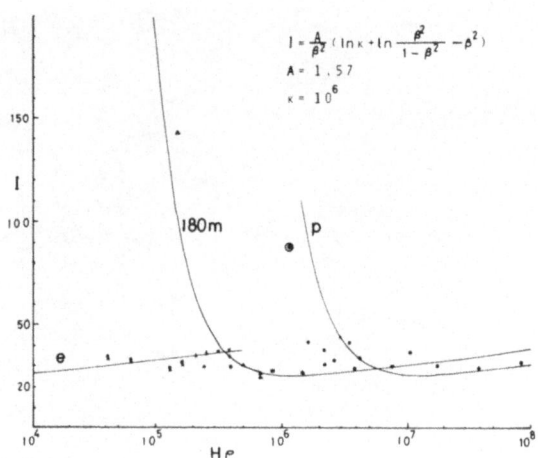

Fig. 4. The Result of observations of momentum and ionization density. The median mass event is shown by a double circle.

Then we decided to use two cloud chambers for further work; the idea was to separate the ionization measurement from the momentum measurement. Actually a large cloud chamber of 40 cm diameter was mounted in a magnet, and a small cloud chamber of 15 cm diameter was placed above the magnet on top of the large chamber. Aiming at improving the accuracy of momentum measurement, the large chamber was expanded immediately when the triggering pulse was accepted to get sharp and long tracks. On the contrary, the expansion of the small chamber was a little delayed to diffuse ions to obtain the accurate ionization densities.

We started this experiment in 1941. However the intensity of cosmic rays which passed simultaneously through both chambers was too low. Before long we realized that it was hopeless to get a sufficient number of tracks with high ionization densities within a reasonable time, and finally we gave up continuing this experiment.

6.

In 1941 we planned to take cloud chamber photographs of cosmic rays at aeroplane altitude; we intended to do this experiment on a military fighter. Since we could not get on board with our apparatus, and since the fighter's cabin was not pressurized at that time, the airborn apparatus should function automatically under the low temperature and low atmospheric pressure. Takeuchi then tried to develop an automatic cloud chamber photographing device. We built a model apparatus, and repeated the test experiment in the low temperature − low pressure room at The Central Research Institute of Aeronautical Science. However, after some time we had to discontinue this experiment owing to the war.

7.

Thus the cloud chamber study of cosmic rays had continued for about ten years in the Nishina Laboratory, and we could accumulate quite a number of photographs of

Fig. 5. Y. Nishina.

high energy cosmic ray particles. Unfortunately these photographs and most of the records of experiments were burned together with the experimental apparatus by the air raid on Tokyo on April 13 in 1945, and only a few documents were left in my hand.

This report was written on refferring to the small amount of documents that escaped damage and to my memory (July, 1980).

REFERENCES

Blackett, P.M.S.: 1937, *Proc. Roy. Soc.* **159**, 1.

Brackett, P.M.S. and Occhialini, G.P.S.: 1933, *Proc. Roy. Soc.* **139**, 699.

Crussard and Leprince-Ringuet, L.: 1937a, *Compt. Rend.* **204**, 240; 1937b, *J. Phys. Rad.* **8**, 213.

Fujioka, Y.: 1973, *The Biography of Hantaro Nagaoka,* (ed. Yoshio Fujioka), Asahi Shinbunsha, Tokyo, pp. 584–595.

Neddermeyer, S.H. and Anderson, C.D.: 1937, *Phys. Rev.* **51**, 884.

Nishina, Y. and Ishii, C.: 1936, *Nature* **138**, 721.

Nishina, Y., Takeuchi, M. and Ichimiya, T.: 1937, *Phys. Rev.* **52**, 1198.

Nishina, Y., Takeuch, M. and Ichimiya, T.: 1938, *Phys. Rev.* **55**, 585.

Street, J.C. and Stevenson, E.C.: 1937, *Phys. Rev.* **51**, 1005.

THE "MAGNET HOUSE" AND THE MUON

1. Introduction

This article is a recollection of work with which I was directly involved during the very short period in which the identification of what is now the muon was established. I came to the 'Magnet House' in Malet Street, London, in September 1936 and continued there until January 1938, when the whole apparatus was dismantled and removed to the Physical Laboratories at Manchester, where P.M.S. Blackett, for whom I was working as a research assistant, had been appointed Head of the Department two or three months earlier. While our own work, which is the basis of this article, concerns only data collected at the Magnet House, there was an inevitable lag in the measurement of plates and the interpretation of the material, and it would be artificial and misleading to exclude material which was not analysed until we were working in Manchester.

2. Background

Work at the 'Magnet House' grew directly from the pioneer experiments of Blackett and Occhialini (1932, 1933) at Cambridge, in which a cloud chamber was interposed in the aperture defined by a counter telescope and triggered by the defining counters discharging in coincidence. This system was engineered to complete expansion with minimum delay after the counter discharge (~ 0.01 s) and operated in a magnetic field of about 3000 gauss. It had a repetition rate of about one expansion every two minutes and it yielded the impressive success rate of roughly 75%.

The potentiality of 'counter control' was to be compared with what was possible making random expansions without any sort of control as demonstrated at the same time by Anderson (1932a). He photographed in a sustained repetition operation of random expansion in a field of 17000 gauss. Of some 3000 photographs, 62 showed cosmic ray tracks long enough and well enough illuminated for energy measurements. Assuming that the tracks mostly represented those of downward-moving particles, (and there were good grounds for this assumption), both positively and negatively charged particles could be identified, and the positive particles were in the first instance identified as protons (that is to say *not* helium or even heavier nuclei).

The advantages and disadvantages of counter-control were set out by Blackett and Occhialini. In favour of control, this made it possible to concentrate on well-placed and well-illuminated tracks, and in particular it was considered much easier to adjust the chamber conditions for a few good track photographs than to maintain high quality over large numbers of photographs the majority of which were abortive. The formidable drawback was that the level of magnetic field reasonably attained by

* Emeritus Professor, University of Leeds, Leeds, U. K.

Y. Sekido and H. Elliot (eds.), Early History of Cosmic Ray Studies, 145–160.
© *1985 by D. Reidel Publishing Company.*

peaking for a regular cycle of random expansions could not be maintained in the steady condition.

At this particular stage the two approaches were nicely balanced in what they had achieved. In the high field available Anderson had been able to deflect the great majority of incident particles, but while Blackett and Occhialini were able only to secure reasonable deflections for a smaller proportion of the 500 tracks which they had secured, the much larger body of data allowed a first attempt to be made at describing an energy spectrum. But broadly the two very differently conceived experiments were notable for the agreement of their conclusions. A substantial proportion of the tracks were identified as of positively charged particles, and in both accounts considerable effort was directed to criteria which excluded any other explanation. Anderson at an early stage satisfied himself that these were the tracks of protons (i.e. singly-charged). Negatively charged particles were exclusively described as 'electrons'. In his second paper (Anderson, 1932b) Anderson identified certain positive tracks, by means of the momentum-ionization relation, as of mass much smaller than that of protons and, indeed, more like that of electrons. Blackett and Occhialini also recognised that many of the positively charged particles could not possibly be protons, and related them to the concept of 'positive electrons', seeing this conclusion as one more of the diverse lines of evidence pointing towards the real physical existence of the Dirac 'positive electrons'.

Work was of course undertaken elsewhere on similar lines, but the activity of the workers I have referred to was dominant in determining the plans from which my work at the 'Magnet House' developed. Inevitably there was some convergence from past experience: counter-control was recognised and adopted immediately, the necessity of higher fields than those available to Blackett and Occhialini was apparent, but the implementation could not be so immediate.

3. The Magnet House

In 1933 Blackett left Cambridge to take the Chair of Physics at Birkbeck College in London, a move which surprised many people. But a base in London had obvious attractions, the teaching load at Cambridge was surprisingly heavy and at Birkbeck College all undergraduate teaching took place in the evening, leaving the whole normal day for other work. Within the crowded complex of old buildings which then housed Birkbeck College there was no possible site for the installation now planned of a large electromagnet, with its associated equipment, and for the cloud chamber to work within its field, and so this came to be sited in Malet Street, a few hundred yards from the London University Senate House, on land already ear-marked for a new Birkbeck College building. (In the event, the new Birkbeck College was erected on a site rather nearer to the Senate House than had first been planned, and does not quite reach to the point at which the 'Magnet House' once stood.) The 'Magnet House' was a quite small hut approached by an unobtrusive doorway through a high hoarding covered with advertisements. The door carried the non-committal notice "Birkbeck College Magnet House" in small letters, together with a letter box. I imagine that the great majority of people passing up and down Malet Street never even noticed that the door was there! But besides being hardly noticable the site happened to be one of the few places in London in which there was still a reliable and readily accessible public D.C. supply.

It was many years later that I was able to read through the two thick files which recorded the major operation of funding this exceptionally expensive installation (roughly £1000) of which most of the cost related to the magnet itself, its cooling system and its controls.

The Magnet House measured about 5m x 3.5m and was divided into two parts, the larger room, say 3.5 m square, housed the magnet and all its equipment, (Fig. 1), and a workbench, and from one corner a minute dark-room had been taken off. The second room was planned for measuring, recording and storage, and for interpretation, and for these purposes would have been relatively roomy. However in December 1936, Lajos Janossy who, after having built a considerable reputation working with Kolhörster in Berlin, was forced to leave Germany, came to us at the Magnet House. His apparatus, with which he was measuring the absorption of penetrating particles in up to 2 m of lead, occupied about a third of the 'measuring room'. Then things were rather cramped, but perhaps good for us, with this reminder of the great penetrating power of the 'hard component' towering above us! From the beginning of the Birkbeck period, Blackett had appointed a young technician, Arthur Chapman, for his own researches. At the Magnet House he was quite invaluable, and his special contribution to effective operation is acknowledged in many papers recording the work there. Chapman went to Manchester when the Magnet House work moved there, and later moved on with Blackett to Imperial College, London.

The Magnet House installation and the first two years are described in full detail in papers by Blackett and with R.B. Brode who came over to London on a Guggenheim Travelling Fellowship (Blackett, 1936; Blackett and Brode, 1936; Blackett, 1937). When I took over the day book of operations, I found that this included occasional notes from the outside world: "George Washington's birthday!" with a neatly drawn American flag. Correctly, I knew that I could not live up to this.

These papers recorded the potentialities of the equipment, the treatment and then the near-elimination of optical distortions and introduced the concept of a 'maximum detectable energy (in fact, momentum)* which was to be the basis of a standard performance through all the later work. The last of these papers (Blackett, 1937) also brought into prominence an optical compensator, an achromatic prism through which a track image could be projected and which was rotated until the image of a track of uniform curvature was judged, viewed at grazing incidence, to be straight. Under these conditions personal judgement of straightness is very sensitive. Over its range of application this instrument was from the outset both more accurate and far less tedious in use than the track coordinate method which it replaced, although the earlier procedure had to continue for tracks of greater curvature than the upper limit of the range of the compensator.

While the principle of both procedures has been described in these papers, certain features concerning application have not been fully developed. Both had to be used on

---

* The curvature of the tracks in a magnetic field actually gives a momentum measurement, and gradually these came to be reported in units eV/c. The earliest work used instead an 'equivalent electron energy', E_e, in units eV.

Fig. 1. The large electromagnet at the Magnet House. Cooling air is drawn into the coil assemblies through the two vertical ducts and is drawn out by an extractor fan which is behind the magnet. Here, the magnet is set up for photography down the exis of the near pole piece, and the end of the camera support frame is on the extreme right of the photograph. The cloud chamber is shown is the servicing position, withdrawn from the magnet gap. On the left hand side of the chamber is the release valve through which expansion is made. Just below the release valve block and just to the right of the lifting lug at the top of the chamber, may be seen two of the three calibrated expansion screws which determine the exact position of the piston before expansion.

Fig. 2. Here, the chamber has been returned to the operating position in the pole gap of the magnet. The slide, and the carriage which carries the chamber and its associated equipment in and out of the magnet, can just be distinguished in their relation to each other at the bottom centre of the photograph.

material of wide ranging quality, from the very high quality which is possible in a clean and well-adjusted chamber to the just-usable condition of a chamber deteriorating to the level at which dismantling and re-assembly had to be undertaken. In both the most prominent features on all tracks were the blobs of secondary ionization, and of these the most prominent were often centered slightly off the general trajectory line. In poor conditions, each point on a coordinate plot presented a problem, for the limited field of view of the eyepiece micrometer offered only a little material, the most prominent feature might or might not be exactly centered on the line of the trajectory, and there might be slight angular uncertainty in the direction of the track upon which the micrometer cross-wires had to be centred. Using the compensator the problem was much easier, for the whole track length was viewed at once, and decisions about the more prominent and potentially misleading features were more confidently resolved. At a late stage in what I am going to describe (in 1938) a more powerful compensator became available, with a compensatory limit about twice that of the earlier instrument. With the new compensator the great majority of the 'hard' particles fell within its range.

4. The Behaviour of Particles Traversing Metal Plates

Blackett and Brode had used the Magnet House installation to establish an energy spectrum of near vertical tracks up to $E_e \sim 10^{10}$eV, and this was based on about 800 tracks. When I came to the Magnet House in September 1936, this work was stopped and I undertook the application of the apparatus to the behaviour of particles traversing metal plates (in the first instance 1 cm of lead) mounted horizontally across the chamber. Quite apart from the occasional gross local distortion illustrated by Blackett and Occhialini and from the motion starting at surfaces in the chamber after expansion, when a considerable temperature difference arises between the chamber wall and the expanded gas (Wilson and Wilson, 1935), references in earlier work emphasised considerable difficulties, and there was no indication that anything had been done to resolve these. At least there seemed to be a great deal to put right!

From the outset, we used the apparatus in the mode in which a 45° mirror sent the image of a full-length strip of the chamber out sideways (parallel to the plane of the chamber disc): the hole down the pole-piece opposite the chamber was plugged, the pole separation was somewhat increased and the result was a rather lower but more uniform field. This choice was not really in question, for the whole operation depended on the provision of useful lengths of track in both halves of the chamber.

Photographs taken in this way were immediately interesting, particularly those well within the range of the compensator and for which there was little apparent change of curvature of tracks traversing the plate, of which there were many. The compensated image, seen at grazing incidence, would be expected to show two straight lines meeting somewhere in the plate. It was at a glance obvious that this was not so, and it was only a short step to the realization that the residual departures from straightness, and the failure of the two images to meet within the plate, represented the distortions which we aimed to reduce. These distortions were not the same from one photograph to another, but it was soon apparent that they were all of the same kind, and that in the plane of photography, in each half of the chamber, they were consistent with a slow circulation of gas (~ 1 cm s^{-1}) in that plane. The motions in the two halves were not simply correlated. Since the time lapse between the setting of the chamber and its triggering by the counter control were very different among the various photographs used, this circulation had to be seen as a (dynamic) steady-state of the chamber gas, and this could only be related to temperature differences existing between various parts of the chamber, derived, perhaps from the temperature changes which were plainly taking place in various parts of its surroundings — the main core of the magnet, the ducting containing the magnet coils and the air circulating to cool them, the room temperature and heat from the lamp which was certainly not negligible.

It did not seem promising to attempt to vary each of these factors in turn, and rather we decided to identify one feature which would most positively control the temperature of the chamber and then as far as possible to exclude the influence of the remainder. The clear choice was the main bulk of the magnet yoke, to which the back plate of the cloud chamber (itself a sheet of magnet steel) settled into the closest possible contact during operation. The 8 ton yoke also was by far the object of greatest heat capacity in the whole installation, and so varied most slowly in temperature whatever the surrounding influences might be.

Accordingly, a fibre-board enclosure was built around the chamber including the forward end of the camera and with a thick glass window through which the light of the illuminating flash entered. A further measure, to introduce a positive element of stability into the gas of the chamber, was to attach a small water-cooled pad to the lowest part of the chamber glass. Such a cooling attachment had in fact been used in earlier work to ensure that any liquid condensed in the chamber would collect at this region and so minimise any clouding of the window through which photographs were taken.

Initially, the thermal contact between the plate across the middle of the chamber and the main metal parts of the chamber structure had been slight, but in the light of what was now being attempted this was increased as far as other requirements allowed.

The success of these measures was dramatic, and the distortions which could so readily be detected and interpreted using the optical compensator were almost completely eliminated. In the first paper describing the use of the chamber with the metal place across the centre (Blackett and Wilson, 1937) we were able to report a precision of measurement in each half of the chamber almost as great as that in the earlier work in the undivided chamber, with twice as great a length of track and with a rather higher magnetic field.

The decision, which had been taken before my connection with the project, to photograph stereo-pairs on plates rather than on film, was now of particular value. Many photographs were, as a matter of routine, developed immediately after exposure, and these were regularly examined within perhaps two hours of exposure. Thus the absence of distortion was constantly monitored, although as I shall now explain, this was not the primary reason for this routine.

The cloud chamber used in all of this work was engineered for a specific purpose, and in operation was used in a way in which some of the measures which were reputedly favourable to good, clean, reproducable operation took second place. It had a large surface-to-volume ratio with rim regions in which clearing fields were not very effective, and it was designed for as rapid expansion as possible. Over and above these features the stabilizing of the gas in a direction perpendicular to the main surface area of the chamber was likely to lead to a saturation gradient down the chamber, while the presence of a plate allowed droplets falling after an expansion to slip below the plate through the gap between the back of the plate and the surface of the piston, while it hindered the diffusion of the condensant into the upper part of the chamber after near-stability had been re-established. Finally, there was some evidence of differential diffusion of the condensant through the rubber of seals and of the piston mounting. Altogether, it was not an easy chamber to work. After cleaning and filling it could be expected to work well for a time and then slowly to deteriorate to the condition when there was no alternative but to open it, clean and refill it. Following this process, and making minor alterations of expansion ratio in an attempt to prolong the useful life of each filling and to follow long term temperature changes, were the primary reasons for the ongoing scrutiny of photographs as they were taken.

Good examples of poor chamber conditions and of distortion are given in Rochester and Wilson (1952) page 8, and are fully described there. Although these illustrations (previously unpublished) are listed as from Manchester, very little work at Manchester used 1 cm plates, and they are probably photographs taken to Manchester from the Magnet House more than ten years previously.

Given time, there is no doubt that the general quality of photographs from this chamber could have been greatly improved, but, as will be clear from the remainder of this account, it would have been quite out of the question to stop the ongoing physical investigation for an indeterminate time for any feature of redesign.

5. The Identification of the Muon

(Note: for convenience, where a *name* was used for the newly-identified particles, e.g. mesotron, meson, the modern form, muon, has been substituted, although it was of course not used at the time.)

It is extremely difficult for physicists of a later age to understand how, until about 1937, the postulate of a new particle was simply not considered by those trying to understand cosmic ray phenomena. In this sense the neutron did not rank as totally new, since its relationship with protons and electrons could be emphasised. The positron also, as the striking confirmation of the Dirac formulation of the theory of electrons was readily acceptable: its experimental recognition was straightforward, even if the detailed interpretation of the theory was not. So these two particles were not thought of as disturbing the essential simplicity of a physical world which still appeared explicable in terms of photons, protons and electrons. As early as 1933, Kunze, working at Rostock, published a photograph of a track which seems certainly to be that of a particle of intermediate mass. He thought that it came from a nuclear explosion, and so it was perhaps a pion. To my knowledge, nobody engaged in the work I am now describing related this single record of something from a nuclear event with the abundant stream of 'hard' particles which then held our attention.

After the work already described on the substantial elimination of distorting motions of the gas in the cloud chamber, it was late in 1936 that our cloud chamber began to yield a useful output, and it was then that the 'state of play' on the energy loss and the scattering of cosmic ray particles as observable in a cloud chamber became important to me.

The local ongoing situation was set out in the Blackett (1937a) paper on the energy spectrum of cosmic ray particles, which had been submitted in mid-December of 1936. This greatly extended the data of the earlier work of Blackett and Brode (1936) to measurements of more than 800 tracks and described the energy spectrum up to $E_e \sim 10^{10}$eV. The spectrum exhibited maxima at $E_e \sim 10^9$eV and also at $E_e \sim 3 \cdot 10^9$eV. Of these the first was readily shown to be instrumental corresponding to the reduction of collection efficiency at low energies. The second was not unquestionably outside the possibility of statistical fluctuation from a smooth spectrum, but its interpretation if real had to be considered. Blackett suggested that this feature might represent a rather sharp change of behaviour of the particles at about $E_e = 2.5 \cdot 10^9$eV. For example it would be possible for about half of all particles to lose energy in a rapid degradation to shower-like particles near this energy, but so crude a model had no unique advantage. It led to another view, that all particles reach shower potentiality at near $E_e = 3 \cdot 10^9$eV, some unquestionably at higher energy, many at lower. It was this apparent abnormality of the cloud chamber spectrum which was related in a second paper (Blackett, 1937b) with the range-energy relationships for the cosmic ray beam which were then emerging from other forms of experiment.

But importantly, of about 1500 near-vertical tracks about 150 had $E_e < 6 \cdot 10^8$eV, a level below which *all* protons should have been recognisable as such in terms of ionization density, and no certain example of a proton was recognisable. There were three tracks, $E_e \sim 6 \cdot 10^8$eV, all positive, which might perhaps have been protons. Thus the fraction of the vertical beam which might be recognisable as protons was about 0.2%, a fraction which was in good enough agreement with two other estimates. Since this recognition was possible over only a small energy range, and the vast proportion of the vertical tracks corresponded to higher energies, Blackett concluded that as much as about 15% of the beam at sea-level might be protons. "The rest must be electrons, about equal numbers being positive and negative". There was no evidence for the existence of negative protons.

Also in the late spring, a paper of critical importance (Bhabha and Heitler, 1937) developed the concept of the electron-photon cascade as applicable to the cosmic ray beam seen intercepted in cloud chambers. Based on the work of Bethe and Heitler (1934) and on the compelling arguments put forward by v.Weizsäcker and by Williams that there was no reason for expecting any change in electron behaviour at $E_e \sim 137$ mc^2 (i.e. near the classical electron radius), they developed an electron-photon model which was postulated to be applicable at all energies, however high. Recognising that a treatment using only average behaviour at each step could not possibly describe electron penetration from the top of the atmosphere, they included a closely-argued treatment of fluctuations, and reached the conclusion, among other features, that primary electrons of $E_e > 3 \cdot 10^{12}$eV should be capable of leading to a reasonable sea-level flux. But they were well-aware of great difficulties in the interpretation of such features as the Rossi transition curve and of the latitude effect, both at sea-level and at high altitude. Above all, the absorption coefficient of the general cosmic ray beam under thick layers of matter had been established to be of the order of 1% of that predicted from their model. They therefore concluded *either* that the extremely hard radiation leading to this value of absorption consists of particles of protonic mass *or* that the quantum theory of radiation indeed breaks shown at the highest energies as far as electrons (and photons) are concerned. Like Blackett, they emphasized here that the proton possibility does not of necessity carry with it the existence of negative protons.

In retrospect, it is interesting to note the total neglect in these papers (and this was a position which I certainly did not question myself) of any role for what we would now describe as the strong interaction of protons.

I have already referred to early work of Anderson (1932) with a randomly-operated cloud chamber, comparing the efficiency of track collection with that of Blackett and Occhialini (1932, 1933) in their earliest reported work with counter-countrol. Among others, Anderson, working with Neddermeyer, was quick to recognise the importance of counter-control, and developed apparatus using this principle but which differed in important respects from that which was being installed at the Magnet House. The Magnet House chamber was particularly developed in a form directed to the establishment of precision measurements of the behaviour of single particles, and indeed relatively rarely, and then by chance, other accompanying particles appeared in measurable conditions. Anderson and Neddermeyer, on the other hand were much more concerned with the phenomena of shower groupings. Their rather smaller chamber was proportionately deeper, and gave in comparison data about such shower groupings which was more complete within its limitations than were possible at the Magnet House. They probably did not put so much stress on gas stability, with a greater counting rate and quicker recycling of the chamber.

In August 1936 Anderson and Neddermeyer published data which came close to establishing the existence of a muon. This paper was primarily a study of the differences of shower records near sea-level and at high altitude on Pike's Peak (4300 m). The proportion of counter-controlled photographs showing features both entering the chamber from outside and also originating in the metal plate within the chamber was much greater at Pike's Peak than at sea-level: not only were there more showers recorded but they also tended to be larger. While this effect was qualitatively clear, Anderson and Neddermeyer were cautious in making numerical estimates, for showers operated the counter control in a way which biased against single traversals and which further favoured large rather than small showers.

They stated that "until the absorption laws of high energy electrons and photons are known ··· the soft component of cosmic rays cannot be identified", and "direct measurement of the energy loss of 'electrons' in 1 cm lead show nuclear impacts involving loss much greater than ionization and secondary electron formation".

Using a thin (0.3 cm lead) plate, the loss of energy by incident electron was like that predicted by Bethe and Heitler (1934) up to energies (E_e) of about $3 \cdot 10^8$ eV, but at 'high energy' (not very well determined) the number of secondaries was an order of magnitude less than the Bethe-Heitler treatment would suggest. Therefore, *either* the theory breaks down somewhere about 10^9 eV *or* these higher energy particles falling on the plate in the cloud chamber are not electrons. Many features seemed to require Bethe-Heitler electrons at much higher energies, and it was accordingly important to identify the penetrating particles. They proceeded to note that protons would suffer an ionisation loss of about $2 \cdot 10^9$ eV in the atmosphere and that geomagnetic deflexion would cut-off protons approaching the earth with momenta $< 5 \cdot 10^9$ eV/c: at this stage they were relating the penetrating particles in the atmosphere directly with protons in geomagnetic trajectories.

In May 1937, critical conclusions were reached from experiments in which the lead plate across the chamber was replaced by a plate of platinum, which is very much denser. They established a distinction between single particles and particles in showers up to about $E_e = 4 \cdot 10^8$ eV, the single particles being characteristically penetrating. Distortions in the chamber tended to be serious, and limited confidence in measurements which appeared to show small energy losses at the plate. Very crude scattering measurements were used to establish that energy attributions were reasonable. No interpretation of scattering (measured to about 5°) in its own right was attempted.

In this paper (Neddermeyer and Anderson, 1937) it was concluded that *either* electrons (positive or negative) can possess some property other than mass and charge and capable of accounting for the absence of numerous larger radiation losses in a *heavy element, or* there exist particles of unit charge but with mass (not necessarily unique) larger than that of a free electron and smaller than that of a proton. They inclined to the second hypothesis, but since such particles seem to be absent from ordinary matter, some effective process for removing them must happen.

In May 1937 also, the first energy loss measurements from the Magnet House installation were reported (Blackett and Wilson, 1937). Roughly the first two months after my arrival there in September had been taken in identifying and then greatly reducing the distortions due to persistent gas movement in the chamber: these have already been discussed earlier in this article. Now, track measurements were undertaken using a 1 cm lead plate and a field of 10^4 gauss; there were also some measurements with a 0.33 cm lead plate with a $3.3 \cdot 10^3$ gauss field (for low-energy incident particles).

The average behaviour derived from these measurements accorded closely with those reported by Neddermeyer and Anderson for $E_e < 3 \cdot 10^8$ eV and also with a recently described measurement by Crussard and Leprince-Ringuet (1937) at roughly $8 \cdot 10^8$ eV. At an intervening energy, $4 \cdot 10^8$ to $5 \cdot 10^8$ eV, for which no comparison material was available, the proportional loss appeared to be even lower than at the rather higher incident energy. At the time of this publication, these measurements appeared definitive, with no data in conflict with them, and it seemed necessary to examine the views of theoreticians (Nordheim, 1936 and others) which concerned the question of a breakdown of the radiation from electron-like particles at high energy.

Nordheim had explored, in a semi-classical model, changes in the radiation behaviour of 'electrons' arising from various changes at the limiting impact parameters of the model, and we drew attention to a variation which would describe reasonably what we had observed over the whole range to which our measurements referred. But views were changing with great rapidity (it will be noted how much was in fact published in May 1937) and, when the American journals of that date reached London, the problems engaging attention were quite different. (Contacts between Europe and America were then slow, with copies of even major journals coming only by slow sea passage.) This time lag had interesting consequences. The data in Blackett and Wilson (1937) had affected planning at the Magnet House perhaps two months earlier, and already work was going forward using a 2 cm plate of copper rather than that of 1 cm of lead. The choice of the thicker plate related to apparently greater loss at about $8 \cdot 10^8$ eV and above which has just been indicated, and any study of this effect in lighter metals seemed to require a greater absorbing layer. This work went forward and was in due course published (Wilson, 1938), but the consequence was that immediately the Neddermeyer and Anderson (1937) paper became available (probably sometime in June) we had already satisfied ourselves that a 2 cm plate could be used in the chamber without leading to serious distortion. Accordingly, with the advantage of a more massive plate manifest in their work with platinum, we were able to plan to use a 2 cm heavy metal plate (in fact gold) as soon as it could be prepared, with confidence that this would not lead to new distortion problems. The 2 cm gold plate was probably fitted some time in July 1937* and was then used continuously until work ceased for the dismantling of the Magnet House, when all its equipment was transported to Manchester in January 1938.

The main work of the evaluation of the Magnet House work with the gold 2 cm plate was not carried out until after the move to Manchester, and then it could only rank as one task of many, but some gold plate data was available late 1937, and part of this was summarized in a paper (Blackett, 1938) which was submitted for publication in December 1937. The purpose of this paper was to re-emphasize and strengthen the point of view brought out in the earlier Magnet House work, that there was little or no overlap between the energy range for which the particles were electron-like and that for which they were penetrating. Blackett used here calculations which I had completed but did not publish until later in the year, (Wilson, 1938), deriving the distribution of probable energy losses in plates of various characteristics, and was able to show that for $E_e < 2 \cdot 10^8$ eV the *spread* of losses as well as the *average* value closely followed what had been predicted for electrons. At this stage the gold measurements were simply used, track by track, to distinguish whether particles were or were not penetrating. This separation, seemed to be stronger than Neddermeyer and Anderson appear to have thought, and led to a discussion which clearly aimed at an understanding of a mechanism by which the penetrating particles could transform in due course into electrons, and it postulated as a necessary third body some target within the atmosphere and *not* an additional product in free decay. (The involvement of a neutrino, which was only then becoming generally understood in its relationship to β-decay, was not of course considered, and thought in general on the problem remained hampered by the persisting belief that the sea-level components were residues of like primary, extraterrestrial components.)

During the later part of 1937, two positive identifications of tracks as those of particles of intermediate mass made any question of the fact of new particles the more difficult to refute but also showed that the particles were relatively heavy. Stevenson and Street (1937) operated a counter-controlled chamber with field but with a time

---

* All records and material were taken to Manchester and worked upon through 1938 and early 1939. During the war some records seem to have been lost, and this particular date is based only on personal recollection.

delay of about one second before the completion of expansion, in order to allow ions to diffuse and condensation upon them to yield countable drops: since the fastest possible expansion was not aimed at, photographs of high quality were achieved. A track identified by curvature and a direct count was estimated to be that of a particle of mass about 130 m_e. At almost the same time Nishina *et al.* (1937) obtained a track in a more conventional manner which corresponded to that of a particle of mass between 1/7 and 1/10 that of a proton (round about 200 m_e). These two photographs while establishing an order of magnitude of the mass were not sufficient to show whether there was a unique mass. With my own experiences of track distortions in mind, the distortion in a time of one second adopted by Stevenson and Street seems to have been unknown, and might have been quite large. The Nishina measurement seemed much the more strongly based, and within a few months measurements reported by Neddermeyer and Anderson (1938) and by Williams and Pickup (1938) also centered around 200 m_e. Most people from this time thought in terms of a unique mass. (The actual Nishina photograph was not published until rather later, and shows features which are not completely understood.)

By the end of 1937 data from the Magnet House had been assembled on the scattering of particles traversing plates (Blackett and Wilson, 1938). It had been pointed out to us by E.J. Williams, in relation to the time when active thought had been directed to mechanisms for the suppression of radiation from electrons at high energy, that any such mechanism was so intimately connected with that leading to Coulomb scattering in the absorbing matter that this also would be greatly reduced. For a singly-charged particle of whatever mass and without such suppression, scattering for example in 1 cm of lead would be multiple, without any significant 'tail' of single scattering. From this situation, the range of measurements required clearly had to extend well above $E_e \sim 3 \cdot 10^8$ eV, beyond which tracks of normal electron behaviour were rare, and the measurements were aimed to be carried out up to several times 10^9 eV.

This was a measurement of particular difficulty, and I do not think any other group of workers was equipped to undertake it. Since the angle of scattering in any plate was expected to be of the same order as the magnetic deflexion, its measurement had effectively to be that between the tangents at the plate surface of the entrant and emergent tracks, and the suppression of distortion close to the plate, where it was always greatest, assumed particular importance. The result of this work was quite clear-cut, even although a relatively small number of tracks were measured. Scattering remained normal up to at least beyond $2 \cdot 10^9$ eV: it was distributed over a normal multiple scattering Gaussian, with no indication of any single scattering 'tail'. Very many of these particles were certainly *not* protons.

From work which will be referred to shortly below, Bhabha (1938a) had concluded, independently of the first two isolated measurements which have been referred to, that the unknown particle was most likely of mass $\sim 100 \ m_e$, rather than $\sim 10 \ m_e$, and had indicated in his calculations how the difference between these postulates would show up in the number of secondary particles (products of 'knock-on' electrons) emerging from a plate. My analysis of the gold material had now gone far enough to allow a test of this conclusion, and in a letter to 'Nature' for July 9th 1938 (Wilson,

1938), I grouped all the available material on primary particles entering the plate according to whether the primary was of energy above or below $3 \cdot 10^9 \, \text{eV}$. I used the estimate for secondaries $E_e > 10^7 \, \text{eV}$, since there was considerable uncertainty about the efficiency with which lower energy secondaries would in fact succeed in emerging from the plate. The results for the two groups were self-consistent and corresponded, in Bhabha's analysis, with a particle mass greater than 100 m_e and certainly consistent with a value 200 m_e.

Through the later part of 1937 and on into 1938, when substantial amounts of cloud chamber data were becoming available, the theoreticians were at no time in a position to reach firm conclusions about the problem of the hard and soft components on cloud chamber data alone. (Our main contacts at the time were with Bhabha, Heitler and E.J. Williams.) Neddermeyer and Anderson (1937) had satisfied themselves about the necessity to postulate an entirely new particle, although in this particular paper without any detail, and this in itself was indeed a bold step in the climate of the time. This postulate acquired as it were flesh as well as bones from the direct measurements of Street and Stevenson and Nishina and soon by others discussed above: it was also reinforced by my measurements on secondary production. The scattering measurements (Blackett and Wilson, 1938) had clarified the field within which the relationship of hard and soft components had to be sought. But the part played by protons was obscure, and all workers were puzzled by the way in which the vast majority of non-protonic hard particles (the new particles) seemed just to vanish at roughly the energy at which electrons appeared in roughly comparable numbers.

In important papers during 1938, (Heitler, 1938; Bhabha, 1938a, b), theoreticians turned to the increasing quantity of data then being brought together about phenomena within the atmosphere and underground, data which had to yield an account consistent with what was now being established in cloud chamber observations. The work which is covered in these papers falls mainly outside the scope of this article, and since they were written in ignorance of the nature of any of the secondary (atmospheric) cosmic radiation in relation to the extraterrestrial true primaries, it is not surprising that these workers had to accept rather forced agreement between predictions and observation. However, it was from Bhabha (1938b) that I first became aware of the work of Yukawa, and to the pervasive role of strong interactions. The problems posed for various phenomena by this approach form the background against which the last two papers derived from the Magnet House material were written. Looking back, I have no doubt that it was this linkage of material we had already obtained and were now examining with the theoretical studies of unfamiliar concepts which were now gaining real momentum which more than anything else excited me, and forced me to understand the importance from both sides of constant dialogue between experimentalists and theoreticians.

I am not going to attempt to relate these two papers with ongoing work elsewhere: in fact I do not think that I was maintaining more than passing concern with such work. Much of my energy was once more concentrated on the engineering and instrumentation for further work, and in the last months other matters became dominant: the second of these papers (Wilson, 1940) was completed early in August of 1939 and submitted for publication about ten days before I left Manchester for work which was to keep me from Manchester and cosmic ray studies until 1945.

These papers differed in type from the earlier work. My arrival in September 1936 had been into an atmosphere in which precision was to be a dominant characteristic, and for the early part of that time this was not what was really effective. We began work (that is to say on particle traversals of metal plates) in a period of great activity with material emerging elsewhere which was, for the immediate purpose, ahead of us. Our very superior position in precision measurement did not avail us in this initial stage. It was only as time went on that the advantages of the quality of our data became manifest: this was so in material which I have already described, but it was most apparent in the two papers which I now turn to. As a matter of detail, these papers use data measured through a much improved curvature-compensating prism system which allowed the range of this mode of curvature measurement to be extended down to energy measurements of roughly $E_e \sim 3 \cdot 10^8$ eV.

The first paper (Wilson, 1939) gave the final conclusions coming from the measurements (completed in December 1937!) of the energy-loss of particles in 2 cm of gold. As is explained in the paper the conditions of use effectively excluded all electrons and so these results refer to the penetrating particles only.

Over the range $2 \cdot 10^8$ eV $< E_e < 7 \cdot 10^8$ eV these showed close agreement of observed energy loss with that computed by Bhabha (1938b) for particles of mass about 200 m_e. Two slow tracks of electron energy (E_e) of 165 MeV and 230 MeV respectively yielded energy losses significantly greater than those for the $2 \cdot 10^8 - 7 \cdot 10^8$ eV band, and since the indication for that band was clear that only ionisation losses were of any importance for these particles in gold, it seemed permissible to interpret this greater energy loss as that of normal ionization theory for particles moving towards the end of their range. The first of these tracks indicated a muon mass value of (170 ± 20) m_e, the second was less informative, and yielded $m_\mu = (250 \pm 50)$ m_e. Of these the first, more precise measurement fell very near to that of the Nishina *et al.* (1937) estimate: of it I wrote "the curvature change is very sensitive to an *increase* of the assumed mass of the particle, and the measurement cannot be reconciled with a mass as great as 220 m_e". The true mass of the muon of course does fall below this limit, and the confidence with which this limit could be sustained became important when detailed calculations of the properties of Yukawa's particle demanded a mass greater than 250 m_e.

These two mass-sensitive measurements were the only ones to be possible with our techniques out of many hundreds of plate traversals. At the level of chamber performance we were achieving, when condensation conditions were often different in the two halves of the chamber, *neither* would have been recognised by its intrinsic ionization even in the sector below the plate! This single fact emphasises the difficulty which had confronted us (and other workers) from the outset in any consideration of what became of the hard particles.

The second paper was more important (Wilson, 1940). It very greatly extended the data on scattering which had been used in earlier work (Blackett and Wilson, 1938) particularly by the inclusion of a substantial number of measurements from data derived using the gold plate. The measurements, under the most stringent conditions possible were estimated to be made to better than $0.1°$. This work seemed particularly necessary in the light of a detailed critical development (Williams, 1939) of the theory of scattering, taking into account the possible short-range interactions between muons and nuclear particles. A fairly sharp angular separation between the Coulomb scattering and any short-range force deflexions was indicated.

The measurements yielded a well-defined Gaussian distribution, and indeed the indicated magnitude of multiple Coulomb scattering distinguished between that for a finite nucleus and that which would have been given by a point nucleus. Only one track showed scattering significantly outside the Gaussian of Coulomb scattering, and for this the probability of occurrence as a fluctuation from the main distribution was less than one in a thousand. Thus, the non-Coulomb scattering of muons was certainly small, and even the single event observed might have been the result of one of the small minority of protons (assumed to be subject to short-range forces).

This work was carried out when the development of the relationship of the muon with the Yukawa particle was unresolved. The short-range scattering established in this paper for muons, which has never been seriously questioned, represents a second critical point in the quantum-mechanical treatment of the behaviour of the free Yukawa particle, for which this measured scattering has proved to be about 100 times too small. These two papers showed that the mass of the muon was significantly too small, and the non-Coulomb scattering perhaps one hundred times too small to meet the properties which were in due course demanded for the Yukawa particle.

Although the analysis of data was to continue for another eighteen months, my thoughts go back to the actual physical end of work at the Magnet House: the day when I lifted the gold bar from the cloud-chamber, wrapped it in a sheet of notepaper, put it in my raincoat pocket and walked down from the Magnet House to Hatton Garden, and to the bullion merchant from whom it had been bought only about six months earlier.

REFERENCES

Anderson, C. D.: 1933a, *Phys. Rev.* **43**, 491.
Anderson, C. D.: 1933b, *Phys. Rev.* **44**, 406.
Anderson, C. D. and Neddermeyer, S. H.: 1936, *Phys. Rev.* **50**, 263.
Bethe, H. A. and Heitler, W.: 1934, *Proc. Roy. Soc. A* **146**, 83.
Bhabha, H. J. and Heitler, W.: 1937, *Proc. Roy. Soc. A* **159**, 432.
Bhabha, H. J.: 1938a, *Proc. Roy. Soc. A* **164**, 257.
Bhabha, H. J.: 1938b, *Proc. Roy. Soc. A* **166**, 501.
Blackett, P. M. S.: 1936, *Proc. Roy. Soc. A* **154**, 564.
Blackett, P. M. S.: 1937a, *Proc. Roy. Soc. A* **159**, 1.
Blackett, P. M. S.: 1937b, *Proc. Roy. Soc. A* **159**, 19.
Blackett, P. M. S.: 1938, *Proc. Roy. Soc. A* **165**, 11.
Blackett, P. M. S. and Brode, R. B.: 1936, *Proc. Roy. Soc. A* **154**, 573.
Blackett, P. M. S. and Occhialini, G. P. S.: 1932, *Nature* **130**, 363.
Blackett, P. M. S. and Occhialini, G. P. S.: 1933, *Proc. Roy. Soc. A* **139**, 699.
Blackett, P. M. S. and Wilson, J. G.: 1937, *Proc. Roy. Soc. A* **160**, 304.
Blackett, P. M. S. and Wilson, J. G.: 1938, *Proc. Roy. Soc. A.* **165**, 209.
Crussard, J. and Leprince-Ringuet, L.: 1937, *Compt. Rend.* **204**, 240.
Heitler, W.: 1938, *Proc. Roy. Soc. A* **166**, 529.
Kunze, P.: 1933, *Z. Phys.* **83**, 1.
Neddermeyer, S. H. and Anderson, C. D.: 1937, *Phys. Rev.* **51**, 884.
Neddermeyer, S. H. and Anderson, C. D.: 1938, *Phys. Rev.* **54**, 88.
Nishina, Y., Takeuchi, M. and Ichimiya, T.: 1937, *Phys. Rev.* **52**, 1198.
Rochester, G. D. and Wilson, J. G.: 1952, *Cloud Chamber Photographs of the Cosmic Radiation*, Pergamon Press, Oxford.
Street, J. C. and Stevenson, E. C.: 1937, *Phys. Rev.* **52**, 1003.
Williams, E. J. and Pickup, E.: 1938, *Nature* **141**, 648.

Williams, E. J.: 1939, *Proc. Roy. Soc. A* **169**, 531.
Wilson, C. T. R. and Wilson, J. G.: 1935, *Proc. Roy. Soc. A.* **148**, 523.
Wilson, J. G.: 1938a, *Proc. Roy. Soc. A* **166**, 482.
Wilson, J. G.: 1938b, *Nature* **142**, 73.
Wilson, J. G.: 1939, *Proc. Roy. Soc. A* **172**, 517.
Wilson, J. G.: 1940, *Proc. Roy. Soc. A* **174**, 73.

WHEN MUONS AND PIONS WERE BORN

Erich R. BAGGE*

1. The Break-down of Rostock's Electricity Power Supply and Kunze's Cosmic Ray Experiments

When in the year 1932 W. Kunze performed his cosmic ray experiments with a Wilson cloud chamber in the university of Rostock the electrification of this town was still a small scale one. For measuring the momenta of the cosmic ray particles Kunze installed a large coil with 1.2 tons of copper to produce the necessary magnetic field in the cloud chamber. For that purpose a strong current of about 1000 Amperes at 500 Volts had to be sent through this coil and the only way to do this was to use the electricity power supply of the town (Kunze, 1933).

At that time the full power level of the electricity station was still so low that always when Kunze operated his cloud chamber – about 500000 Joules in the time of a second, normally in some minutes distance – the whole electricity net of the town went down for about half of a second: The copper coil produced practically a short circuit for the electricity generator.

Soon all people of Rostock knew when the electricity lights darkened in regular intervals: "Aha, Prof. Kunze now is making his cosmic ray experiments!"

The story was reported later on in all German papers and people smiled about this amusing situation. Especially in the physics laboratories of the universities and in lectures the story was told to the younger students to demonstrate them the short circuit effects.

But there was nobody who could confirm what was added humorously by the elder physicists: Namely, that always the nails have been drawn out from the walls of Kunze's laboratory and were flying like fired bullets through the room when during an expansion of the Wilson cloud chamber the magnetic field of the copper coil was on the peak of excitation at about 18500 Gauss.

True or false, the story still is told from time to time, when elder physicists – the former students of that time – come together in physics conferences and always the same smile is on their faces.

Nevertheless another thing is certainly true: Prof. Kunze was the man who observed and realized in those experiments for the first time positively electrical charged particles in cosmic radiation of an intermediate mass "between that ones of electron and proton" (Fig. 1).

He did it 1932, three years before H. Yukawa (1935) published his famous and fundamental ideas on the nature of nuclear forces and as a consequence of them the hypothesis of the possible existence of unstable particles, now called pions. It was also already four years before C.D. Anderson and H. Neddermeyer (1936) made their

* University of Kiel (emeritus), Kiel, West-Germany.

Y. Sekido and H. Elliot (eds.), Early History of Cosmic Ray Studies, 161–164.
© *1985 by D. Reidel Publishing Company.*

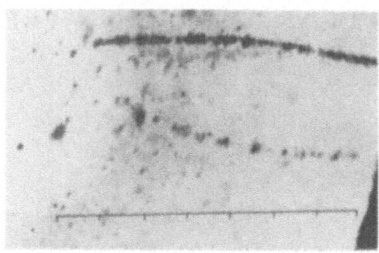

Fig. 1. One of Kunze's cosmic ray pictures of Rostock which shows a relatively fine electron track of 37 MeV and a second track of a positively charged particle much more ionizing than an electron but obviously lower than a proton. Kunze is speaking in the text about a particle of unknown character (1933).

Fig. 2. The professors: Niels Bohr, Werner Heisenberg and P.M. Dirac at a discussion during the 1962 – meeting of the physics – Nobel prize-winners at Lindau (W-Germany).

Fig. 3. Prof. Erich R. Bagge (left) with his co-workers discussing a Kiel-neonhodoscope picture of an extensive air shower (1980).

splendid experiments on the "mesotrons" by which they established the real existence of these particles in a way convincing all physicists.

The picture we see in Fig. 1, is a copy of one of Kunze's original cloud chamber photos in the "Zeitschrift für Physik" of the year 1933. There is no doubt that his interpretation of the phenomenon was the correct one. But at that time theoretical aspects of elementary particle physics have not been ripe for a serious discussion of Kunze's results. So his observations were forgotten and only by Yukawa's theory and Anderson's and Neddermeyer's extensive and very careful investigations the particles of "intermediate mass" came fully to the conscience of the physicists all over the world.

2. Kulenkampff's Ideas on the Penetrating Particles in the Atmosphere

At the same time an X-ray-physicist, H. Kulenkampff at Munich became very much interested in cosmic radiation. Especially he was fascinated by the unintelligible high penetrating power of the cosmic rays on their way through the earth's atmosphere on the one side and the relatively high intensity decrease on the other. His own experiences in the field of X- and γ-rays and of electrons in gases lead him to the conclusion that all these quanta and particles cannot be able to reach the ground level of the atmosphere. So he discussed several ideas for the understanding of this curious phenomenon.

He also made investigations to find out if there are intermediate steps by neutral particles to explain this astonishing effect of cosmic radiation.

It must have been a highlight in his life, when he heard about Yukawa's theory of mesons. He realized at once that such particles could be those, which are responsible for the observed penetrating power of cosmic rays in the atmosphere. Their masses, about 200 times bigger than the electron mass, are large enough to avoid higher radiation losses on their way through the air. Further on Yukawa's idea of the instability of these particles presented the reason, why they cannot be observed as members of stable atoms but it can explain the big absorption effects of the cosmic radiation on the way through the atmosphere. When he gave a lecture on this new interpretation during a physics conference at Jena (Kulenkampff, 1938) a young theoretical physicist from Heisenberg's institute at Leipzig was there. Just at the same time he had made theoretical investigations on electron-photon-cascades on the basis of the theories of Bethe and Heitler (1934a, b) and Oppenheimer (1936) for pair production using also the Bethe-Bloch-formula (see for example, Heisenberg, 1953) for the electron stopping power in gases. So he was well informed on the difficulties of cosmic ray theory concerning the penetration of the atmosphere. At once after his return to Leipzig he reported with great enthusiasm to Heisenberg about Kulenkampff's lecture on the possible role of Yukawa's mesons.

3. Heisenberg and Euler's Disappointment about the Meson-Lifetime

Only a few days later Heisenberg and Euler had developed a theory on the basis of Kulenkampff's ideas. The interesting result was, that they could explain the observed particle spectra of the so interpreted "muons" on the ground level assuming a muon-lifetime of $2.7 \cdot 10^{-6}$ sec. This was nearly the same lifetime as that one of Yukawa's particles calculated in one of his first papers ($\sim 10^{-6}$ sec).

Very happy about this situation they formulated at once a new chapter in their already finished report on cosmic radiation for publishing in "Ergebnisse der exakten Naturwissenschaften" (Heisenberg and Euler. 1938).

Heisenberg was so highly excited that he wrote also a letter to Yukawa informing him on this amazing result hoping that Yukawa would be as happy as he himself about this application and – in a certain sense – wounderful confirmation of his theory.

Several weeks later nearly at the same time, when the galley proof of their publication in "Ergebnisse der exakten Naturwissenschaften" arrived at Leipzig, Heisenberg also received a letter from Yukawa indicating that there was a mistake in his (Yukawa's) paper just in that caluclation for the lifetime of the decaying particle which was found in a different way by Heisenberg and Euler. Yukawa's theory really doesn't give a lifetime of about 10^{-6} sec. The correct value instead is $2 \cdot 10^{-8}$ sec.

For the first moment the younger Euler was disappointed very much. In his eyes this was a full breakdown of the splendidly operating picture they had performed. But Heisenberg did not agree. He said to his pupil, as Euler told me still on the same day: "Look, our calculations show that a lifetime of $2.7 \cdot 10^{-6}$ sec allows us to understand how it is possible that cosmic ray particles are able to penetrate the atmosphere and how they can produce nearly the correct decrease of cosmic ray intensity in the atmosphere. Let us stay at our result! The lifetime of 10^{-8} sec is far too short to explain the observed effects. We should wait and see what the future will bring."

So they published their paper in the originally intended form and the results in this report later on played a very fruitful role in the discussions of the cosmic ray community.

Today we know that in this case like in many others Heisenberg's highly cultivated physical sensitiveness and his steadiness collaborated together very well in an admirable manner. In fact, there are engaged in this connection really two sorts of particles:

The muons with a lifetime of $2 \cdot 10^{-6}$ sec and the pions which have been detected later on by Gardner et al. in accelerator experiments (Gardner and Lattes, 1948) with a lifetime of $2 \cdot 10^{-8}$ sec. They belong together very closely. The pions are generated in nucleon-nucleon collisions and in decaying they produce those muons which penetrate the whole atmosphere and in many cases still several hundred meters of the ground.

REFERENCES

Anderson, C. D. and Neddermeyer, S. H.: 1936, *Phys. Rev.* **50**, 263; 1938, *Rev. Mod. Phys.* **11**, 191.
Bethe, H. and Heitler, W.: 1934a, *Proc. Roy. Soc. London, A* **146**, 83; 1934b, *Proc. Cambridge Phil. Soc.* **30**, 524.
Gardner, E. and Lattes, C.: 1948, *Science* 1.
Heisenberg, W.: 1953, *Kosmische Strahlung,* Springer, Berlin.
Heisenberg, W. and Euler H.: 1938, *Ergeb. Exakten Naturwiss.* **17**, 1.
Kulenkampff, H.: 1938, *Verh. Dtsch. Phys. Ges.* **19**, 92.
Kunze, P.: 1933, *Z. Physik.* **83**, 1.
Oppenheimer, J. R.: 1936, *Phys. Rev.* **50**, 389.
Yukawa, H.: 1935, *Proc. Phys.-Math. Soc. Japan,* **17**, 48; 1937, **19**, 717, 1087; 1938, **20**, 319.

PART 4   SEA, MOUNTAIN AND UNDERGROUND

ULTA-PENETRATING AND EXTRA-TERRESTRIAL ASPECTS

SOME RECOLLECTIONS OF EXPERIENCES ASSOCIATED WITH COSMIC-RAY INVESTIGATIONS

Scott FORBUSH*

By 1930 a few investigators had begun to make continuous registrations of cosmic-ray intensity, using ionization chambers. Studies of any observed variations were naturally expected to provide some clue about the source of cosmic-rays.

Fig. 1. S. E. Forbush

Among the first continuous registrations were those obtained by Axel Corlin in northern Sweden in June, 1930 and from September, 1932 to March, 1933. These data together with their comprehensive analyses were published by Corlin in his academical dissertation *"Cosmic Ultra-Radiation in Northern Sweden,"* published in the Annals of the Observatory of Lund, Nr. 4, 1934. This publication contains excellent summaries of the status of cosmic-ray investigations based on research by other investigators. In spite of excellent observational work and careful analyses Corlin found no variations that were ascribable to causes other than those associated with barometric variations. The limited sensitivity of his equipment and the large erratic pressure variations in Northern Sweden plus large variations due to μ-meson decay (then unknown) from variable heights, probably obscured more interesting effects. In addition these registrations were made near a sunspot minimum with less chance of finding magnetic storm effects.

My own first contact with cosmic-ray observations involved assisting with eye-readings from a Kolhörster ionization-chamber on board the Carnegie Institution of Washington non-magnetic yacht, The Carnegie, from September, 1929 until its destruction by explosion and fire in November, 1929 in Samoa. From this I was extremely lucky to have escaped alive.

* Dept. of Terrestrial Magnetism (emeritus), Carnegie Institute of Washington, Washington, D.C., U. S. A.

Y. Sekido and H. Elliot (eds.), Early History of Cosmic Ray Studies, 167–169.
© *1985 by D. Reidel Publishing Company.*

About 1933 or so the Carnegie Instituion of Washington Committee for coordination of cosmic-ray research sponsored the construction of seven model C cosmic-ray ionization chambers designed by Professor Arthur A. Compton and Dr. Ralph D. Bennett. These chambers of 19.3 liters capacity were filled with very pure argon at 50 atmospheres pressure. The ionization current produced by cosmic rays in the main chamber could be balanced by an adjustable ionization current in a small chamber (inside the main chamber) by β rays from metallic uranium. Thus, it was possible to record intensity variations on photographic rolls, 6 cm wide, with ample sensitivity. Unfortunately, the balance ionization current was subject to drifts, probably due to radioactive decays of unknown origin. These disturbing drifts also changed when the balance was readjusted. Although these slow drifts did not interfere with finding the world-wide changes associated with magnetic storms, they nearly obfuscated the demonstration of significant world wide changes over a period of several years.

Before the model C meters were sent by the Chairman of the Cosmic-Ray Committee of the Carnegie Institution of Washington, to cooperating observatories in widely separated world wide locations, the seven meters had been operated simultaneously at the University of Chicago. The Committee designated Forbush to be responsible for the operation and maintenance of the meter installed at Cheltenham, Maryland in June, 1936. About this time a paper "Fluctuations in cosmic-ray ionization as given by several recording meters at the same station" appeared in the Physical Review (R.L. Doan, *Phys. Rev.* **49**, 107–122, 1936).

This responsibility for the meter at Cheltenham and more especially because of his interest in statistical analysis in geophysical problems led Forbush to scrutinize this paper very carefully. His interest in statistics generally had been greatly influenced by the original outstanding contributions (see for example: Cosmical and Geophysical Periodicities, *Terr. Mag. Elect.*, 1–60, March, 1935) of Professor S. J. Bartels to statistical procedures in geophysics and by fortunate personal association with him. The paper referred to was found to contain several erroneous conclusions due to common statistical pitfalls and some other mistakes. Consequently, the paper provided no valid useful comparison of the meters. This stimulated Forbush with a lasting endeavor to try to derive from cosmic-ray data, conclusions that would stand the scrutiny of rigourous statistical tests. Naturally this was essential to determine what apparent variations of intensity could be regarded as real. Since the meters had already been sent to Godhaven, Cheltenham, Mexico City, Huancayo, and Christchurch, the reliability and defects of performance had to be assessed from data obtained from the meters in these separated places. Had there been data from only a single station, it is doubtful that very much could have been learned about variations in intensity.

The interest of Forbush in the value of statistical procedures together with considerable serendipity that was not Corlin's good fortune, such as the occurrence of magnetic storm effects, solar flare increases and world-wide changes provided a continuing interest in such investigations. Statistically invalid conclusions claiming reality for apparent diurnal variations often appeared in print. Hence, Forbush started measuring and reducing records from Cheltenham in 1937. After applying the statistical procedures of Bartels, Forbush was quite surprised to find a statistically significant diurnal variation from 273 complete days of data. About this time Professor P.M.S. Blackett showed that the varying altitude at which μ-mesons are generated would, because of their short life time, result in unwanted diurnal and seasonal variations, which could not be reliably calculated and removed. For this reason further investigation of diurnal variation was avoided for about 30 years. Not until the mid 1950's were data available from neutron monitors from which reliable determinations (without the unfortunate μ-meson decay effect) of the diurnal variation were made by several other investigators. This led to the accepted corotational theory for the diurnal variation which accounted for the observed maximum in the asymptotic direction 90° East of the sun.

In the 1970's data from the CIW network of ionization chambers were available for a much longer period than that from neutron monitors. This stimulated an analysis of the diurnal variation from data in spite of the distressing meteorological perturbations due to μ-meson decay from variable altitudes. These perturbations were luckily found to be well eliminated in the yearly means of the diurnal variation by proceeding on the assumption that, at a given station, the yearly means of the unwanted contribution to the diurnal variations were the same for every year. That this proved to be the case was due to the kind cooperation of nature and to the use of statistical procedures. This led to the finding that the observed yearly means of the diurnal variation were due to two components. The first was assumed to have its maximum in the asymptotic direction 90° E of the sun that had been established by other investigators using neutron data. From this the additional component was found to have its maximum in the asymptotic directions about 128° E of the sun, along the Archimedian spiral stream. Together these two components provided a thoroughly satisfactory statistical account of the annual mean diurnal variation. In addition the amplitude of the component with maximum in the asymptotic direction 128° E of the sun was shown to vary with a "period" of two sunspot cycles and reversals of sign at the time of reversals in the solar general magnetic field as determined at the Hale Observatory of the Carnegie Institution of Washington. The amplitude of the corotational component was found to increase with magnetic activity varying somewhat like the sunspot cycle. Thus, in one way or another phenomena originating on the sun were directly or indirectly responsible for all the established variations. Since no siderial diurnal variation was established, the source of primary particles is evidently isotropic. There is thus some satisfaction in concluding that most of the more interesting effects derivable from the Carnegie Institution of Washington ionization chambers, may now have been obtained. A much greater satisfaction has come from the meeting and making of so many lasting friendships with outstanding investigators from all parts of the world.

HIGH ALTITUDE OBSERVATORIES FOR COSMIC RAYS AND OTHER PURPOSES

Serge A. KORFF[*]

1. Introduction

The original proof that cosmic radiation was of extra-terrestrial origin came from balloon flights made by Hess and others, in the early years of this century. Since the cosmic radiation was attenuated by the atmosphere, it was clear that to study this radiation one should go up as high as feasible. From the first, balloon flights were used, but these can carry only limited loads and have hard-to-control trajectories and flight characteristics. We discuss these below. Aircraft could not attain high altitudes in the early years and only since about 1940 have been much used for high altitude studies. For example, a flight made by the present author in 1936, in the original DC-2, broke the then South American altitude record, reaching over 30000 feet, and this record stood for a number of years.

Stations on high mountains therefore early became of importance, and immediately provided several important advantages to high altitude research. On a mountain one can, in principle, stay as long as one wishes, and bring up heavy equipment. In the early 1920's, Millikan and his colleagues made studies, using equipment that could be lowered into Arrowhead Lake, a mountain lake in Southern California. The lake is accessible by road. From the lake, the summit of Mt. San Gorgonio at 11500 feet is accessible on foot without the aid of ropes or ice-climbing equipment.

Access is clearly the principal problem. The maximum altitudes are determined by the snow and ice carried by the particular peak. When a road exists, heavy equipment can be brought up, but loads carried by porters are strictly limited. The food, fuel and living equipment for the observers is also limited. The maximum altitudes reachable on mountains are in the neighborhood of 20000 feet. Above this altitude the problems very rapidly become considerable. This altitude, about 6100 meters, represents about half the atmosphere. Actually the halfway level in the standard atmosphere is 18600 feet, but this varies with latitude and other parameters. At this altitude half the atmosphere is below the observer. Since in this work, precise altitudes are generally unimportant, I will give these in rounded numbers.

Historically other disciplines had already stimulated high altitude work. Such fields of work included meteorology, geology, glaciology, high altitude biology and physiology, studies of geomagnetism, the zodiacal light and many others. Cosmic ray studies came along and in the early years made use of stations established for other purposes.

Whereas it is not now feasible to report on the majority of high altitude sites occupied over the twenty-five years from 1930 to 1955, we can summarize information about some in actual use today. We will subdivide these by the continent on which they are located. We arbitrarily define high altitude as above about 8000 feet or

[*]New York University (emeritus), New York, U.S.A.

Y. Sekido and H. Elliot (eds.), Early History of Cosmic Ray Studies, 171–179.
© *1985 by D. Reidel Publishing Company.*

about 2500 meters, below which there are many stations. A more complete list exists (see reference, The World's High Altitude Research Station).

2. Balloons

Since high altitude balloon flights also comprise a platform from which cosmic rays have been observed, we will also mention these briefly. In the very earliest experiments, in which the extra-terrestrial nature of the radiation was established, in the years around 1910, Hess made use of such flights. In the decade of the 1920's, Millikan and his co-workers made many such flights. As balloon fabrics were improved and operating techniques developed, progressively greater altitudes were attained. Controls of rates of ascent and flight trajectories, tracking and recovery procedures were all improved. Transmission of flight data by short-wave radiosonding was developed.

Most of these flights were unmanned. The inclusion of observers meant a great increase in the weight of the life-support systems and consequent greater cost. Man-carrying flights were piloted by the Piccards, by T.G.W. Settle, Kepner, Stevens and Anderson, in the 1930's. Probably the most publicized flight was launched by the National Geographic Society and the U.S. Army Air Corps. On November 11, 1935, this flight, launched from a natural amphitheater in South Dakota, attained an altitude of 72400 feet, or about 23 kms, which remained for years the altitude record for manned flights. It was piloted by (then Major) Albert W. Stevens and (then Captain) Orville Anderson. It carried among other instruments several cosmic ray experiments, and also the photographic plates on which neutrons were recorded. Stevens later went on to command the Aerial Photographic School in Denver, and Anderson became a Major General during World War II.

After the war, in 1956, Morton Lewis and Malcolm Rose took a manned balloon to 96000 feet, and the following year Major D.G. Simons ascended to 102000 feet.

3. North American Stations

Probably the first station, established with mainly cosmic ray work in mind, is that on Mt. Evans, Colorado, at 14000 feet or 4300 meters. In the mid 1930's, the present author, together with Professor J.D. Stearns of the University of Denver, visited John Evans, a banker in Denver, after whose grandfather the mountain was named, and persuaded him to join in placing an observing station on the peak. Since this mountain is in the "front range" of the Rockies, it has a line of sight to Denver, permitting short wave communication with low power. A structure consisting of two rooms, one for equipment and one for personnel was erected, and has been in use ever since. It is accessible by road, and power is provided by a generator. A summit-house is nearby, which provides snacks for tourists.

While snow closes the summit road during the winter, another facility is located at 10700 feet on the same peak, where some quite substantial buildings, again including a chalet for tourists, have been built. A power line reaches this station, and the road is kept open all year by the Highway Department plows.

Fig. 1. Mount Evans, Colorado, Cosmic Ray Laboratory, 14100 feet (4350 m) Altitude.

Many other facilities exist in the U.S., at altitudes of 9000 feet and below, including coronagraph stations and other observatories at Climax, Colorado, Sacramento Peak, N.M., Kitt Peak, Arizona, Sunspot, N.M. and White Mountain, California.

4. Alaskan Stations

On the summit of Mt. Wrangell, Alaska, two Jamesway huts were erected in 1952–3. One houses the power supply, a small generator, the other the personnel, the buildings being separated as precaution against fire. Access is by air, with ski-wheel combination aircraft. The 14000 feet summit has a large snowfield which is saddle-shaped, so that one can land uphill and take off downhill with almost any wind direction. The first thirty four landings on this peak were made in 1952–3 by the present author in a small plane flown by Dr. Terris Moore, then President of the University of Alaska. The huts and equipment were airdropped.

Fig. 2. Mt. Wrangell Cosmic Ray Observatory, Alaska. Upper structure, dormitory; lower, generator hut and apparatus. Elevation about 14000 feet.

In Mexico, the City is at about 7500 feet, and the University, in a southern suburb, holds a neutron monitor. Various roads lead up over 11000 feet. No cosmic ray observing sites are at present occupied at high altitudes, but easily could be, and have been from time to time.

In Canada, a major cosmic ray observing station is located on top of Sulphur Mountain, about 9000 feet above Banff, Alberta. This station is accessible by road and has a powerline. About half a mile away there is a ski-lodge, at almost the same height, which is reached by a telepherique. This station has been in operation since the 1950's and houses a neutron monitor and much cosmic ray equipment.

5. South American Stations

In South America, many high altitude sites are available, and have been used, some since the 1920's. Much of Colombia, Ecuador and western Venezuela lie at high altitudes, but the presently operating stations lie in Peru, Bolivia and Chile. Going from north to south, the first one is at Huancayo, Peru, at about 11000 feet, located some miles north of the town of Huancayo and is accessible by road. Huancayo was established in the 1920's as the Huancayo Magnetic Observatory, and was operated by the Carnegie Institution of Washington. It is close to the geomagnetic equator. About a decade later the observatory was ceded to the Peruvian Government, and now houses not only the original Compton Model C ion chamber, but much other equipment for both cosmic rays and other disciplines. The station is on the alti-plano, with good housing for personnel, and has a power line. The town itself is the terminus of the Central Railway of Peru, which crosses the Andes at Ticlio, 15600 feet (4700 meters). The Central Railway is the highest standard gauge railway in the world. The author and others have made observations at Ticlio. There is no town here, only a modest stone station-house, so that food and sleeping equipment must be brought by the observer.

Continuing south, the Southern Railway of Peru crosses the Andes at Crucero Alto, at 14,688 feet. Again only a modest station house exists here, but it has been occupied by H. Victor Neher and later by the author, with cosmic ray equipment in the early 1930's.

Adjacent to Arequipa, (7500 feet), rises the dormant cone of the volcano El Misti, 19,166 feet or 5843 meters. This mountain is almost unique in that the summit is occasionally free of snow. It is reached by mule from Arequipa. Many years ago the Harvard Observatory had a summit station for astronomical work, but the advantages of height were not great enough to over-balance the increased difficulties in access and maintenance and the station no longer exists. The present author operated an electroscope on the summit for a week in 1934, coming down each evening to 16000 feet to sleep. Professor Norman Hilberry operated an extended air shower array on El Misti in 1940. There is nothing on the summit now except a large iron cross set in concrete, the storms having obliterated all traces of the summit installation.

In Bolivia, the world's highest operational station is situated on Mt. Chacaltaya, at 17000 feet or 5200 meters, above La Paz. A road leads to the station as does a power line. Substantial buildings are there, both for personnel and for equipment and for a maintenance staff. Cylinders of oxygen are available for personnel emergency needs.

Fig. 3. Millikan-Neher type cosmic ray electroscope, in 10 cm lead shield, on top of El Misti, 19200 feet (about 5850 meters) in southern Peru.

Fig. 4. World's highest station, Chacaltaya, Bolivia. Altitude 17000 feet, about 5200 m.

The site has been occupied for years but the buildings were erected in the 1950's. This station is largely due to the initiative of Professor Ismael Escobar of the Universidad Mayor de San Andres, in La Paz. Measurements of the zodiacal light were made there by a British team in the 1950's, and other disciplines are also represented.

Still further south, above Santiago in Chile and accessible by road, lies a station at Infernillo, 4320 meters. Here several pieces of cosmic ray equipment are in operation, and several buildings exist. This station was constructed through the initiative of Professor Gabriel Alvial of the University of Chile.

6. European Stations

In Europe, two ranges, the Alps and the Pyrenees, each have seen cosmic ray observing activity since before 1940. The most elaborate and in many ways the finest observing station in the world is at the Jungfraujoch in Switzerland. The Swiss had built a chalet for skiers and mountaineers and a railway that led up through a tunnel through the Eiger to the 11000 feet or 3570 meter level. The research station is adjacent. The rail access permits heavy equipment and electrical power to be brought in, and observers can find food and lodging at the chalet. A small meteorological station atop the Mönch is accessible by elevator.

The Jungfraujoch station is truly international. Austria, Belgium, France, Germany, Great Britain and Switzerland all contributed to its establishment. Administration is from the Hochalpine Forschungs station, Buhlplatz 5, Bern. Located as it is at the top of the longest glacier in Europe, it is an excellent starting point for a long ski-run. When the present author first visited it in 1936, it had already been operating as an important facility for serveral disciplines, and much biological work as well as cosmic ray observing was in progress.

There are also a number of other high stations in the Alps. There is a French station of the Aiguille du Midi, at 3600 meters above Chamnix, accessible by telepherique and a small shelter on Mt. Blanc at 4353 meters, accessible on foot. An Italian station exists at Testa Grigia, above Breuil, at 3380 meters and accessible by telepherique. A Swiss for snow and glacier research exists at 2670 meters at Davos, and several other stations are at lower altitudes. A major set of buildings on the Zugspitze, at 2960 meters in the Tirol are accessible by road and telepherique from both the German and Austrian side. This station has seen important cosmic ray work since the mid 1930's. Again it is not possible to give a meaningful date since this and other stations started as mountain shelters for alpinists and grew into scientific stations over the years.

Fig. 5. Jungfraujoch High Altitude Laboratory, Switzerland. The laboratory is the second building, on the right. The foreground building with the triangular roof is the tourist hotel. On top of the "Sphinx", left top, is the meteorological station. The main railway station is inside the mountain, behind the hotel. The Sphinx station is reached by an elevator.
Altitude about 11000 feet or 3400 meters approx.

Fig. 6. High Altitude Laboratory, Zugspitze, Germany **Altitude 2960 m.** started in 1900.

Fig. 7. Pic du Midi Observatory, Pyrenees, France. Altitude about 9300 feet.

Finally there is the station of the Pic du Midi in the French Pyrenees, at 2857 meters, originally started as an astronomical observatory with a substantial telescope, and now housing also cosmic ray equipment, accessible by road and telepherique. It also has a coronagraph.

7. Stations in Asia

Although the world's highest mountains are located in Asia, there are few cosmic ray stations there except in India and Japan. Gulmarg, Kashmir, located at about 9000

Fig. 8. Gulmarg Research station, India. Altitude about 9000 feet (Photo courtesy Information office, Jammu & Kashmir Govt.)

feet, has seen cosmic ray activity, and Mt. Norikura, in the Japan Alps has a substantial station at 2840 meters. This station has several good buildings, much equipment, and an ancillary station housing a coronagraph on the summit. Access is by road.

At the time the survey referred to below was made, we had no data on stations in the USSR or in China. Letters of inquiry went unanswered. There are today some stations in the USSR.

8. Stations in Hawaii

Summit stations exist, both on Mt. Haleakala at about 11000 feet on the island of Maui and Mt. Mauna Kea, at 13800 feet on the island of Hawaii. The first was the scene of Grote Reber's classical radio astronomy experiments many years ago, and the latter is now the site of a major astronomical telescope, built with the aid of Canadian funds. Both stations are accessible by road and both stations have housed or still house, cosmic ray equipment.

REFERENCES

The World's High Altitude Research Stations, by S. A. Korff, Editor. 105 pages. List and description of stations, prepared for the Joint Commission on High Altitude Research stations, of the IUBS-ICSU-UNESCO. Published by the Research Division of the College of Engineering of New York University, with some funding from UNESCO, and the office of Naval Research, March, 1954.

Supply long since exhausted, but we have on had some two dozen sets of unbound pages, almost complete, which will gladly be supplied on request as long as available. Write Prof. Korff, Dept. of Physics, New York University, 4 Washington Place, New York, N.Y., 10003, USA.

See also: The Geographical Aspects of Cosmic Ray Studies, by Serge A. Korff, *The Geographical Review*, Vol. 50, pp. 504—522, (1960).

HOW IT STARTED IN BUDAPEST

Antal J. SOMOGYI*

1. It Started Twice

The first start dates back to the early thirties, about 1931. One or two years before that date a young assistant professor came to the Institute of Experimental Physics of the Pázmány Péter (now: Eötvös Lóránd**) University of Sciences, Budapest. The then professor of experimental physics, a former pupil of Eötvös, was an open-minded man who gave his assistants the liberty to choose the field of research they wanted to work in, if he saw ingenuity and spirit in the young people. In such cases he encouraged and helped them on their new, audacious paths – albeit he did not join them.

So it happened in this way, when the new assistant, Miss Magdolna Forró, after having obtained her Ph.D. degree on the basis of a Thesis written in the subject of piezoelectricity, turned to the promising and exciting field of cosmic rays. Soon she was joined by another young assistant, Mr. Jenö Barnóthy who just came from Kassa (Kosice), where he had graduated from the Technical High School. While he was still an assistant and already doing succesful research in cosmic rays, he studied physics at the same University and obtained his Ph.D. in Physics (subject: cosmic rays) in 1935. He married Miss Forró in 1940.

It was a gradual and difficult start. The financial means and technical possibilities of the Institute of Experimental Physics were very moderate in those times. Notwithstanding the circumstances, Barnóthy and Forró arrived at outstanding results many of which are quoted in classical textbooks of cosmic rays, like Jánossy: *Cosmic Rays,* Oxford at the Clarendon Press, 2nd Ed., 1950; Heisenberg: *Kosmische Strahlung,* Springer, Berlin, 2nd Ed., 1953; Dauvillier: *Les Rayons Cosmiques,* Dunod, Paris, 1954; Dorman: *Variatsii Kosmicheskikh Luchei i Issledovanie Kosmosa,* Izdat. Akad. Nauk SSSR, Moskva, 1963 and others.

They used the techniques of Geiger-Müller counters which they developed into a highly stable device in the early thirties, and also introduced coincidence techniques first applied by Bothe and Kolhörster in 1929.

The GM counters were first put into operation at the top plateau of the tower of the Institute of Experimental Physics, in the centre of the town. Some thirty counters formed a telescope with variable inclination to the zenith: the zenith angle dependence of the barometer coefficient was first measured with this device.

---

* Central Research Institute of Physics, Hungarian Acad. of Sci., Budapest, Hungary.

**Since 1948, the University of Sciences in Budapest has been carrying the name of the great experimental physicist, Eötvös Lóránd (1848 – 1919), inventor of the torsion balance as a device for earth resource prospecting, the man who formulated the basic law of surface tension, who gave the first and surprisingly precise experimental proof of the proportionality of gravitating and inertial mass – to mention only a few of his achievements.

181

Y. Sekido and H. Elliot (eds.), Early History of Cosmic Ray Studies, 181–185.
© *1985 by D. Reidel Publishing Company.*

Barnóthy and Forró's most cited results are those obtained in their underground measurements. In the coal mines at Dorog (40 km north-west from Budapest) and Tatabánya (60 km west from Budapest) they measured the absorption of cosmic rays down to 1000 m water equivalent depths, as well as the temperature effect and the Rossi transition effect.

Barnóthy and Forró had considerable influence on the development of research in physics in Hungary, and on the development of cosmic ray research in general. They took part in the activities of the First International Cosmic Ray Conference at Kraków in 1947, where they read a paper on the results of their underground measurements. In 1948 they left Hungary to live in the United States*. Although they had educated several young scientists in cosmic rays physics – two of them, E. Fenyves and O. Haiman continued and finished the measurement of the transition effect underground – the group was too small to survive without them.

2. The Second Start

Years following the second world war saw an unparallelled boom of scientific research in many places in the world. Unparallelled also in the sense that "basic" research got its due emphasis with respect to the "applied" branches. Physics was in the first line of interest due to its splendid, or catastrophic, but by every standard prodigious results which promised (or threatened) to transform the mode of human life on earth.

The People's Republic of Hungary, born in 1946, promptly recognized the new role of the new science and acted accordingly. The Hungarian Academy of Sciences, founded in 1827 as a learned society with no executive power, was entrusted with organizing and controlling basic research in the country. Large scale projects were worked out in many branches of science, especially in physics. A huge institute, the Central Research Institute of Physics (its acronym in Hungarian: KFKI) of the Hungarian Academy of Sciences was founded in 1950 with the aim to develop up to date techniques and methods of modern physics. When founded, KFKI consisted of three departments only: Department of Cosmic Rays, Department of Optical Spectroscopy, and Department of Electromagnetic Waves. The staff numbered some hundred people in all. Nowadays KFKI (i.e., the Central Research Institute of Physics of the Hungarian Academy of Sciences) is an agglomerate of five almost independent research institutes each of which emerged as the product of development of certain sections of KFKI throughout its history. The staff today consists of something like 2300 people.

The second start in cosmic ray research in Budapest was thus completely different from the first one. Plans were ambitious, means were provided for to the extent which could only be found in the country and, last but not least, there was an expert leader of international reputation, Lajos Jánossy, who returned to Hungary, his home

---

* Barnóthy and Forró published only a few papers on cosmic rays after 1948. They lectured at Barat College, Illinois and University of Illinois, respectively, from where they retired in the early seventies.

country, in 1950* to become not only Head of the Department of Cosmic Rays in KFKI, but also a Member of the Hungarian Academy of Sciences and Professor of Physics at the Eötvös Lóránd University, Budapest. The research activity of the Department was conceived on a large scale: the first program envisaged working out most of the up to date experimental techniques in cosmic ray physics, such as the building of discharge counters and a triggered, multiplate cloud chamber in a magnetic field, learning the techniques of nuclear emulsions, and construction of up to date electronic circuits underlying all these methods.

The greatest difficulty was, of course, in the insufficient number of experts. In addition to Jánossy and two coworkers of Barnóthy and Forró, there was practically nobody who had any experience in the field of cosmic ray research. Enthusiasm together with Jánossy's expertise and energy helped to overcome the difficulties.

As early as 1952, the first experimental devices were available. They were the GM counters, the production technique of which had to be worked out again. The number of GM counters produced increased rapidly: about thousand counters were produced − and checked − each year by a small group consisting of a technician and four laboratory assistants under the supervision of a scientist, for a period of almost two decades.

The first measurements were thus carried out with GM counters. After repetition of a few classic measurements like the lifetime of the muon and the meteorological effects of extensive air showers (EAS), there followed systematic studies of the Rossi transition curve (the second maximum was still haunting in the early fifties) and the composition of EAS. The International Geophysical Year (1957−59) prompted the idea of installing a pair of big semicubical muon telescopes in an underground laboratory, under a layer of 18 m rock (about 40 m water equivalent). This pair of telescopes, with a total sensitive area of 3 m^2 and a total counting rate of more than 100000 muons per hour, is still running and has been the source of a number of important discoveries such as the existence of Forbush decreases and 27 day variations at primary energies as high as 10^{11} eV, and others.

It took a longer time to build the triggered cloud chamber. The first useful pictures were obtained around 1956. The diameter of the chamber was 30 cm, its depth 15 cm, and contained seven lead plates. The chamber was triggered by the EAS setup. Interesting results were obtained on the photon/electron ratio, the zenith angle distribution (i.e., absorption mean free path) of EAS, and the density spectrum of EAS at large densities. The latter measurement showed that, in contrast to results of other measurements, there was no "breakdown" of the density spectrum around densities higher than 800 particles/m^2. The magnet, however, was never built. This was already

---

* Lajos Jánossy was born in Budapest in 1912. His stepfather, György Lukács, the renowned marxist philosopher, played an eminent role in the short lived Hungarian Soviet Republic in 1919, and, after its subversion, had to emigrate together with his family. Lajos Jánossy was educated in Vienna and in Berlin, where he began his career in cosmic ray physics in Kolhörster's group. In 1937 he joined Blackett's famous school of cosmic rays at the University of Manchester, from where he was called to be professor at the Dublin Institute for Advanced Studies in 1947. From 1950 to 1956 he was Head of the Department of Cosmic Rays in KFKI, Budapest, and from 1956 to 1970 Director of KFKI. From 1970 till his untimely death in 1978 he acted as scientific counsellor to the same institute and concentrated his efforts on research in general relativity and basic problems of manybody systems.

the sign of times: particle interaction studies in energy regions where magnetic deflection in cloud chambers was still perceptible no longer belonged to cosmic ray physics.

Pushed to rapidly rising energy regions in the field of high energy interactions, cosmic ray physics developed more and more to be a tool of space physics and a tool to investigate the structure of the universe. This was recognized early by the Budapest group and accordingly, projects like penetrating shower investigation and extensive air shower research were closed down in the mid sixties, whilst space research programs, begun in the late fifties, were extended both in size and scope. EAS were used as a technique for exploration of galactic structure via measuring galactic angular distribution of high energy cosmic rays: measurements were carried out, in cooperation with the Sofia group*, at the peak Musala, Bulgaria, from 1969 till 1976. Evaluation of satellite measurements and investigations into low energy particle propagation in interplanetary space, begun in the early seventies, have formed a rapidly increasing part of the program in the last years, together with continued activity in the field of high energy modulation and galactic angular distribution, with a strongly growing interest in theoretical problems.

It cannot be the goal of this short review even to outline the history of the Department of Cosmic Rays of KFKI. Instead of giving more details, I should like to finish with a few personal reminiscences.

I was so fortunate as to be in the first group formed around Jánossy in 1950, and begin my work in cosmic ray physics as the leader of the group in charge of working out the process and technology of GM counter production, later (1952) also of the EAS group and various other groups too, and to become second successor to Jánossy at his post as head of the Department of Cosmic Rays, in 1964.

The grandiosity of the conception of KFKI, and Jánossy's personality were my two most remarkable experiences.

I ask the reader to imagine a forest with an area of 40 hectares on the hilly outskirts of Budapest, with a single spacious new building, some others under construction, given to a handful of people to work there on cosmic ray physics: that was the site of KFKI in 1951. A wooden bungalow with a light roof and practically no metallic component parts, built specially for cosmic ray measurements, was made ready in 1952, and it did not take long before the underground laboratory — a shaft with altogether six tunnels at three levels underground — was given at our disposal. Of course, new departments were established at rather frequent intervals and they were all given a place on the same territory. So it happened that this part of land is now crowded with more than twenty five buildings housing some two thousand and three hundred people. One thing, however, is still in common with the times of thirty years ago: namely, that construction (or lately — again a sign of times — rather reconstruction and enlarging) of buildings is still going on.

Working with Jánossy made, as on all his co-workers, even deeper impressions. I should like to mention two things which I found very characteristic of his style both as a researcher and a leader.

* Department of Cosmic Rays, Institute of Nuclear Physics and Nuclear Energy, Bulgarian Academy of Sciences, Sofia.

Jánossy had a strong sense for facts, both in general, and in science. This was strongly reflected in the severity he required in handling experimental data. "Threefold statistical error" was made a cardinal point in drawing conclusions in his group and, even more important, the meaning of the concept of error had to be scrutinized, the way of finding its value had to be optimized. There was a constant warning against schematism, the usual pitfall in applying statistical methods. Although the maximum likelihood method has generally been applied, its limits had always to be kept in mind, and other methods had to be preferred in cases of contraindication.

As a leader and organizer, he was strongly against "amateurism" in research. In his mode of thinking, research was a collective game in which each post had to be filled in by a person specially trained for the post. I clearly remember a case, when – in the early fifties – one of us, a physicist, proudly demonstrated to Jánossy a newly built automatic device for taking photos of hodoscope bulbs – and was rebuked by him for having done something outside of his profession. Jánossy carried through this principle, as far as circumstances allowed, consequently and right from the beginning. Later, during the 14 years of his directorship, he developed it, together with other leading experts of the Institute, to a policy not restricted to individuals, but extended to sections and tasks of the Institute. This policy led not only to a general high level of experimental techniques in KFKI, but also provided the basis of a technological activity reaching far out of the Institute to influence development in certain fields of technology on a national scale, and to gain reputation for KFKI also abroad.

It was somehow in this way that the collective game of doing cosmic ray research started in Budapest. It has been a game worthwhile playing and worthwhile the efforts spent in it. Not only for the results, but also because it has been collective. It has brought together many people inside and from outside of national boundaries, and joined them in working for a common goal. Surely, this was so in the case of other groups, perhaps all cosmic ray groups, too. Is not this as important as the results themselves?

INTENSITY AND ANISOTROPY OF COSMIC RAYS

Yataro SEKIDO*

These are my personal reminiscences. Since 1932, I was a student of Hokkaido University, Sapporo. Through frequent contact with Professor Ukichiro Nakaya (1900-60), I was charmed by many essays of his teacher Torahiko Terada (1878-1935) bearing his wonderful ability of finding many new fields in physics. Nakaya was working with Fumio Yamasaki operating a cloud chamber to study ion formation preceeding to spark discharge. It was suggested in 1929 by Terada after he and Nakaya worked since 1924 on the morphology of a spark to study the growth of an inanimate object (Sekido, 1981) which was pure but not yet a current problem of physics. In 1934, Nakaya let me read C.Bosch's paper on Elekronenzählrohr as an exercise before graduation telling me his interest in the growth of a discharge from only one pair of ions. I surveyed every available paper in which the Geiger counter was concerned and found unexpectedly the existence of cosmic ray physics. Nakaya's favourite book, J.J.Thomson's Conduction of Electricity through Gas (Fig.1), was also useful for knowing the beginning of cosmic ray physics. I tried to make a Geiger-Müller counter to detect cosmic rays with batteries and Shimizu type fiber electroscope but could not succeed before it was interupted by the season of snow. In November, I visited Terada's home at Tokyo wishing personal contact with him. Accidentally on the next day, a monthly meeting of the Physico-Mathematical Society was held, which was my first chance to peep at a scientific meeting. The chairman was the famous Dr. Nishina, but I did not know the lecturer. When the lecture was over, the chairman stood up and said, "It is a matter for congratulation that such an excellent work came out from a member of this Society". He was Hideki Yukawa and the lecture was his first paper of mesotron theory.

In every winter season, Nakaya was working on another growth phenomenon, a snow crystal, and I participated in it during that winter. Early in 1935, Yamasaki moved to Tokyo to join Nishina's cosmic ray work. I did not know that Terada, who wrote an essay on a possible biological effect of cosmic rays, was a member of the Cosmic Ray Committee of Gakushin (Japan Society for Promotion of Science) which was encouraging Nishina's work. Then, Nakaya concentrated his interest on snow crystals, which was likened by him to "a letter sent from heaven". I was involved in it with fair interest and was appointed his assistant after my graduation in March. In that November, when I felt my first love to be one-side and was painful to live in Sapporo, accidentally he told me, "Dr. Nishina wants a worker to expand his research on cosmic rays. Will you go?" I answered yes impulsively. My last work under Nakaya's kind guidance was in March 1936, when I fortunately succeeded in the artificial production of a snow crystal, however, I did not become aware that cosmic rays were also a letter sent from heaven nor that it was a product of a growth phenomenon of energy. On 1 April, I attended Nishina's

* Nagoya University (emeritus), Nagoya, Japan.

Y. Sekido and H. Elliot (eds.), Early History of Cosmic Ray Studies, 187–206.
© 1985 by D. Reidel Publishing Company.

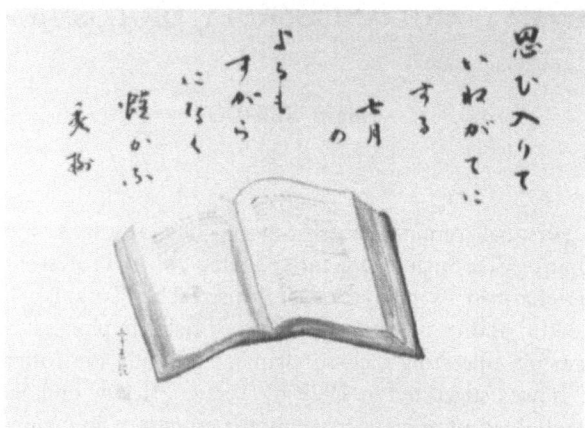

Fig. 1. Picture by U. Nakaya, Waka (Japanese 31 syllable poem) composed by H. Yukawa and written
by him when he stayed at Nakaya's home in about 1938.

room with an empty mind, when Nishina said briefly to me "You work with Mr. Ishii".
That gave the orientation of the four decades of my work.

1. Beginning of Cosmic Ray Intensity Observation in Japan

Yoshio Nishina (1890-1951), who observed Compton electrons from X rays with a
Geiger point counter at the Cavendish Laboratory in 1921-1922 and found the Klein-
Nishina formula in 1928 at Copenhagen, came back to IPCR (Institute of Physical and
Chemical Research), Hongo, Tokyo in December 1928. On his suggestion, his teacher
Hantaro Nagaoka (1865-1950) invited P.A.M.Dirac and W.Heisenberg to give lectures at
Tokyo in September 1929 (Fig. 2*). The lectures stimulated H. Yukawa and Shin-ichiro
Tomonaga but no reference was made of cosmic rays as far as the note (Lectures of
Heisenberg and Dirac, 1932) interpreted by Nishina states. In this year, Suekichi
Kinoshita (1877-1935), who detected a particles by photographic emulsion under
E. Rutherford at Manchester in 1909, introduced Bothe-Kolhörster's paper (1929) at
Tokyo University and let his student, Chihiro Ishii, read it for exercise in that autumn.
In December 1930, Mankichi Hasegawa (1894-1970) came back to the Geophysical
Institute, Kyoto University, from Potsdam with a Kolhörster's ion chamber (20 cm ϕ X
40 cm) for durchdringende Strahlung.
    Since January 1931, M.Hasegawa tried to detect cosmic rays by various experiments.
He told me one of them. He made an antenna coil (Fig.3) designed by him to detect the
transient magnetic field due to a cosmic ray "charged particle" but got no response.
According to Yuichi Tamura, who did this experiment, photographic emulsion plates
were also examined and the main effort was to make Geiger counters to participate in the
Japanese cooperative works in the 2nd Polar Year (1932-33) with particular item of
cosmic ray intensity observation. The counts decreased by lead shielding, but they could

* <Ed.> Figure 2 is displaced to page 38 (Old Photographs).

$\longleftarrow 5\,cm \longrightarrow$

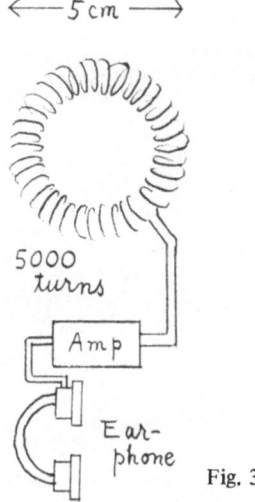

Fig. 3

not get stable counters in that year, while Nishina, who had experience with Geiger point counters at the Cavendish Laboratory, started cosmic ray experiments with Ryokichi Sagane (1905-69), a son of Nagaoka, since July 1931. According to Masa Takeuchi, their counter could trace a well known cosmic ray barometer effect on the occasion of a typhoon. Thus, the initial cosmic ray works in Japan was concerned with intensity observation. However in 1932, Hasegawa changed his interest to geomagnetism while Nishina turned to cloud chamber work on cosmic rays.

On the occasion of the 2nd Polar Year, V.F.Hess planned a network ranging from Abisco to Cape Town, and E.G.Steinke made standard cosmic ray meters for them. It is not clear at present when Nishina asked Steinke for one of his standard meters. However, a field work was probably discussed in the autumn of 1933, when Masao Kinoshita, a nephew of Suekichi, was expected to be a member of the Cosmic Ray Committee of Gakushin. He looked for a worker in his laboratory of IPCR. One of his assistants, C.Ishii, showed interest. According to Ishii, after Ishii had once hesitated, M.Kinoshita encouraged him saying "Desk work for Tomonaga, field work for you". Ishii made up his mind in November 1933, cooperated with Nishina in preparing an ion chamber of the Compton type (7 cm ϕ)at the IPCR workshop since June 1934, and was appointed the first Nishina-Gakushin personnel in December 1934. The chairman of the Cosmic Ray Committee was Takematsu Okada (1874-1956), Director of the Central Meteorological Observatory, and Masami Kawano (1934, 1935) of the Local Observatory at Mishima made observations on the east-west effect of cosmic ray at the top of Mt. Fuji in August 1934 after preparations since the summer of 1933 directed by Okada.

According to Ishii's memory, F.Yamasaki and Yoshihiro Asano came to IPCR in 1935 to form Ishii's team under Nishina. Their first observation was done with the Compton type ion chamber by visual reading night and day at Mt. Hakone in April (Fig.4) and by a recording camera at Mt. Fuji in August 1935 to confirm the intensity-altitude curve. The Steinke's standard ion chamber (20 cm ϕ x 50 cm) arrived at Tokyo in March and began operation in May. In October, Nishina planned with Ishii to make five ion chambers

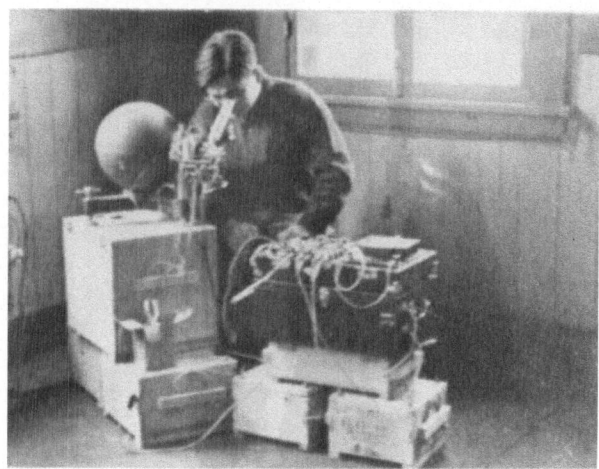

Fig. 4. C. Ishii at Mt. Hakone (1935).

(20 cm $\phi \times$ 50 cm), called, afterwards, the Nishina type, to form his own network ranging from Saghalien to Palau. Meanwhile, A.H.Compton made several new ion chambers called Model C meters with uranium compensation. Ishii amended his design in the course of its construction. In November, when H.V.Neher visited Japan on his way to Phillipine with his portable ion chamber (15 cm ϕ), Nishina asked him for a chamber of the same type. Yamasaki moved to the group of nuclear experiments under Nishina. In December, Nishina decided with Ishii's team to participate in the Japanese cooperative observation during the solar eclipse on 19 June 1936.

2. Early Ion Chamber Observation of Cosmic Ray Intensity

When I joined Ishii's team in April 1936, our urgent program was the observation during a solar eclipse. For Nishina had desired to do it near the top of Mt. Syari (1545 m), Hokkaido, which was the highest point in the band of the total eclipse, Ishii inspected the place to built a small hut (1260 m). We hurried to make one of the five chambers available. Ishii and I climbed up the snow-laden mountain with equipments and lead blocks on our backs together with 200 young men of the village and started observation from 22 May (Fig. 5). Just in time for the eclipse, Nishina came to our hut (Fig. 6) with the Neher's chamber without any lead shield to observe at the summit. On 27 June, when we finished the observation, there was neither snow nor road to carry the equipments back. The Forest Office cut a lane for us. The result (Nishina et al., 1937) showed no effect of the eclipse with quite a better accuracy than those of our predecessors.

Neher's ion chamber was carried into Shimizu tunnel in August 1936 by Nishina and Ishii to determine its background current. A cosmic ray burst was first found even at a depth of 800 mwe (Nishina and Ishii, 1936) (Fig.7). In December 1936, the chamber was mounted without shield on a Navy airplane, and an altitude curve up to 9 km was obtained. Considering the discrepancies between Compton's and R.A.Millikan's world maps of cosmic ray intensities, Nishina planned repeated observations along a particular course, and asked the steamship company NYK to carry Neher's chamber on board. The

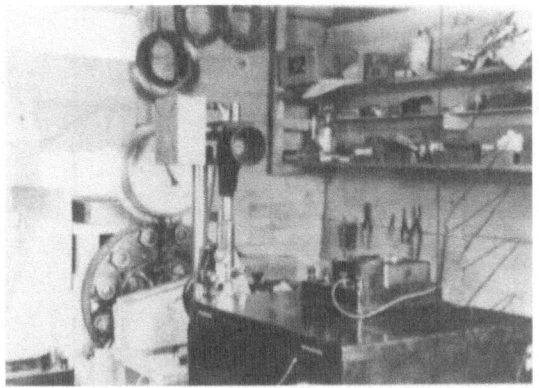

Fig. 5. Mt. Syari (1936).

Fig. 6. Mt. Syari. From left, Y. Sekido, Y. Nishina and C. Ishii. (1936).

Fig. 7.   Y. Nishina seeing the film of cosmic ray burst at Minakami Spa near Shimizu tunnel (1936).

observations were done on 8 trips between Yokohama and Melbourne for one year from April 1937, and then on 8 trips between Kobe and Vancouver until November 1938. This was extended as described in Section 4, and the result (Sekido *et al.*, 1943) showed that the discrepancies were due to time variation of cosmic rays.

In 1937, a young astronomer Hukutaro Shimamura joined us and made comprehensive analysis on the time variation of the data obtained by our Steinke's chamber during the period January 1937 - July 1938, and we (Ishii *et al.*, 1939) published the results in 1939. A remarkable negative correlation with solar activities was shown in their 27 day periodicity, as W. Kolhörster and Miczaika found in the same year. Both the events of April 1937 and January 1938, from which S.E.Forbush discovered a worldwide decrease of cosmic rays, were also observed at Tokyo as reported in our paper, and also recorded by our Neher's chamber on board a ship. Operations of Nishina's ion chambers were started one by one according to their preparation, even though Ishii was called for military service in July 1937. The parallel running of the five chambers was done in the campus of Tokyo Astronomical Observatory at Mamiana, Tokyo, since 1941 by Ishii who came back in August 1940. Shimamura processed the data to study Forbush decreases and atmospheric modulations. But, the five sites network was not realized.

3. Cosmic Rays in the Stratosphere and Deep Underground

One day in 1936, Nishina was reading Comptes Rendus when I opened the door of his room. He told me excitedly, "Auger is saying that there are soft and hard components in cosmic rays." In December 1936, I read at our colloquium G.Pfotzer's paper on his observation in the stratosphere. Nishina said instantaneously, "This is interesting. Let us try it." I did not know that a balloon experiment was once planned by Nishina and Ishii in October 1935 but interupted for a long time. Our new plan was to observe Pfotzer's curve first, and then vertical intensities of hard and soft components separately. Ishii began its preparation from January 1937, but was soon called for military service. Shin Iio joined us in 1938 and made preparations throughout this work. In 1940, M. Schein *et al.* observed hard components by balloon flight; however, our discussion together with Tomonaga and Mituo Taketani suggested that a different kind of meson, the fluff meson,

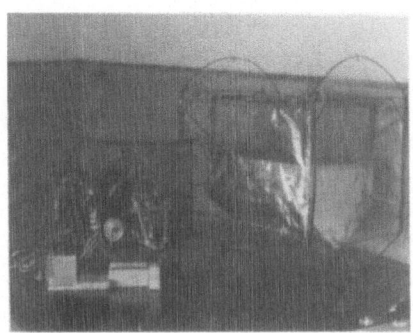

Fig. 8. BAlloon flight (1942-43).

might exist at high altitudes. Ishii came back, and a friend of Nishina made a very small primary battery, 400 volt in about 25 cc. We sent up a balloon flight with radio transmission in December 1942 and two more flights in January 1943 (Fig.8). To absorb the decay electrons of fluff mesons, insertion of 3 cm Pb was planned for the 3rd flight, but Iio requested one more repetition. Thus, three Pfotzer curves were obtained. The maximum intensity was shown in my book (Sekido, 1944, Fig. 9), and the curves were published later (Ishii, 1959).

In the spring of 1937, I read at our colloquium J.Barnothy and M.Forro's paper (1936, 1937). They reported that most of the charged particles at 732 mwe underground were so soft that half of them were absorbed by 1.5 cm Pb and thought the cosmic rays penetrating down to deep underground to be some neutral particles, probably neutrinos. When I finished my talk, Nishina said, "This is interesting. Let us try it at Shimizu tunnel." I did not know Nishina had tried theoretically to attribute the ultra-penetrating power of cosmic rays to neutrons just after the neutron was discovered, though it was not promising. We had been testing G.M. counters with argon gas and iodine vapour for the balloon experiments but did not know any reliable method to produce many big counters which would be stable for a year. Ishii was called soon, Asano moved away to a lineac group, and I prepared the underground equipment with a young technician. We constructed a cathode ray oscillograph to see the counter pulse by using a primitive Braun tube which was not in the market in those days. The characteristic of the counter was much improved by using alcohol vapour after Trost. Long life was obtained by adopting a

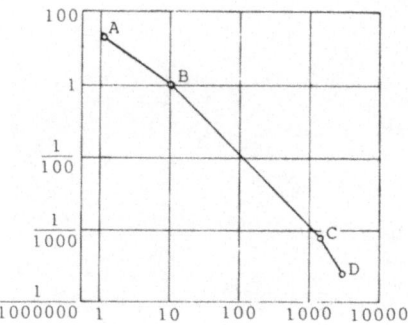

Fig. 9. Cosmic ray intensity vs depth (mH$_2$O) in Japan from the reference (Sekido, 1944). A: Pfotzer max, Tokyo, B: Tokyo. C, D: Shimizu tunnel.

Fig. 10. Shimizu tunnel, 3000 mwe underground. Y. Sekido (left) and Y. Miyazaki (1942).

glass sealed type, though it took a long time to prepare 20 available counters of big size. I designed a counter arrangement to observe single and shower particles with absorbers up to 30 cm Pb, which might absorb all of the charged particles underground. The whole equipment was completed at the end of 1938.

In January 1939, Yukio Miyazaki joined us, and made a test running of this equipment through a laboratory experiment with K.Birus (Nishina *et al.*, 1941a). This was directed by Nishina to find the charge exchange of neutral mesons by using the pole pieces of a cyclotron as a producer and an absorber. In August 1939, we set the equipment in Shimizu tunnel at a site of 1400 mwe, and it was carried out by Miyazaki and Tokio Masuda for one year. Most of the charged particles underground penetrated 30 cm Pb (Nishina *et al.*, 1941b). Meanwhile, Barnothy and Forro (1939, 1940) did further experiments and corrected their previous report, i.e. a major part of the charged particles penetrated 50 cm Pb but about 90% of them were absorbed by 80 cm Pb. We changed our counter arrangement to use 100 cm Pb, and set it at the deepest site, 3000 mwe (Fig. 10) in September 1940. By making preliminary experiments (Sekido, 1944) at both sites, arranging 4 counter trays vertical or horizontal, I believed that the coincidences at

3000 mwe were not accidental but due to cosmic rays. The vertical coincidence was about one per week, but it decreased to less than one per month when one of the trays was shifted off line or 100 cm Pb was inserted between trays. The flux at 3000 mwe was about one tenth of that at 1400 mwe, as shown in Fig.9. The observation was maintained by Miyazaki and Masuda and continued even in wartime since man power for the maintenance was estimated to be neglisibly small, though it was finally discontinued owing to a fire accident. The paper was published by Miyazaki in 1949, including the flux at the deepest point in the world up to that time. At the site of 1400 mwe (1939-40), drosophila were breeded for Nishina and D.Moriwaki's study on mutation by cosmic rays. The result was mentioned by Nishina *et al.* (1941).

4. Cosmic Ray Temperature Effect and a Strange Event

In 1936, the seasonal variation and so called out-door temperature effect were still in doubt. We were operating Steinke's ion chamber keeping the room temperature constant. In 1937-38, the observed mass of a mesotron was nearly the same as that predicted by H.Yukawa's theory, while the improvement of the theory seemed to show a very short life time. I was convinced of the real seasonal variation in November 1938, when Shimamura showed me the beautiful result (Ishii *et al.,* 1939) obtained from our own observation. Next day, I told Nishina the result in a car going to the steamship company NYK to express our thanks. At the company, Nishina asked for 6 additional trips to cover all seasons instead of expressing our final thanks. In December 1938, P.M.S.Blackett inferred that the seasonal variation observed by Compton and Turner should be a temperature effect due to the decay of mesotrons in the atmosphere (Blackett, 1938).

   Early in 1939, when it was difficult to exclude some other possible explanations for the seasonal variation, we started to examine the temperature effect from various sides. Direct comparison with the temperature observed by a meteorological sounding balloon was not realized for some time notwithstanding Nishina's effort. The seasonal variations at 10 districts ranging from Melbourne to Vancouver, obtained by our observation on board a ship, suggested the effect of high altitude temperature because the intensity minimum was at July or January even in districts where the sea level temperature delayed one or two months. The seasonal ranges as well as the apparent barometer coefficients (Sekido, 1943) were widely different at each of the 10 districts. In June 1939, Hidetoshi Arakawa, a meteorologist, showed interest in our results. The close cooperation with him enabled us to compare the cosmic ray intensities at Tokyo with the temperature at high altitude supposed by weather charts. This method was applied to various phenomena (Nishina *et al.*, 1940a,b,c, 1941), such as air masses, cyclones, anti-cyclones and typhoons. Thus, we were convinced of the temperature effect with the life time of the hard component of cosmic rays to be $\tau_0 = 3 \times 10^{-6}$ sec in 1940, whereas Shoichi Sakata obtained $\tau_0 < 1.5 \times 10^{-8}$ sec for the Yukawa particle.

   Our work in 1939-40 on the temperature effect was done by averaging many events, however, the conclusion suggested us the possibility of watching the variation of tropospheric temperature by observing sea level cosmic ray intensities with a sufficiently large area detector. In 1941, Nishina considered that this watching would be useful in the meteorological service for aviation. For this reason, this project started in 1942 and existed in wartime helped by many technicians and mobilized high school students

Fig. 11. A large area counter telescope at Hongo, Tokyo (March, 1945).

including Masami Wada, Koichi Kamata, Ichiro Kondo and Kiyoshi Niu. The method of watching the temperatures of upper and lower half of the troposphere by observing hard and soft components of cosmic rays was also studied. The scale up of two kinds of detectors was started in parallel, i.e. a long ion chamber (20 cm ϕ x 5 m) by Ishii, who moved to the Army Meteorological Department at Suginami, Tokyo, and a large area counter telescope by our team at Nishina Lab., Hongo, Tokyo. The latter was composed of some detector units of wide angle telescope, $0.1 \, m^2$ each. Since October 1944, the telescope with 6 detector units (2.2×10^5 counts/h) was operated with a direct reading pen recorder. In November, the variation of the temperature presumed by this telescope showed a good agreement with that of the harmonic mean temperature observed by meteorological soundings. Before that, cosmic ray data over one year were necessary to get one significant curve of the daily variation. But it was clearly seen on a monthly basis by this large area counter telescope (Fig. 11).

On 16 December 1944, the keeper of the telescope informed me of the sudden occurrence of a very large diurnal variation. It appeared the next day also. We knew that a magnetic storm occurred on 16 December but did not know any instance of such a cosmic ray variation. In those days, a large cyclotron at Nishina Lab. was just starting its test operation. I thought of some radiations from it, which might be operated in day time only. I compared the operation chart with our observations for a few days. They were somewhat in parallel, but not clear. I asked Nishina to stop the cyclotron for a few days. The cyclotron was stopped and the strange diurnal variation disappeared. However, the cyclotron did not start soon for they were improving some parts. Now, I regret that I did not examined the films of the ion chambers at Mamiana distant from the cyclotron and am afraid that the cyclotron program might have suffered some disturbance from our cosmic ray telescope.

In 1945, the telescope was improved to have 8 detector units, and another telescope of the same design was just completed by some mobilized students at their high school. All these equipments and the five ion chambers of Nishina type at Mamiama were carried to Kanazawa in March 1945 to avoid air raids. Ishii, who operated his long ion chamber with Isao Miura since January 1945 under a shielding of earth at Suginami, moved to Manchurian Meteorological Observatory in June 1945.

5. Formation of Postwar Cosmic Ray Group in Japan

In 1945, after the war ended in August, our team under Nishina returned from Kanazawa to Hongo, Tokyo, where the buildings were much damaged, and we settled in June 1946 in a building at Itabashi, Tokyo, which had been used by the Army. Meanwhile, Yuzuru Watase made a big cosmic ray laboratory at Osaka as described in this book by M. Oda. The Research Section of the Central Meteorological Observatory at Kanda, Tokyo, where Osamu Minakawa had been making cosmic ray intensity observation with a counter telescope since 1944 and had been helped by mobilized students Satio Hayakawa and others since 1945, moved after the War to Suginami taking over the site of the Army Meteorological Department and has been renamed as the Meteorological Research Institute since 1946. In August 1946, I contacted Minakawa to hold a joint colloquium and symposium once a month, where Tomonaga and Hidehiko Tamaki were also invited and I made a review talk on meson production in November 1946. Meantime, Nishina assumed the directorship of IPCR, which met with a very hard time, and it seemed to be difficult to keep a big group for pure research. Takeuchi gave up his cloud chamber work. Invited by Sakata, I moved to Nagoya University in April 1947 together with three others from Itabashi, while Miyazaki maintained the Itabashi Branch though IPCR was changed to a limited company called Science Research Institute. Shigeo Nakagawa, who studied accelerator physics at the Nishikawa Laboratory of IPCR, moved to Rikkyo (St.Paul)

Fig. 12. Mt. Norikura (1962). Up right: Corona Obs., Univ. of Tokyo. Below left: Cosmic Ray
Obs., Univ. of Tokyo. Below right: Cosmic Ray hut given by
Press Asahi in 1950.

University, Tokyo, and prepared a balloon flight from 1951 by organizing an emulsion
group. As the Meteorological Research Institute also had to decrease its members in 1950,
Minakawa moved to Kobe University organized a group of cloud chamber and hodoscope
experiments and then started a balloon flight in 1952 having organized an emulsion
group. Ishii came back from China in 1953 to Suginami. He dug out his long ion chamber
with the modulation group there and helped balloon flights in Japan as a veteran.

    In 1950, the Press Asahi gave one million yen to encourage cosmic ray work
nominating Watase, Minakawa, Miyazaki and myself. The four built a hut of 50 m^2
including a power source and some beds at a site 2770 m above sea level near the top of
Mt. Norikura. It was opened in September 1950 for cosmic ray researchers in Japan. The
needs overflowed soon, and the plan for a bigger observatory was discussed in the Science
Council of Japan by the Special Committee for Nuclear Physics. The chairman was
Tomonaga since Nishina had died in January 1951. Our proposal of an inter-university
observatory was the first case discussed in the Council. According to its recommendation,
the Mt. Norikura Observatory was established in 1953 as one of the new kind of research
institutes for inter-university use, and was formally attached to the University of Tokyo.
The first director was Morizo Hirata. The observatory (Fig.12) was equipped with a power
source, a big magnet, some instruments and several bedrooms, but had no research staff
of its own. Minakawa, who helped to build the Asahi Hut, was also active in preparing

Fig. 13. Mt. Loche (1957).

these equipments. Meanwhile, a democratic organization of cosmic ray researchers in Japan, called CRC, was formed in 1953. Watase was elected as the first chairman by vote of about a hundred members.

Then, the Tomonaga Committee described above discussed the idea of an institute for nuclear studies (INS), and opinions of CRC were conveyed to the committee. It was hoped that institute would have divisions of accelerator and cosmic rays, and that the research staffs would take care of the inter-university use of the accelerator and cosmic ray facilities. INS was established in 1955 at Tanashi, Tokyo, as a research institute for inter-university use, and formally attached to the University of Tokyo. According to the discussions in CRC, the cosmic ray facilities were an air shower array and balloon-emulsion equipments, for these were considered to be utilized by the largest possible number of cosmic ray researchers. Seishi Kikuchi moved to Tokyo to assume the directorship. After that, he assumed in addition the directorship of Mt. Norikura Observatory. When the Institute for Cosmic Ray Research (ICRR) of the University of Tokyo was established in 1976 as an expansion of Mt. Norikura Observatory, the cosmic ray division of INS attached itself to ICRR. In 1955, I proposed that CRC should issue a journal in Japanese for prompt report of cosmic ray studies. This journal "Uchusen-kenkyu" was edited at Nagoya Univ. since 1956 and taken over by INS since 1961 and then by ICRR since 1976. In 1957, China invited four of our group to visit, and we saw a multiplate cloud chamber operating with a magnet at Mt. Loche Observatory (220 m^2 at 3184 m, Fig.13) established in 1953.

6. Formation of Cosmic Ray Modulation Group in Japan

The continuous observation of cosmic ray intensities was revived in 1946 with Neher's ion chamber at Hongo and then moved to Itabashi, where observations with the large area counter telescope (Miyazaki *et al.*, 1951) and a Nishina type ion chamber also revived in 1947 and 1948 respectively. At Nagoya University, a small narrow vertical telescope (Sekido and Kodama, 1948) was operated since 1948, and a small neutron monitor was operated since 1954. The latter was moved in 1955 to Mt. Norikura Observatory

(Kodama *et al.*, 1957). Since our participation in IGY, a cubic meson monitor with cross coincidences (Kawasaki *et al.*, 1957) was operated at Itabashi from January 1957, and a neutron monitor of IGY standard was operated at Mt. Norikura. A neutron monitor was operated also at the Showa Base in the Antarctic observation of IGY. Most of the continuous observations were done by the Itabashi group, and Wada took care of the projects throughout its long existence.

In the spring of 1946, an eminent astronomer Yusuke Hagihara (1897-1979) organized an interdisciplinary study group ranging from astronomy, geophysics, physics to radio communication. It was called the Ionosphere Research Committee, however, their interest was what is now called solar terrestrial physics. Suggested by Nishina, I joined them from the beginning to attend their monthly meetings and continued even after I moved to Nagoya. Hot data of every discipline were exchanged and discussed at every monthly meeting. I proposed a long chart with time scale as abscissa to write down the events in the row of corresponding discipline. This chart was filled in at every meeting and continued for more than 20 years. This Committee, which was very effective to help the studies on cosmic ray modulation, acted as one of the initiatives of IGY and WDC through activities of Takeshi Nagata and others.

Since 1948, I sent our intensity data to cosmic ray stations throughout the world asking for data exchange. I dreamed of a worldwide network instead of Nishina's five sites network. Since I organized a working group of cosmic ray modulation in 1951, the data exchange was done as one of the activities of the group. Before the Bagneres Conference of 1953, M.S.Vallarta invited proposals for standardization of counter telescope and format of data sheet. Our modulation group sent proposals to him. In 1954, Vallarta established SCRIV (Special Committee for Cosmic Ray Intensity Variation) in the Cosmic Ray Commission of IUPAP, and discussed such proposals at the Mexico Conference of 1955. To my surprize at Guanajato, Mexico, the proposed solid angles of the telescope were on a monotone function of the latitudes of the proposers, i.e. wide angle proposed from high latitude and narrow angle from low latitude, and it seemed to be difficult to come to a conclusion. At the moment, V.Sarabhai, who had been helping Vallarta since before 1954, expressed his agreement to the cubic monitor as standard, proposing a very narrow angle ($10°$) as a special program. By this, SCRIV meeting could arrive at a conclusion. I was entrusted to write a draft of data format, which was used in time for IGY. After that, the Itabashi Branch of IPCR served as one of the world data centers, WDC-C2, since July 1957 and published a series of cosmic ray data books containing useful data prepared after some basic data processings (WDC-C2, 1959).

I was deeply impressed at Guanajato by active discussions on various problems in cosmic ray physics, and desired to hold this kind of conference in Japan. Blackett's feeling on my idea encouraged me, and I thought the nearest possibility for it to be 1961. At the general meeting of CRC in October 1955 I proposed to hold such a conference in 1961. About the same month, at a committee of geophysics in Science Council of Japan, I agreed positively with Nagata who talked about an idea holding an international symposium on the earth storm in about 1961 when the results of IGY observations would be ready for discussion. These two ideas were jointly realised as the Kyoto Conference in 1961.

Fig. 14a, b, c. From Mexico to Kyoto. a: Forbush and Alfvén at Guannajato (1955). b: Sarabhai at Ahmedabad (1959). c: Alfvén at Pearl Island (1961).

7. Anisotropic Aspect of Cosmic Ray Storms, and Near-by
Acceleration of Galactic Cosmic Rays

Cosmic rays were said to be isotropic, and even the nature of their solar time variation was not clear for a long time. As described in the following paragraphs, the modulation group in Japan discovered the anisotropic aspect of cosmic ray storms, and revealed that the nature of this temporary anisotropy was a near-by evidence of the accerelation of galactic cosmic rays. In other words, the electromagnetic mode of generating cosmic rays was shown practically by observing this one step of the growth of cosmic ray energy. They found the method of locating the source point of the responsible solar blast, which gave evidence to the momentum transfer from the blast.

Compared with 1944, the cosmic ray diurnal variation, the sun spot number and the frequency of magnetic storm were larger in 1947 when we four moved from Itabashi to Nagoya. One of us, Sekiko Yoshida, who was a keeper of the large area counter telescope at Hongo and experienced in 1944 the coincidence of the strange diurnal variation and a magnetic storm (Section 4), examined cosmic ray diurnal variations at the time of magnetic storms by using the cosmic ray data revived after the War, and found in 1947-48 that the amplitude of cosmic ray diurnal variation increased at the time of magnetic storm (Meidai Uchusen-Kenkyushitsu Kiji, 1947, 1948). I wanted to examine the phase shift also, and we found that the time of maximum intensity became earlier in such a case (Meidai Uchusen-Kenkyushitsu Kiji, 1947, 1948). I wrote the result to Forbush, however, he seemed to be doubtful at first. We published it in English in 1950 (Sekido and Yoshida, 1950). This discovery impressed me, for it was my first experience of finding a strange but real cosmic ray phenomenon after many years of observation and many collaborator's effort to build the large counter telescope. The event of 16 December 1944 was a rare case which favored the discovery of the strange diurnal variation. The Forbush decrease in that case did not recover for many days, while the diurnal variation alone revealed clearly its characteristics. In most cases, the recovery and the enhanced diurnal variation were mixed up in so far as they were observed at one station or stations in a limited range of longitude. The above result was obtained statistically by averaging many events happened at various local time. But in 1952, we, including Yoshiko Kamiya, could show a worldwide Forbush decrease and the associated anisotropic aspect separately for each event, by using cosmic ray data observed simultaneously in the lucky distribution of longitudes: Europe, America and Japan (Sekido et al.,1952). Forbush decrease had been recognized by referring to a magnetic storm. But in 1954, our modulation group (Sekido,1954) named the simultaneous occurrence of the worldwide decrease and this characteristic change of anisotropy as "cosmic ray storm", after finding that the direction of the cosmic ray flow responsible for the additional storm time anisotropy was almost constant throughout many observations.

Since the discovery of Forbush decrease by the network of Model C meters, the ring current around the Earth, which caused the main phase of the magnetic storm, was considered to be responsible also for the cosmic ray decrease. But in 1949, two young physicists, Hayakawa at Meteo. Res. Inst. and Jun Nishimura at Itabashi, and two geophysicists, Nagata et al., showed theoretically that the ring current could not block off cosmic rays (Hayakawa et al., 1950). H.Alfvén's early idea of electric deceleration was also difficult to explain the Forbush decrease though we did not know at that time that the difficulty was not due to his idea of deceleration itself but came from the solar stream model adopted by him. Apart from models, Kazuo Nagashima, who was a student of M.Hasegawa and interested in the old ion chamber of Kolhörster mentioned in Section 1, examined the observed modulation spectrum and showed in 1950 that the Forbush decrease seemed to be better explained by considering electric deceleration (Nagashima, 1951) than by the magnetic blocking effect only. After the morphological studies on the anisotropic aspect of cosmic ray storms, done in 1954 by our modulation group (Working Assoc. Prim. C.R. Jpn. 1954/1957), he showed (Nagashima, 1955) that the observed modulation spectrum of the additional anisotropy during storm time was well explained by assuming an electric accerelation from the direction of the Sun.

This implied a practical step towards the experimental study on the mode of generating the energy of galactic cosmic rays. As a model in accordance with this mode of acceleration, I supposed that a "magnetic cloud", i.e. a plasma cloud containing electric current, was coming from the Sun when a cosmic ray storm took place. In the same year, 1955, we (Sekido *et al.*, 1955) classified magnetic storms into two types: those associated with cosmic ray storm (S-type) and the others (M-type). Then, we found that S-type happened sporadically though statistically related with solar activities, while M-type showed a 27 day recurrence period irrespective of any solar event, thus showing the existence of two kinds of magnetic storms with different origins. S-type was supposed to be caused by a blast from a solar eruption, while M-type by a stable solar corpuscular stream. The model adopted by Alfvén in 1941 mentioned above would be this M-type stream, while the above described accerelation from the direction of the Sun was presumed to be a momentum transfer from this S-type blast.

About the S-type storm, however, there were too many kinds and too frequent occurrences of solar flares to specify one event responsible for a storm on the Earth. A hint for solving this problem was already provided in 1950 from the study of the unusual increase of cosmic rays. In 1937, Nagaoka, who was the chairman of Cosmic Ray Committee of Gakushin, came to me frequently with hot data of the Dellinger effect, urging me to examine their relation to cosmic ray time variation. I regret that I did not do so. Forbush's paper on his discovery of the unusual increase in 1942 came to Japan after the war, when Ishii said "I knew it." We saw the old plot of intensities observed by Nishina's ion chambers, which showed a remarkable spike shaped increase (Sekido and Yagi, 1949). He omitted it from data sheet for he thought it was not of meteorological origin, when the study of temperature effect was the supreme order as the wartime project. In 1950, a Nishina's ion chamber was operated at Asahi Hut of Mt. Norikura as soon as it opened in September. An unusual increase was found on 20 September and reported at the October meeting of the Ionosphere Research Committee, when an astronomer Takeo Hatanaka reported a strange solar radio outburst which happened in a very wide range of frequencies (type IV by later classification). The coincidence of these two rare events suggested a close relation between them, as published (Hatanaka *et al.*, 1951). Using comprehensive data obtained since IGY, Kamiya *et al.*, (Kamiya and Wada, 1959; Kamiya, 1961) found the one to one correspondence betweeen type IV solar radio outburst and cosmic ray storm. By this, the solar event responsible for each cosmic ray storm was specified, thus enabling in 1961 to locate (Kamiya, 1961) the source of the blast precisely by using the associated optical flare. This location indicated the direction of the initial momentum of each blast, and enabled us (Sekido, 1963) to show that the direction of each blast was the same as that of the corresponding additional anisotropy during storm time, thus giving evidence to momentum transfer from the blast. When I talked of this relation between the source point and the additional anisotropy in 1963 at Jaipur, A.Ehmert said "I knew it". He did not apply for a lecture on it because he could give no explanation of.

8. An Air Čerenkov Telescope

In 1949, just after I believed that the strange diurnal variation was a real anisotropy, I ventured to start a new experiment seeking for another new fact. Until that time, sidereal anisotropy of cosmic rays was not examined by any telescope of narrow response. In 1950, I made a narrow angle telescope with alt-azimuth mounting to observe cosmic rays of higher energy (about 400 GeV) little deflected by the geomagnetic field, when neither the galactic nor interplanetary magnetic field was known. A narrow anisotropy, say a point source, which appeared in 1952-56, encouraged me to dream of cosmic ray astronomy, and I made the 3rd telescope $(20 m^2)$ with many collaborators at Nagoya Univ. After preliminary experiments from 1956 and construction in 1958-60, the observation started from 1961. But the narrow anisotropy had almost disappeared (Sekido *et al.*, 1961a), and I could not get a convincing proof.

The 3rd telescope was an air Čerenkov telescope (Sekido *et al.*, 1961b) with alt-azimuth mounting. Air Čerenkov radiation produced in a closed tube were reflected by a parabolic mirror (Fig.15) to make an image on the focal plane. The optical image was detected by a mesh of photomultiplier tubes. Therefore, this telescope was nothing but an optical telescope enclosed in a dark room. When this telescope was constructed, we surveyed by changing the zenith distance z of the telescope from vertical to horizontal. The flux was proportional to $\cos^2 z$, which was the same as the

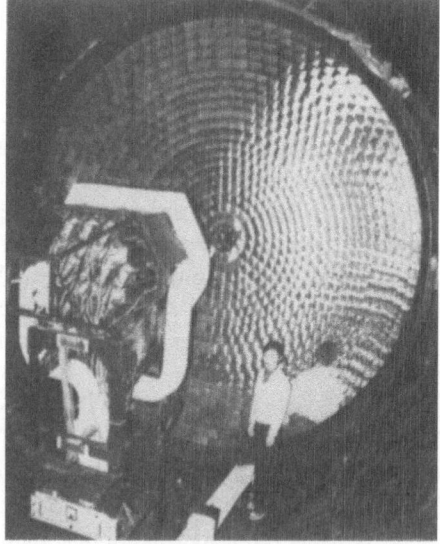

Fig. 15. An air Cerenkov telescope.

characteristic of cosmic rays. If somebody were to do this simple experiment without knowing anything about cosmic rays, he might discover cosmic rays by the dark current of light.

9. Worldwide Network of Cosmic Ray Intensity Observation

I am now recalling Hess' big influence upon my life, including my first work (Nishina *et al.*, 1937) on cosmic rays (1936) and my last publication (Sekido *et al.*, 1975) before my retirement (1975). The former was the observation during a solar eclipse, first observed by Hess (1912) on 17 April 1912, and the latter was the data book of the time variation, first observed by Hess (1912) on 20 May 1912, both in the year of my birth. Cosmic ray intensity observation initiated by Hess was succeeded by many workers in the world. Their cooperation was remarkably accelerated since about 1948 through two channels: the voluntary publication of data books and the international coordinaion with data exchange. As mentioned in Section 6, the preparation of the international data format for WDC had been done by SCRIV, and it was taken over by IQSY Committee, where M.Pomerantz took care of the cosmic ray format. After the retirement of Vallarta as chairman from SCRIV, I agreed with Sarabhai's proposal to recommend H.Carmichael as the chairman, but Sarabhai died before it was officially proposed. In the period of IASY, I proposed a post IASY observation network of solar terrestrial disciplines at the Leningrad Meeting of IUCSTP, and it was realized as MONSEE, while an instrumentation catalog was compiled by M. Shea (1972). At the London Meeting of SCOSTEP (1973), M. Dryer proposed a study group STIP, in which I suggested two working groups of cosmic rays, low and high energy regions, and it was realized just as I suggested. Thus, SCOSTEP and WDC maintain Vallarta's intention of SCRIV.

After the pioneering set-out of the networks of Hess (1913 (Benndorf *et al.*, 1913) and 1932) and Forbush (1936-37), the data book of cosmic ray intensities was first published by Forbush (Forbush and Lange, 1948). I was deeply impressed by those data accumulated over many years and harboured a long-cherished desire of similar contribution. When I moved to Nagoya in 1947, I had my own plan, the directional observation for the study of cosmic ray anisotropy, which started with the small narrow vertical counter telescope operated since 1948. Its development into a multi-directional telescope with high counting rate was much delayed because of my preference for high energy narrow telescope, and then realized by the efforts of Nagashima thus enabling me to publish the data book in 1975 as mentioned above, while Forbush continued his observation even after his double discovery, Forbush decrease and unusual increase, and continued his publications in 1957, 1961 and 1969. In spite of the existence of many improved equipments in the world, his data obtained with his classical instruments by his persistent spirit over four decades have been sparkling with an unique colour through the ever increasing use of these data by many of the younger generation.

REFERENCES

Barnothy, J. and Forro, M.: 1936, *Nature* **138**, 325; 1937, *Zeits. Physik.* **104**, 744.

Barnothy, J. and Forro, M.: 1939, *Phys. Rev.* **55**, 870; 1940, *ibid* **58**, 744.

Benndorf, Dorno, Hess, v.Schweidler and Wulf: 1913, *physik. Zeits.* **14**, 1141; See also page 8 of this book.

Blackett, P.M.S.: 1938, *Phys. Rev.* **54**, 973.

Forbush, S.E. and Lange, I.: 1948, *Researches D.T.M. Carnegie Inst. Washington* **14**.

Hatanaka, T., Sekido, Y., Miyazaki, Y. and Wada, M.: 1951, *Rep. Ionos. Res. Jpn.* **5**, 48.

Hayakawa, S., Nishimura, J., Nagata, T. and Sugiura, M.: 1950, *J. Sci. Res. Inst. Jpn.* **44**, 121.

Hess, V.F.: 1912, *Physik. Zeits.* **13**, 1084.

Ishii, C.: 1959, *Genshijidai no Kagaku*, Dainihontosho, Tokyo, p.153.

Ishii, C., Asano, Y., Sekido, Y. and Shimamura, H.: 1939, *Bull. IPCR.* **18**, 1066.

Kamiya, Y.: 1961, *Proc. Kyoto Conf.* **II**, 391.

Kamiya, Y. and Wada, M.: 1959, *Proc. Moscow Conf.* **IV**, 199.

Kawano, M.: 1934, *J. Meteor. Soc. Jpn.* **12**, 445; 1935, *ibid* **13**, 1.

Kawasaki, S., Kondo, I., WAda, M. and Miyazaki, Y.: 1957, *J. Sci. Res. Inst.* **51**, 107.

Kodama, M., Murakami, K. and Sekido, Y.: 1957, *J. Phys. Soc. Jpn.* **12**, 122.

Lectures of Heisenberg and Dirac, translated into Japanese by Nishina, Y.: 1932, Keimeikai Kiyo No.11, Keimeikai, Tokyo.

Meidai Uchusen-kenkyushitsu Kiji (*Proc. Cos. Ray Res. Lab. Nagoya Univ.*): 1947, **1**, No.4, 15, 18; 1948, *ibid* **2**, No.3, 5, 43.

Miyazaki, Y.: 1949, *Phys. Rev.* **76**, 1733.

Miyazaki, Y., Wada, M., Kawasaki, S. and Kondo, I.: 1951, *Rep. Sci. Res. Inst.* **27**, 173.

Nagashima, K.: 1951, *J. Geomag. Geoelectr.* **3**, 100.

Nagashima, K.: 1955, *J. Geomag. Geoelectr.* **7**, 51.

Nishina, Y. and Ishii, C.: 1936, *Nature* **138**, 721.

Nishina, Y., Birus, K., Sekido, Y. and Miyazaki, Y.: 1941a, *Sci. Pap. IPCR.* **38**, 353.

Nishina, Y., Ishii, C., Asano, Y. and Sekido, Y.: 1937, *Jap. J. Astro. Geophys.* **14**, 265.

Nishina, Y., Sekido, Y., Miyazaki, Y. and Masuda, T.: 1941b, *Phys. Rev.* **59**, 401.

Nishina, Y., Sekido, Y., Shimamura, H. and Arakawa, H.: 1940a, *Nature* **145**, 703; 1940b, *ibid* **146**, 95; 1940c, *Phys. Rev.* **57**, 1050; 1941, *ibid* **59**, 679.

Nishina, Y., Sekido, Y., Takeuchi, M. and Ichimiya, T.: 1941, *Uchusen*, Iwanami Koza Butsurigaku XI.B, Iwanami Shoten, Tokyo.

Sekido, Y.: 1943, *Sci. Pap. IPCR.* **40**, 456.

Sekido, Y.: 1944, *Uchusen*, Kawade Shobo, Tokyo.

Sekido, Y.: 1954, *Proc. Int. Conf. Theoret. Phys. 1953 Kyoto—Tokyo*, Sci. Coun. Jpn., p.69.

Sekido, Y.: 1963, *Proc. Jaipur Conf.* **II**, 199.

Sekido, Y.: 1985, Hess' Old Predecessors, in this book.

Sekido, Y. and Kodama, M.: 1948, *Solar Eclipse Com. Res. Council. Jpn.* p.50.

Sekido, Y. and Yagi, T.: 1949, *J. Phys. Soc. Jpn.* **4**, 353.

Sekido, Y. and Yoshida, S.: 1950, *Rep. Ionos. Res. Jpn.* **4**, 37.

Sekido, Y., Asano, Y. and Masuda, T.: 1943, *Sci. Pap. IPCR.* **40**, 439.

Sekido, Y., Yoshida, S. and Kamiya, Y.: 1952, *Rep. Ionos. Res. Jpn.* **6**, 195.

Sekido, Y., Nagashima, K. *et al.*,: 1975, *Rep. CRRL Nagoya Univ.*, No.1, No.2.

Sekido, Y., Wada, M., Kondo, I. and Kawabata, K.: 1955, *Rep. Ionos. Res. Jpn.* **9**, 174.

Sekido, Y. *et al.*: 1961a, *Proc. Kyoto Conf.* **III**, 131.

Sekido, Y. *et al.*: 1961b, *Proc. Kyoto Conf.* **III**, 139.

Shea, M.A.: 1972, *Ground based Cos. Roy. Instr. Catalog*, AFCRL-72-0411.

WDC-C2: 1959, *Cosmic Ray Intensity during IGY*, Sci. Council Jap.

Working Assoc. Prim. C.R. Jpn.: 1954, *Comm. IATME*, Rome; 1957, *IAGA Bulletin* **15**, 386(abstract).

PART 5 SHOWERS: LOCAL AND EXTENSIVE

TERRESTRIAL GENERATION OF PENETRATING RADIATIONS

PERSONAL RECOLLECTIONS OF EARLY THEORETICAL COSMIC RAY WORK

Walter Heinrich HEITLER[*]

My work on cosmic ray theory began with the paper on Bremsstrahlung and pair production by high energy electrons and photons, published jointly with Bethe. The cross sections for these processes proved to be unexpectedly high. A few years later Blackett and Occhialini discovered the cosmic ray showers in cloud chamber experiments. Blackett thought at first that these were nuclear explosions caused by very high energy particles. I saw at once that in a not too thick layer of a preferentially heavy material an incident electron must cause repeated processes of Bremsstrahlung which in turn led to pair production and so forth. If the energy of the incident electron was high enough, this inevitably must lead to a great number of electrons emerging from the bottom of the layer of material. Thus the showers would find a very natural explanation without introducing new physical processes. I discussed the matter with Bhabha. If I am not mistaken, the discussion took place in England. We settled down to work and very soon had a theory of "cascade showers". It is remarkable that at the same time Carlsson and Oppenheimer published the same results obtained by a very different mathematical method.

Fig. 1. W. Heitler.

Blackett conceded only after an animated conference in Copenhagen. He had in fact an argument. Besides the shower producing particles there were "penetrating particles" which passed through the material without leaving a trace of any interaction. Bhabha and I thought that these were protons. This was a mistake. The ionisation of protons would have been much higher with the momentum shown by the deflection in a

[*]University of Zurich (emeritus), Zurich, Switzerland.

Y. Sekido and H. Elliot (eds.), Early History of Cosmic Ray Studies, 209–211.
© 1985 by D. Reidel Publishing Company.

magnetic field. The only conclusion left was that these were hitherto unknown particles with a mass intermediate between electron and proton.

The time was ripe to remember the short paper of Yukawa's who had predicted in 1935 the very existence of such particles from an almost qualitative consideration of nuclear forces. This paper remained practically unknown, until the problem of the penetrating particles arose. I do not remember who drew the attention of people to this paper, I only remember that it was again at a Copenhagen conference. Then, practically at the same time, three sets of physicists worked out Yukawa's theory into a vector-meson theory, namely: Yukawa, Sakata and Taketani; Bhabha; Fröhlich, Heitler and Kemmer. I omit preliminary notes, in which part of the problems were handled.

In these papers the problem of nuclear forces has found at least a qualitative solution. Moreover: The same was true of the anomalous magnetic moments of proton and neutron. Since the nuclear forces showed charge symmetry F.H.K. could predict the existence of a neutral meson. This was in fact discovered later through its decay into two photons and their subsequent pair production.

Nearly everybody believed that the penetrating particles were the particles predicted by Yukawa. Their mass turned out indeed to be about 200 electron masses. Yet, there was a difficulty. If these mesons were to be responsible for the strong nuclear interaction, it was unlikely that they could emerge from a thick layer of material without hitting a nucleus and thus causing some event that must have been observed.

The solution came only several years later, after the war by a new experimental technique. About at that time, I suppose it was in 1938, I discovered in the library of the physics Institute in Bristol a paper by Blau and Wambacher. In this paper, simple experiments were described, in which ordinary photographic plates were exposed to cosmic rays for some months. Later, when developed, they showed under the microscope tracks due to protons (judging the ionisation by the density of the grains).

I suggested to carry on these very simple experiments in Bristol. I carried a stack of plates up to Jungfraujoch where they were exposed under varying conditions (layers of heavy and light materials between them, etc.). A few months later they were recovered. They showed a wealth of tracks and events partly also due to fast, heavily ionising, nuclei.

Then war broke out. After the war, Occhialini and Powell resumed these experiments and developed the technique. In collaboration with a photographic firm a much more sensitive emulsion was produced in which even the thinly ionising tracks of electrons could be seen. Then Occhialini and Powell made the following discovery: the true Yukawa-particles (hence called pi-mesons) were not very stable but decayed rather quickly into what was in fact identical with the penetrating particles of cosmic radiation, hence called mu-mesons. They also decay, but much more slowly than the pi-mesons and can penetrate the whole atmosphere.

One more discovery remains to be discussed. At about the same time, Janossy and coworkers discovered that also showers of penetrating particles occurred. I believe it was in counter experiments. Undoubtedly they were caused by particles strongly interacting with nuclei, nucleons or pi-mesons, giving rise to several pi-mesons and hence mu-mesons in a collision with a single nucleus.

The interpretation gave rise to a controversy. Janossy and I believed that a strongly interacting particle would collide with several nucleons in the same (heavy) nucleus, thus giving rise to the production of several pi-mesons. In fact, it could easily be seen that this must inevitably happen if the interaction is so strong. Heisenberg believed that, again owing to the strong interaction, several pi-mesons are produced in a single act of a collision with one nucleon. Probably in a penetrating shower both kinds of events occur. The first, analogous to a cascade shower, was named plural production, the second multiple production.

I have not followed up the experimental outcome of this situation. I did not continue theoretical work on cosmic radiation. The major problems were all solved. I shall therefore not describe either the later discoveries of numerous unstable particles with masses between pi-mesons and nucleons or heavier than nucleons. Undoubtedly this work will be described in other contributions to this book.

EXPERIMENTAL WORK ON COSMIC RAYS PROOF OF THE VERY HIGH ENERGIES CARRIED BY SOME OF THE PRIMARY PARTICLES

Pierre AUGER[*]

My first contact with cosmic rays dates from the year 1929 when I did some work with a cloud chamber built by D. Skobelzyne in the laboratory of Madame Curie, the Institute of Radium. The cloud chamber was placed in a magnetic field, not very strong, but which was sufficient for the study of the energy of radio active rays: This was, in fact, the aim of Skobelzyne, and enabled him to publish a number of results in this field. However, from time to time, straight tracks were present in addition to the curved tracks, and we concluded that they were due to electrons of very high energy. As they were always rather close to a vertical direction we recognized in these high energy electrons the so-called cosmic radiation or "high altitude radiation" discovered by Hess with an ionization chamber (Auger and Skobelzyne, 1929).

One of the experiments demonstrating that the nature of at least a large fraction of the primary cosmic radiation was that of electrically charged particles, consisted in measuring directly the change of the number of such particles striking the high atmosphere at different latitudes. Clay had already shown that the so-called "residual current" in an ionization chamber, current due to the ionizing power of the cosmic radiation, showed a marked decrease at the equatorial latitudes, compared to the same measurement at higher latitudes. Louis Leprince Ringuet and I performed the direct measurement of the number of incoming particles, during a trip from the latitude of 50° North to 50° South, using a system of three counters in vertical coincidences. Our result confirmed the corpuscular nature of the primary cosmic rays, through their deflection by the magnetic field of the Earth (Auger and Leprince, 1933, 1934).

I went on studying cosmic rays with different arrangements of counters and cloud chamber, and was soon convinced that there were two components of different nature in the corpuscular radiation (Auger, 1935a, b, c, 1936a, b, 1938; Auger and Ehrenfest, 1934). I studied their properties at sea level, underground, and at high altitude with a number of students using the coincidence method with counters in vertical disposition and screens of lead of increasing thickness, and also in horizontal arrangements to study the "showers" produced in the material placed above them. It was soon clear that the showers were associated with the soft group of cosmic ray particles, the relative proportion of the soft and hard group changing with the thickness of the screen interposed between the high atmosphere and the apparatures I measured this relative proportion at high altitude, at sea level and at 30 meters underground. I could conclude that the penetrating particles had a higher mass than the electron which composed the soft group. Later, when the real nature of the penetrating particles — then called "mesotrons" — was established I could measure their lifetime by a very direct method, in collaboration with R. Maze and Chaminade (Auger *et al.*, 1942). We observed horizontal coincidences between counters even without a "shower producing"

* 12 Rue Emile Faguet, 75014 Paris, France.

Y. Sekido and H. Elliot (eds.), Early History of Cosmic Ray Studies, 213–218.
© *1985 by D. Reidel Publishing Company.*

screen above them, but such studies were made difficult by the "spurious" coincidences due to the fact that the individual counting rate of each counter was high, and the resolving power of the coincidence system unsufficient to eliminate such chance coincidences masking any real phenomenon.

So, with one of my alumni Roland Maze, very able in electronics, we built new coincidence systems of counters with a resolving power of 10^{-6} seconds, much greater than those in current use at that time in cosmic ray research. This progress made it possible to reduce drastically the number of background (or accidental) coincidences which were very frequent with our counters which had a large counting rate, as is often the case in laboratories. This new coincidence apparatus made it possible to study rare events, such as showers with a large spatial extension or the so-called "explosive" phenomena, such as Hoffmann ionization shocks. A number of showers with an horizontal extension of one meter or more were observed by us in the laboratory rooms. They could have been explained as normal showers issuing from the building above the apparatus, but the number seemed too high and I suspected the existence of a phenomenon of greater extension, necessitating the use of counters situated at much greater distance. So I decided to put one of the counters in another building, the horizontal distance being then about 150 meters. The number of coincidences remained quite important, and we could announce, in 1938, the existence of what we called "Grandes Gerbes" that is "Extensive Showers". They are now frequently called Extensive Atmospheric Showers or E.A.S.*, their origin being evidently quite high in the atmosphere above the laboratory.

Having made sure of the existence of these new showers, I wanted to analyse some of their properties: composition, corpuscular and photonic, energy of the components, and the density of impacts per square meters. The result could then be made use of for an evaluation of the total energy of the shower. As all the component arrived simultaneously at sea level, this total energy could be interpreted as the energy of the primary particle arriving from the space outside the atmosphere. The density of tracks per square meter could be evaluated by taking into account the surface of the counters, and the variation of the number of showers observed when additional coincidence counters were added at a few meters distance to the two first counters responsible for the detection of a shower. Densities of the order of 10 to 100 per square meter were thus ascertained.

I then thought necessary to extend as much as I could the distance between the counters. In order to get a reasonable number of coincidences and considering the well known increase of intensity of cosmic rays with altitude, I thought that we would have a better chance in a high mountain laboratory. So the whole instrumentation was transported on two sites where good laboratory conditions could be found. One was at the observatory of the Pic du Midi, in the Pyrénées (2990 m altitude) the other at the Jungfraujoch, in Switzerland, (3500 m altitude). The factor of increase of the number of showers per hour was found to be of the order of ten between sea level and 3500 m (Auger *et al.*, 1939b, 1948, 1949). We could thus observe a rough value of 60 showers per hour at sea level, and 600 at high altitude.

* This abbreviation: E.A.S. can also be interpreted, as some authors do, as meaning Extensive Auger Shower!

In order to extend as much as possible the distance of the counters, and considering the local conditions at the Jungfraujoch, we had to set one or two of them in the laboratory, and a third one at 300 m distance on the glacier, in a small cavity carved out of the ice: We could have increased the distance even more, the glacier being very large, but as we had already to apply a correction of time to the kiks of the laboratory counters in order to bring them in coincidence with those coming from the distant counter, this time lag would have to be corrected for very large showers even slightly oblique, because of a difference of time arrival of particles in different part of the shower.

The particles of the whole shower are in fact contained in a kind of "plaque", perpendicular to the direction of the axis of the shower and a few tens of centimeters thick. With more than 300 m distance, the different parts of an oblique shower touch ground with a difference of time arrival of hundreds of nanoseconds, approaching our coincidence time. In fact this means that we were observing only showers with a small angle with the vertical.

About the composition of the showers, we did a number of experiments with three counters, two being in a vertical plane and separated by a variable lead screen up to 20 cm thick and protected against side effects, the third one placed at a few meters of horizontal distance. The absorption curve obtained could be explained partly by the existence of bundles of electrons of very high energy, but proved also the presence of a small proportion of penetrating particles, the "muons" which at that time were called "mesotrons". This was the first proof of the secondary nature of these particles as I stated in my 1938 papers (Auger et al., 1938).

The presence of narrow bundles of high energy electrons was confirmed by a number of cloud chamber photographs (see Fig. 1), the expansion being triggered by triple coincidences. Some photographs showed the presence of penetrating particles (see Fig. 2), and the existence of nuclear disintegrations was also observed, in particular by the frequent presence of highly ionizing tracks, about 12 tracks per 100

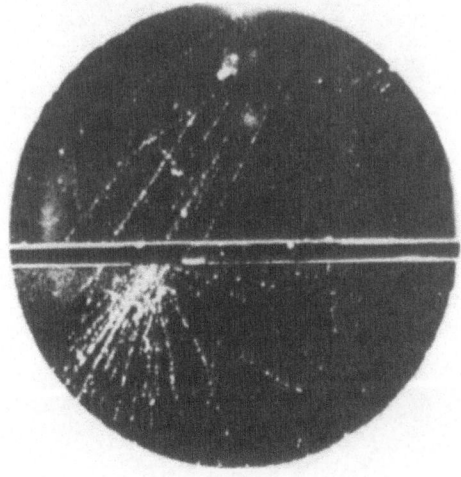

Fig. 1

photographs, due probably to slow protons, and suggesting the presence of neutrons in the beam. But the striking feature was the density of electron tracks, sometimes so thick that the cloud chamber looked like filled with a heavy rainfall (see Fig. 3): in this case we thought that the chamber had been striken by the axial part of the shower, probably the direction of the incoming primary particle. Some photographs showed much less dense tracks, characteristic of the periphery of the showers. We did also some work with two chambers, situated vertically above one another and separated by screen of lead, and could make visible the multiplication effect of the screen.

But the main surprise came when I tried to evaluate the energy of the incoming primary particles. For that purpose I had to evaluate the total number of electrons, and also the mean value of their energy. The total number could be deduced from the density of tracks per square meter and the distance at which the coincidences could be obtained. For that last variation of number of showers with distance, it was clear that showers of much more extension did exist. Anyway, with a total surface of between 10^4 and 10^5 square meters this meant a munimum of 10^6 particles in the showers of large extension. Now as to the energy of the electron, not having at my disposal at that time a cloud chamber in a magnetic field, I had to admit that they had at least the critical energy in air, that is 10^8 eV. By another method, the consideration of the total screen of air between the origin of the showers and the instruments, the same value can be derived from the Bhabha-Heitler calculations.

Taking that mean value of 10^8 eV and the total number of 10^6 electrons and taking into account the loss of energy by crossing the atmosphere, I came to the conclusion, rather astonishing at that time, that particles of at least 10^{15} eV arrived at the top of the atmosphere (Auger *et al.,* 1938, 1939a, b, 1948, 1949). The number per square meter of these primary particles could also be evaluated, and taking only the particles

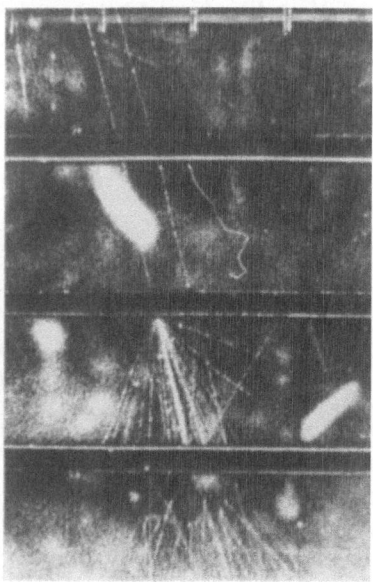

Fig. 2. Au sol (Labo.).

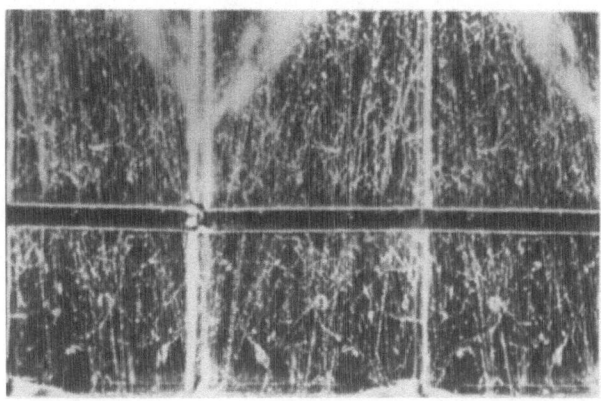

Fig. 3

with an energy higher than 10^{15} eV their number was found of the ordered one per day on 10 square meters. Of course the number of particles with smaller energy, say 10^{12} eV, was more than 10 times larger.

As for the energy spectrum of the primaries responsible for the extensive showers we tried to deduce it from the number of showers obtained at different altitudes. Having done some experiments with three to nine coincidence counters installed in an airplane with a total horizontal extension of 15 meters, we found at the altitude of 7250 meters a fifty fold increase. We found also a strong increase of the density of the tracks in the big showers. At first sight, the spectrum seemed to be of an exponential form, the number of showers with an energy higher than E being roughly proportional to E^{-2}.

Anyway the existence of a large number of particles striking our atmosphere with an energy of that magnitude could not be easily explained. Since the time of our discovery, in 1938, the existence of these high energy particles has been confirmed and even much higher energies have been measured by different authors, 10^{18} and perhaps even 10^{20} in very exceptional cases. How particles with such energies are produced is not fully understood even if some interesting theories have been proposed, for instance by Fermi, involving large and extended magnetic fields in space. It is presently admitted that the primary particles responsible for the EAS are nuclei, mostly protons, and some heavier ones. The initiation of the shower takes place at very high altitude, say 20 km above ground, and consists in a first collision between the high energy primary particle and the nucleus of an atom of the air. All sorts of particles are then produced, but the main part of the shower after a few km of atmosphere, are electrons and positrons, with a certain proportion of muons, – those which are counted under 10–20 cm of lead –, of pions, of light nuclei and hadrons. So the total analysis of the phenomenon tends to be rather complicated, and has been studied by a large number of physicists. Many publications have been made since our first observation of long distance coincidences leading to the discovery of the extensive showers.

REFERENCES

Auger, P.: 1935a, *Nature* **135**, 820
Auger, P.: 1935b, *Compt. Rend. Acad. Sci. (Paris)* **200**, 1022.
Auger, P.: 1935c, *J. Phys.* **6**, 226.
Auger, P.: 1936a, *Kernphysik,* 95.
Auger, P.: 1936b, *J. Phys.* **7**, 65.
Auger, P.: 1938, *Compt. Rend. Acad. Sci. (Paris)* **206**, 346.
Auger, P. and Ehrenfest, P.: 1934, *Compt. Rend. Acad. Sci. (Paris)* **199**, 1609.
Auger, P. and Leprince-Ringuet, R. L.: 1933, *Comp. Rend. Acad. Sci. (Paris)* **197**, 1242; 1934, *J. Phys.* **5**, 143.
Auger, P. and Skobelzyne, M.: 1929, *Compt. Rend. Acad. Sci. (Paris)* **189**, 55.
Auger, P. *et al.*: 1938, *Compt. Rend. Acad. Sci. (Paris)* **206**, 1721
Auger, P. *et al.*: 1939a, *J. Phys.* **10**, 39.
Auger, P. *et al.*: 1939b, *Rev. Mod. Phys.* **11**, 288.
Auger, P. *et al.*: 1942, *Phys. Rev.* **62**, 307.
Auger, P. *et al.*: 1948, *Phys. Rev.* **73**, 418.
Auger, P. *et al.*: 1949, *Rev. Mod. Phys.* **21**, 14.

COSMIC RAY SCHOOL OF OSAKA UNIVERSITY OSAKA CITY UNIVERSITY – FOR THE MEMORY OF LATE PROFESSOR YUZURU WATASE –

Minoru ODA*

Prof. Watase will be remembered among Japan's scientific community as a great leader in cosmic ray physics. He was one of the leaders who revitalized activities in Japan's physics, which was deeply damaged by the war, and helped to build it up to the present level.

I will describe the historical developments up to around 1960 of Watase's School to which a number of scientists belonged. Some of them are still in the discipline of cosmic ray physics, some are in astrophysics, some are in elementary particle physics with accelerators and some are working in industries. My emphasis here is primarily on historical episodes and not on the details of physics.

Watase's first contact with cosmic ray physics may be traced back to the mid-1930's when he poineered cosmic ray research in Japan with Seishi Kikuchi. Prof. Kikuchi guided the laboratory for nuclear physics at the Osaka Impereal University from 1933 to 1955: From the start of the laboratory Kikuchi had been interested in cosmic rays and encouraged Watase, who had joined the lab in 1933, to initiate research in this field.

The mechanism of cosmic ray shower production was investigated in the period prior to 1936 when it had been debated whether the shower particles are produced successively or simultaneously in the material. Figure 1 shows the counter arrangements and the experimental results (e.g. *Nature* **39**, 671 (1937)).

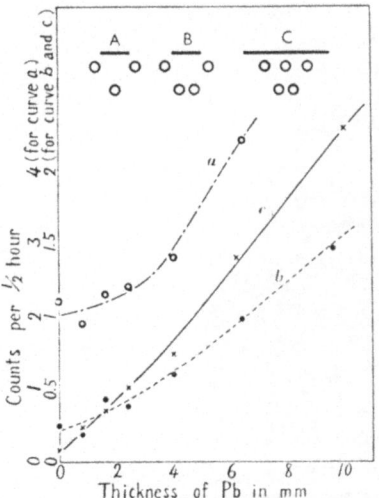

Fig. 1. From Watase's paper on "Cosmic Ray Showers": *Nature* **39** 671 (1937).

* Institute of Space and Astronautical Science, Tokyo, Japan.

Y. Sekido and H. Elliot (eds.), Early History of Cosmic Ray Studies, 219–224.
© *1985 by D. Reidel Publishing Company.*

They found that "the quadratic character becomes less prominent as the number of counters of which simultaneous discharges are counted increases.", and concluded that "The present results do not exclude the existence of showers produced by a succession of elementary processes, but show clearly that the latter is not necessary for the production of a shower containing a large number of particles; or, in other words, there exists an elementary process in which many particles are produced simultaneously, as suggested by Heisenberg." They were very much encouraged by discussions with Prof. N. Bohr who then visited Japan.

Watase discontinued cosmic ray research in ~1938 when he joined the group under the direction of Kikuchi which started to design and construct Japan's first cyclotron. Years later he regretted that he had not pushed the research further: he confessed that it was difficult psychologically to continue cosmic ray study practically alone in a laboratory where the main stream of research was nuclear science, where accelerators had been designed and constructed, and activities in the experimental studies had been almost explosive. He wrote "Of course then we did not know μ-mesons, but we sensed some future for the penetrating component ... We should not have closed the activity regardless of how non-productive it appeared superficially as far as the number of publications is concerned." Thus Japan's activity in cosmic ray research transferred to the hands of the group at the Nishina Laboratory of the Physical and Chemical Research Institute. There, Miyazaki, Sekido, Takeuchi ..... continued the research through the war time.

After the war in 1947, Watase was appointed as a professor of physics at Osaka University, and also at Osaka City University in 1949. By this time, even though the effect of the War still remained strongly in the chaotic social condition a deep desire for research was arising among scientists. A number of young physicists gathered to

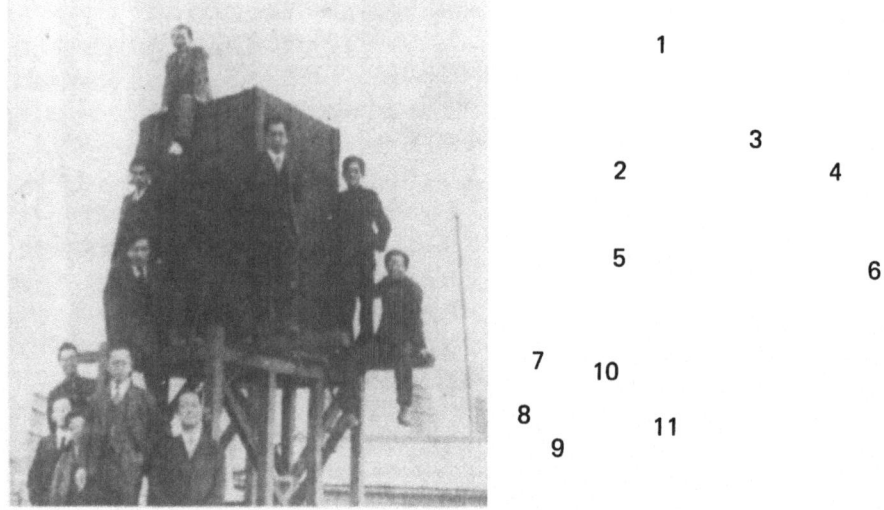

Fig. 2. Kikuchi's group aside the water tank for the cosmic ray experiment.

1, Fushimi ; 2, Kumagai (Aoki) ; 3, Watase ; 4, Itoh ; 5, Kikuchi ; 6, Kawamura ; 7, Ichimiya ; 8, Sakata ; 9, Yamaguchi ; 10, Oka ; 11, Yukawa

participate in Watase's Lab., some were his or Kikuchi's students, some were introduced to him by his friends, most of them not being in the payroll of the University.

While he conducted the research of microwaves as a continuation of his work during the war time for the Navy, and its various applications to feed his starving colleagues, and also solar radio astronomy which I helped him to develop in the lab, he declared that cosmic ray research should be the main target for the future. We planned the program so that within a few years the fabrication techniques of a number of large G-M counters and ionization chambers and the techniques of the large cloud chamber would be developed; but, under the economical and social circumstances in the few years after the end of the War this was not simple.

The City of Osaka, in the meantime, decided to take a far more progressive policy for the newly founded Osaka City University in forming research groups than in any of the other national universities. As a result Watase managed, in addition to the lab in Osaka University, a laboratory three to four times larger in size and in the number of young researchers than the average physics laboratory in the national universities by 1950.

In 1949 the City of Osaka gave approval to Watase to build a high altitude laboratory at Mt. Norikura for cosmic ray research and for high altitude medical science. By this time, several other cosmic ray research groups started to become active. They were Minagawa's group at Kobe University, and Nakagawa's group at Rikkyo (St. Paul) University: also Miyazaki succeeded the director of Nishina's lab at the Physical & Chemical Research Institute and Sekido moved from the Nishina group to Nagoya University to build a new group. We could also count a few more groups of smaller size. Watase and these post-war leaders acquired a fund from the Asahi Press being helped by the strong support of Mitsuo Taketani and built a mountain observatory also at Mr. Norikura. The role of these two high altitude observatories was later in 1953 taken over by the Norikura Cosmic Ray Observatory, University of Tokyo which was established as "inter-institutional" among researchers of universities and institutions from all over Japan. This was the realization of a new concept which had emerged in the Science Council of Japan and had been crystalized by the Ministry of Education.

Watase recognized the importance of having a strong theoretical group in close contact with the experimental group. He made all efforts to realize this idea and, thus, for some time Y. Nambu, S. Hayakawa, Y. Yamaguchi, K. Nishijima and T. Nakano gathered at the Osaka City University and the experimentalists enjoyed the luxury of every-day contact with these theorists: They later scattered and became the most active theoreticians in various places.

Among the several subjects of research, which were then performed in the laboratory, on the roof of the laboratory, and also in the mountain laboratory, Watase put specific emphasis in proving the multiple pion production in a nucleon-nucleon collision: we recall that in the late 1940's whether pions are produced in the target nucleus "plurally" or "multiply" was the important question which might have crucial bearings on elementary particle physics.

Watase encouraged and supported Saburo Miyake and his colleagues to construct a high pressure cloud chamber with hydrogen gas to resolve this question. The photograph of an event that occurred in the chamber was the conclusive evidence in favor of the multiple production in the collision of a proton with a hydrogen target. Although by that time there was little doubt about the existence of multiple production by means of extensive studies with the nuclear emulsion, this was the first direct and clear proof of the multiple pion production. Koichi Suga and his colleagues from the Watase Lab also attacked this problem using a cloud chamber located underneath a high pressure hydrogen tank.

Simultaneously among the Watase group special interests in cosmic rays underground had been emerging. By this time there was the pioneering work by Miyazaki *et al.* of Nishina's Lab.: the intensity-depth curve had been obtained upto a depth of 3000 meter water equivalent using the Shimizu Tunnel indicating that the energy spectrum of muons extends up to 3000 GeV and probably even further. The London subway station had been used by the British group to study the nuclear interaction of the earth penetrating component, presumably muons, and the results was analysed by means of the Weiczächer-Williams method; i.e. the muon-nucleus interaction was interpreted in terms of the photon-nucleus interaction. The muon-nucleus interaction was then studied by the scattering experiment represented by e.g. that of Amaldi *et al.*

Among the members of the group there was a strong feeling, partly supported by suggestions of our theorists, that the study of muons or the weak interaction would be related to something more fundamental in physics and to a deeper extent than the study of penetrating showers produced by the strong interaction: the study of penetrating showers, at least with the counter or cloud chamber techniques, appeared to be coming to a dead-end or being replaced by experiments with acclerators. Watase's philosophy, then, was that in the field of muon physics, using cosmic rays penetrating to the underground, we would suffer from the competition with accelerator physics to a much less extent. This was particularly so since under the circumstances we were still not free from damage by the War and only half way to developing the laboratory techniques. At the time the average laboratory could not purchase a synchroscope and had to construct it from the oscilloscope itself. High frequency transmission cable was not yet known.

In 1951, the abandoned tunnel of the Japanese National Railroad at Yaizu in Shizuoka Prefecture was chosen to be the location of the underground laboratory for cosmic rays and the group for this study was formed. This tunnel was not very deep but was sufficiently deep to filter the cosmic ray muons: so, the subject was almost exclusively restricted to muon interaction events. I will describe a couple of episodes by which our living conditions in the field work there may be understood. A silkworm hut of a farmer-house near the tunnel was rented as a small laboratory for the observing station where we had only poor photographic facilities: Without having a dark room clock we sang songs in the dark to measure the length of time for processing the photographic film; on another occasion, using a small stream in the tunnel as a sink, the gelatin of a nuclear emulsion plate was eaten by a frog etc. etc. Starting from GM counter hodoscopes which were used for the study of muon scattering, a large cloud chamber and GM counter array were introduced to investigate the nuclear interaction of muons.

The Yaizu Laboratory operated until 1962 when it was moved to the Mitsuishi Tunnel in Okayama Prefecture. We can recall the names of Oshio, Shibata, Higashi,

Higashino, Kitamura, Fukui, Watanabe and many others as authors of a number of papers published in this period. In the meantime (1962-71) the experiment of ion-chamber bursts and the deep sea experiment at a depth of 1500 m to produce a precise intensity-depth relation of muons had been performed as correlated programs. Ozaki who had worked with Kitamura, Fukui and Murata on muon scattering and the momentum spectrum of muons joined K. Greisen's group at Cornell to work on novel techniques for EAS (Extensive Air Shower) detection and then worked for Keuffel of Utah for the neutrino experiment.

After working for Rossi of MIT for the construction for MIT's EAS facility, in 1956 I moved from Watase's Lab to the newly established Institute for Nuclear Study, University of Tokyo, to construct a new EAS facility with Suga who was from the same group and Miura from Sekido's group at Nagoya. Fukui and Miyamoto of Osaka University contributed their efforts towards the construction of the neon-hodoscope as a part of the EAS facility. During the course of developing the instrument, in 1956 they invented a technique which they then called the discharge-chamber but which really was nothing but the "spark chamber": this of course became a widely used technique in high energy physics later.

Miyake, after his work with the two generations of the high pressure hydrogen could chamber, constructed a large cloud chamber (at that time the World's largest) and

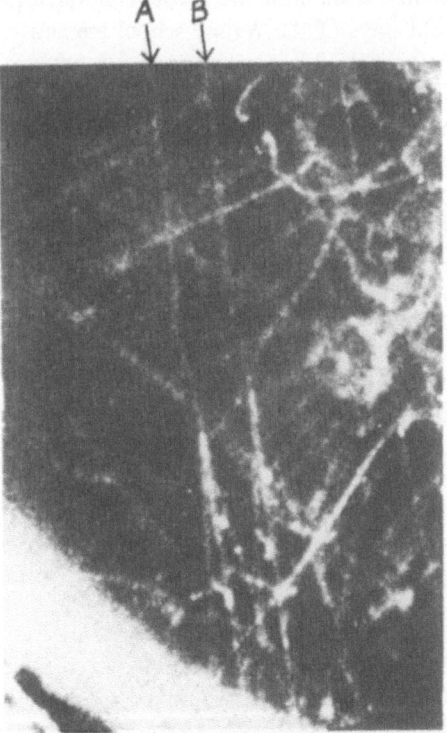

Fig. 3. A picture, taken by the high pressure cloud chamber, proves the multiple production. A and B are positive and negative ion tracks separated from one cosmic ray track by sweeping electric field and show multiple production of secondaries in the
high pressure hydrogen gas.

Fig. 4. Watase near the entrance of the Yaizu Tunnel where the laboratory was located. Detectors above ground, the silk worm hut and the pig hut may be seen.

formed a group for the study of EAS. In 1960, following Watase's visit to India, Miyake visited the Tata Institute of Fundamental Research, Bombay. Since then the fruitful Indo-Japanese collaboration first in the study of EAS and soon after with the underground experiment at the Kola Gold Mine has been initiated. From the collaboration emerged a new field of muon and neutrino physics in cosmic rays.

As of 1980, over twenty years after the above-mentioned period and two years after the death of Watase, old boys of the Watase school are active in various institutions in various fields. The Osaka City University group continues to be active in fields of cosmic ray physics, accelerator physics and plasma physics being led by Oshio, Ozaki, Higashi, Kusumoto and others. Miyake directs the Cosmic Ray Laboratory, University of Tokyo which grew up from the Norikura Cosmic Ray Observatory: Kitamura also joined the Laboratory. A number of Watase's former students including Fukui and Kaneko moved to the field of high energy particle physics. Suga who had been at the Institute for Nuclear Study, University of Tokyo and moved to the Tokyo Institute of Technology has been actively leading the air shower experiment at Mt. Chacaltaya, Bolivia which has continued from BASJE (Bolivian Air Shower Joint Experiment) which was originally established by the collaboration between MIT (Rossi, Clark, Bradt and others), the Bolivian's (Escobar) and the Institute for Nuclear Study (Oda, Suga and others). Tanaka and I have been working in the field of X-ray astronomy since mid-1960's at the Institute of Space and Aeronautical Science, University of Tokyo.

BEGINNING OF HIGH-ENERGY PHYSICS IN COSMIC RAYS

Nicolai A. DOBROTIN*, Georgi T. ZATSEPIN*,
Sergei I. NIKOLSKY* and Sergei A. SLAVATINSKY*

This essay presents a brief history of certain research into the nuclear-physical aspect of cosmic rays which was performed, on the whole, before 1960 by the scientific school of Academician D.V.Skobeltzyn (in the main, at the Lebedev Physical Institute of the Soviet Academy of Sciences on high mountains). Other such research is touched upon briefly, only so far as it has been stimulated by its inkage to the described work.

In the twenties, the investigation of cosmic rays in the Soviet Union, as in other countries, was of a general, phenomenological nature. A.B.Verigo measured cosmic-ray flux at various altitudes using a variety of filters. L.V.Mysovsky and his collaborators studied the absorption of cosmic rays in water and the relation of their intensity to meteorological factors — pressure and temperature. Thus, in 1925, i.e., even earlier than R.Millikan's well-known experiments on cosmic-ray absorption in mountain lakes, Mysovsky (1925) showed that the absorption coefficient of such rays in water amounts to $\sim 2.10^{-3}$ cm^2/g, i.e. about one-tenth of the absorption coefficient of γ-rays from radioactive substances. He observed also that with increasing atmospheric pressure the intensity of cosmic rays decreases: for $\Delta p = 1$ mm of Hg, $\Delta J/J = -0.345\%$ (Mysovasky and Tuvim, 1928). Thys, Mysovsky's experiments demonstrated again that this radiation was of extraterrestrial origin.

A very important step in the general study of primary cosmic rays was made by Vernov, who first used for this purpose unmanned balloons with radio transmission of instrument readings. The application of such a method acquired subsequently tremendous significance. In particular, this enables Vernov (1934) and his collaborators not only to obtain complete altitude curves of cosmic-ray intensity and thereby establish certain characteristics of particle interaction at such high energies (Vernov, 1936), but also to investigate geo-magnetic effects for primary cosmic-radiation and on this basis determine its nature. Later, when investigations of the space began to be conducted beyond the atmosphere by means of artificial satellites, radiation belts around the Earth were discovered by Van Allen in the United States and Vernov, Chudakov and their collaborators in the Soviet Union. But we shall not consider the space-physical aspect of the investigations. We shall deal with the history of those cosmic-ray investigations performed in the Soviet Union which are connected, in the main, with the study of high-energy particle interactions, particularly at mountain altitudes.

As is well known, high-energy particle physics was begun in Skobeltzyn's classic experiments with a cloud chamber in a magnetic field (Skobeltzyn, 1927, 1929). By means of such equipment, Skobeltzyn was studying the Compton effect. In 1927, he observed that in the cloud chamber there appeared tracks of particles having energies of at least tens of MeV. And in 1929, he established that these cosmic-ray particle tracks

* P.N. Lebedev Physical Institute, Acad. Sci., Moscow, U.S.S.R.

Y. Sekido and H. Elliot (eds.), Early History of Cosmic Ray Studies, 225–237.
© 1985 by D. Reidel Publishing Company.

appear comparatively frequently in groups. In this manner, he discovered the phenomenon of formation of cosmic-ray particle showers, i.e., the first specific effect related to the physics of high energies. This discovery transformed the study of cosmic rays from a branch of geo-physics into a new branch of physics — the physics of high-energy particles. A more detailed history of the birth of this new branch of physics is described by Skobeltzyn in his article.

After Skobeltzyn's experiments there followed experiments by Kunze (1933a,b) and Anderson *et al.* (1934), also using a cloud chamber in a magnetic field, in which new photographs of cosmic-ray particle tracks were obtained. In addition, Anderson (1933) succeeded in observing the positron. Particularly convincing were the pictures of positron-electronic showers obtained by Blackett and Occialini (1933).

Separately from experiments with cloud chambers for studying cosmic-ray particles, experiments using coincidence counters began to be performed. A particularly large number of such investigations was undertaken by Rossi (1930, 1933).

Counter experiments made it possible to separate cosmic rays at sea level and low altitudes into hard (penetrating) and soft (strongly absorbing) components. After the discovery in 1938 by Anderson and Neddermeyer of mesons (μ-mesons), this division was no longer merely phenomenological, but acquired a physical basis.

Subsequently in a number of experiments for studying the soft-component at sea level, performed in the main by Azimov by means of a counter "telescope" and later in more detail (Azimov, 1949) by means of a circular installation at various heights, it was shown that even at sea level not all of the soft component was due to the decay of μ-mesons and the formation by them of δ-electrons. This additional, "non-equilibrium" part of the soft component is formed, in the main, in the upper layers of the atmosphere in the processes of interaction of cosmic-ray primary particles with nuclei of air atoms. Subsequently, the mechanism of its production (decay of π°-mesons) was established.

In the middle of the forties Alikhanov and Alikhanyan (1944, 1945) and their collaborators already noted that even at sea level the composition of the penetrating component includes a definite number of protons, the number of such protons in the atmosphere increasing rapidly with altitude. In many works of that period, the flux of nucleons in cosmic rays began to be called their third component.

Systematic investigations of cosmic rays on high mountains began in the Soviet Union in 1934 as part of a wide-range program of expeditions to Mt. Elbruz. In the first experiments, performed by Frank, Cherenkov and one of the authors of this essay (Dobrotin *et al.*, 1936), it was shown by means of a small cloud chamber that at an altitude of 3 to 4 thousand meters above sea level there existed a comparatively large number of low-energy electrons in cosmic rays. Radioactivity was excluded by setting the apparatus on the thick, solid glacial cover of Mt.Elbruz. In that period, before the advent of the cascade theory and, in general, any conception of the production mechanism of cosmic-ray secondary particles, this result was unexpected and went un-noticed.

Moreover, even the first variants of the cascade theory proposed by Bhabha and Heitler (1937), Landau and Rumer (1938), Carlson and Oppenheimer (1937) and others were restricted to the consideration of only relatively high-energy particles. The energy spectra of secondary electrons obtained by them were in sharp contradiction with the experimental data obtained by means of ionization chambers (in particular by Vavilov, 1949). Only subsequent development of the cascade theory, first on a semi-quantitative level by Skobeltzyn and Vernov (1940) and then more rigorously by Tamm and Belenky (1939) showed that there was indeed a large number of low-energy particles in the soft component of cosmic rays.

Beginning in 1937, cosmic-ray work on Mt.Elbruz was conducted under the leadership of Veksler. It was established by means of proportional counters that the number of strongly ionizing heavy particles in cosmic rays exceeded the expected if one considered the penetrating component as consisting of protons. From this it was concluded that secondary, strongly ionizing particles (apparently, secondary slow mesons) were produced at an altitude of 3 to 5 km above sea level (Veksler and Dobrotin, 1941).

The scattering of cosmic-ray particles in various materials ("backward showers") and certain other problems were also studied on Mt.Elbruz (Dobrotin *et al.*, 1938).

One can say that the pre-war expeditions to Mt.Elbruz were the culmination of the "heroic" stage of cosmic ray research in the Soviet Union. After the war, the work was conducted at an entirely different level as regards the availability of apparatus and living conditions. Before the war, donkeys loaded with packing cases served as the main means of transportation on Mt.Elbruz. Thus, during the first 1934 expedition, together with Frank and Cherenkov we packed a cloud chamber with solenoid and storage battery for creating the magnetic field, a heliostat for solar illuminating and other devices into three cases and loaded these, as well as sleeping bags and provisions, onto two donkeys. It was interesting to observe how carefully they sought their footing at each step on rocks and ice, but how difficult it was for them in snow where they sunk to their bellies. And on the saddle of Mt.Elbruz, at an altitude of 5300 m, where even donkeys could not pass, scientific equipment and provisions were carried by person. A man, motivated by a sense of duty, can endure more than a beast of burden, inspite of the fact that on Mt. Elbruz the height at which even trained people begin to feel mountain-sickness lies below an altitude of 5000 m ...

After an interruption caused by the beginning of the war, the work on cosmic rays at mountain levels was resumed in 1944 on the Pamirs and somewhat earlier on Aragatz (in Armenia). In the forties, permanent, high-altitude stations for studying cosmic rays were

Fig. 1.   The general view of the Pamir Expedition Camp (about 4000 m above sea level); 1946.

Fig. 2. The investigation of penetrating component of extensive air showers at Pamir expedition (1946).

established on the Pamirs (altitude 3860 m; under the overall direction of Skobeltzyn, during the initial period − Veksler, and then the authors of this essay); in Armenia (alltitute 3250 m; supervisors Alikhanov and Alikhanyan); and a little later in Georgia (Tskhat-Tskaro; altitude 2800 m; supervisors Andronikashvili and Chikovani).

On the Pamirs, investigations were channeled into two main directions: the study of the mechanism of secondary particle production and the formation and growth in the atmosphere of extensive air showers, discovered before the war by Auger (1938) and Kohlhörster (1938).

The basic experiment on the mechanism of secondary particle production on the Pamirs was arranged as follows: a group of counters connected in a multiple-coincidence scheme was surrounded by thick layers of lead, excluding ordinary photon-electronic showers. This installation was exposed at various altitudes. It turned out that the number of showers due to penetrating particles in lead increased with the observation altitude significantly faster than the particle flux of the hard component of cosmic rays (Zhdanov and Lubimov, 1947).

It became clear that these showers were due to nuclear-active particles, namely, hadrons.

Many experiments were performed with the aim of studying the nature of shower particles themselves. A variety of filters were placed between counters inside a lead "oven" and the pictures of showers were photographed by means of a cloud chamber with plates in the working region. It turned out that these showers had a complex, hybrid composition: in addition to the photon-electronic component, there were nuclear-active particles (mesons, nucleons) in the showers (Birger *et al.*, 1949).

In the pre-war years and at the beginning of the forties, extensive air showers were considered to be pure photon-electronic avalanches, produced in the atmosphere by photons or electrons of ultrahigh energy (Euler, 1940). However, even during the war

Fig. 3. The preparation of the Pamir experiment with air shower detectors placed at the distance 1 km between them (1946) (G.T. Zatsepin with a cable on his neck).

Fig. 4. The cloud chamber in a magnetic field. (Tien-Shangstation; 1960).

years, Skobeltzyn (1942a,b) drew attention to the discrepancy between experimental results and calculations based on such assumptions, especially on the important role of penetrating particles.

Decisive experiments making it possible to form our modern view of basic processes during the passage of cosmic rays through the atmosphere were performed on the Pamirs in the second half of the forties. Thanks to these studies, not only electron-nuclear showers were detected, but nuclear-cascade process was discovered (Zatsepin, 1949). A step-by-step analysis of all the experimental data in the light of the achieved understanding of overall cosmic-ray processes made it possible to draw very important conclusions: the conservation by a nucleon in nuclear collisions of a considerable share of the energy, the lack of dependence of the shape of spectrum of secondary particles produced in nuclear interactions on primary energy in the $10^{10}-10^{12}$ eV range, assuming their energy is expressed in fractions of primary energy, i.e., energy scaling exists.

It was shown that when this conception was applied to extensive air showers, the main characteristics of the showers, such as altitude dependence, the fraction of muons in a shower etc., depended significantly on the characteristics of the elementary act of nuclear interactions, e.g., multiplicity, fraction of energy going over to π-mesons and transverse momentum. Models of the Heisenberg or Fermi type were shown to be incompatible with the experimental data.

The nuclear-cascade scheme connected the study of extensive air showers with the investigations of interactions involving nucleons, pions and ultrahigh-energy nuclei and, at the same time, made it necessary to reject an a priori view of showers. For further investigations of extensive air showers, there was proposed a method of complex installations in which registration of a shower and the initial stage of experimental data processing were based only on the qualitative assumption of axial symmetry of the shower. Complex installations consisting of 1000–2000 hodoscopic counters, and also other particle detectors, turned out to be feasible in the "pre-transistor" age thanks to an ingenious invention by Korablev, an engineer who used miniature gas-discharge neon bulbs to amplify pulses, select coincidences and achieve memory. The creation and operation of complex installations required that groups of 20 to 30 experimenters had to be formed rather than the customary small group consisting of several people around a single experiment. The problem was solved using students. But as for reliability ... Data from a complex installation could be obtained if only, at all twenty data registering points, nothing disturbed the operation of counters, electronic equipment (the hodoscope functioned only if regimes were held to within several percent accuracy), automatic photography, and good development of film. In the course of the experiment, there was elaborated a whole system of tests and signalling of faults, equipment monitoring: and a list of mandatory measures to be taken was compiled.

However, in the Pamir expeditions, some very capable students participated. Many of them are now engaged in important work as leading scientists in various scientific institution. Work on the Pamirs with starting physicists played a very important role in training specialists in the Soviet Union in high energies (and, for that matter, not only in high energies).

Detailed investigations of secondary particles in electron-nuclear showers were conducted by means of cloud chambers. In these experiments on the Pamirs, it was first shown that the main products of multiple production were the lightest hadrons, namely, pions. They exceeded 60% of the total. Comparatively infrequently, "strange" particles — kaons and hyperons — were observed in the chamber.

A number of interesting events occurred during this period. Thus, in one of the photographs in the cloud chamber, there was found the track of a strongly- ionizing particle with negative electric charge. Unfortunately, the point of production of this particle could not be determined since it was not accompanied by other shower particles. Therefore, one could not assert categorically that this track was not due to a proton entering the chamber from the bottom in the upward direction (although the target, naturally, was located above the chamber). It is now clear that this was an antiproton, making its appearance 10 years before the discovery of antiprotons in accelerator experiments. In another case, there was observed in the chamber a weakly-ionizing particle of high energy, produced in a beryllium target above the chamber. The particle momentum was 3 ± 0.5 GeV/c and the ionization was only 40% of that of a relativistic particle. At first glance, this was the track of a quark with a charge of 0.7 of that of an electron. Excitement among the physicists who had obtained this photograph was at a high pitch while careful measurements of the ionization and curvature of all the adjacent tracks on this and neighboring photographs were conducted until, alas, it became clear that one of the four pulse lamps illuminating the chamber during this period was not functioning properly. In both of the described cases, the events were not published, although in the first case we had clearly been over-cautions.

In parallel with the investigations conducted on the Pamirs, the analogous work was performed by the Georgian physicists Chikovani, Mandzhavidze, Roinishvili *et al.*, under the overall direction of Andronikashvili. They constructed the largest precision cloud chambers of the time at the Mt.Elbruz Scientific Station, located 4000 m above sea level, and then at the Tskhat-Tskharo Pass. The aim of this work was the quantitative study of the mode of decay and lifetime of "strange" particles. In all, about 200 V-typed events were registered. This is the most thorough statistical material on the decay of "strange" particles registered in cosmic-ray work. The data obtained on the lifetime of Λ-hyperons were then confirmed in experiments on accelerators.

Georgian physicists (Roinishvili *et al.*, 1961) first determined $<p_\perp>$ for strange particles in the experiments with cloud chambers in a magnetic field on Mt.Elbruz. It turned out that $<p_\perp>$ was somewhat larger for these particles than for π-mesons.

Of the studies undertaken on the Pamirs in the early fifties, one should also mention the experiment by Zhdanov (1949) on ascertaining the mechanism of μ-meson decay. This experiment was proposed by M.A.Markov. If a μ-meson decays into two particles (electron and neutrino), the energy spectrum of decay electrons from stopped μ-mesons should be a monoenergetic line. However, Zhdanov's experiments, performed by the method of "delayed coincidences" simultaneously but independently of Steinberger's experiments (1948), showed that the spectrum of decay electrons from stopped μ-mesons corresponded to a decay into at least 3 particles.

Beginning in the early fifties, a study of the interactions of cosmic-ray particles was undertaken by the Physical Institute of the Soviet Academy of Sciences at altitudes of 9000 to 12000 m above sea level by means of an airplane laboratory (Smorodin's group). In the plane, there was mounted a small cylindrical cloud chamber in a magnetic field of induction 10 kGs, operating alternately with a beryllium or lead plate in the effective region of the chamber (Baradzei *et al.*, 1955).

On obtained photographs of interactions at primary particle energies > 10 GeV, there was observed a process of nucleon-produced showers consisting of a small number of particles (2 or 3) with small escape angles of secondaries and low value of primary-nucleon coefficient of inelasticity.

Later, it was found that this phenomenon was caused by the diffractional production of pions. The momentum (energy) spectrum of hadrons in the stratosphere was determined by Smorodis's group and, in accordance with data from several other sources, it was shown that it reproduced the spectrum of primary radiation ("similarity of energy spectra"). When this phenomenon was re-discovered in accelerator experiments, it was called scale invariance or scalling.

At the end of the fifties, the cloud chamber in the airplane laboratory was replaced by an ionization chamber and counter installations (Baradzei et al., 1964).

Experiments with a 20 m hodoscopic counter installation mounted on an airplane showed that at altitudes of 9 to 12 km the flux of extensive air showers significantly exceeds the expected one for scalling conservation up to primary energies of $\gtrsim 10^{15}$ eV.

Subsequently, their conclusion that the degree of energy dissipation increases when going over to energies $> 10^{15}$ eV was confirmed in extensive air shower experiments performed by the Physical Institute of the Soviet Academy of Sciences, Moscow University, and also several other laboratories (Nikolsky, 1970).

The main weakness in the investigations conducted in the fifties was the absence of reliable information on the energy of the primary hadron, E_0. Therefore, much attention was then concentrated on measuring the energy of hadrons producing showers.

A way out was found by measuring the energy using a "calorimetric" method, i.e., determining the flux energy absorption curve. A detailed elaboration of this idea and its realization in a working device was achieved in 1955 in the work headed by Grigorov. The device for measuring the energy of high- and ultrahigh-energy particles was called an ionization calorimeter. Many difficulties had to be overcome in creating the first calorimeter. Such difficulties involved the gathering of a large volume of information from very many ionization chamber probes, the development of a method of measuring the amplitudes of a large number of signals etc.

In 1957, the cosmic-ray laboratories of the Physical Institute of the Soviet Academy of Sciences and the Scientific Research Institute of Nuclear Physics of Moscow University began work on a unique, complex installation consisting of a large cloud chamber in a magnetic field and an ionization calorimeter. The first such installation was constructed at the Pamir High-Altitude Station of the Physical Institute of the Soviet Academy of Sciences in 1957–58.

It should be particularly noted that the new and abundant experimental interaction data gathered with this complex installation made it possible to obtain infromation on the multiplicity and angular distribution of produced particles, fluctuations of these quantities for known energy of primary particles in the range $10^{11} - 10^{12}$ eV, coefficients of inelasticity etc.

An important result was the presence of a significant propotion of interaction at energies $\gtrsim 10^{11}$ eV with asymmetric (in the center-of-mass system) divergence of secondary particles. Such asymmetrical showers were interpreted as being the result of the decay of a mesonic mouving fireball (cluster) having a mass of several GeV/c^2, moving slowly in the center-of-mass system.

The momentum spectrum of the decay products of a cluster was determined in the system of coordinates in which the cluster is at rest. It turned out to be close to the Bose-Plank distribution for radiation of a black body at a temperature of $kT = 0.8\ m_\pi$, where m_π is the mass of a pion (Grigorov et al., 1960; Dobrotin et al., 1962).

This result helped to consolidate the view advanced somewhat earlier by Miesowicz (1958), Niu (1958), and Cocconi (1958) regarding secondary particle production as a decay of mesonic fireball-clusters.

On the basis of photoemulsion experiments, the hypothesis was advanced that, for interactions due to particles with energies $\gtrsim 10^{12}$ eV, there were produced two short-lived fireballs which flied apart in opposite directions in the center-of-mass system of the colliding particles. From the data of the Pamir work it followed that for energies of $10^{11} - 10^{12}$ eV one heavy fireball-cluster was produced in most cases in the pionization region.

The obtained results had a great influence on subsequent research in cosmic-ray physics. In particular, a large contribution in this direction was the work performed in the sixties on the Pamirs and then at the Tyan-Shang High-Altitude Station near Alma-Ata (Slavatinsky et al., 1970). These investigations significantly expanded our knowledge of the characteristics of fireball-clusters themselves. It turned out that, in addition to purely mesonic clusters moving in the center-of-mass system with a comparatively small Lorentz factor (in modern terminology, pionization clusters), clusters were formed that were joined to the bombarding nucleon or target nucleon (isobars – excited nucleons in the old terminology or fragmentation clusters in the modern).

In recent times, ideas on the production of secondary particles via the decay of clusters are finding wider and wider acceptance among accelerator specialists as well. Thus, in the Annual Report of CERN for 1973, on the basis of data obtained at CERN and FNAL, practically the same thing is stated in an article by the General Director, W.Jentschke, on the mechanism of secondary particle production via the decay of pionization and fragmentation clusters as in a summary article on Pamir and Tyan-Shan experiments, published in the mid-sixties (Dobrotin et al., 1965).

It is appropriate at this point to note the contribution of Soviet scientists to the development of the nuclear photo-emulsion method. As is well known, the use of this method yielded results that were of paramount importance to the discovery of new particles (in particular, π-mesons by C.Powell) as well as in the study of interactions at high and ultrahigh energies. Its successful application continues to the present in work performed with cosmic-ray particles and particles obtained from accelerators.

This method became widely used after 1927, when Mysovsky proposed that special thick-layer nuclear photoemulsion be fabricated (Mysovsky, 1927). He and his assistants (especially A.P.Zhdanov) succeeded in making such plates and using them to obtain visual evidence of nuclear disintegrations caused by cosmic-ray particles, mainly in the substance of the photoemulsion itself. This method acquired particularly great importance when in the early post-war years the British firm Ilford and subsequently laboratories in the United States, Japan and the Soviet Union developed emulsions that could register the tracks of individual, singly-charged relativistic particles.

At the time, great interest was shown in an event registered in a stack of nuclear photoemulsions exposed in the stratosphere by Schein and his co-workers (Lord et al.,

1950). It very clearly demonstrated the production of a large number of high-energy secondary particles in an act of elementary interaction. An analogous event was found and studied in 1957 by physicists from Kazakhstan, headed by Takibayev, in an emulsion stack that was also irradiated in the stratosphere (Boos *et al.*, 1958). This shower was produced in the emulsion substance by, apparently, a proton having an energy $\sim 5.10^{12}$ eV. An analysis performed by the authors of this work showed that all 16 shower particles were produced in a single act of proton-nucleon interaction. The characteristic feature of this event is that its coefficient of inelasticity is only $0.1 - 0.15$.

In emulsion stacks irradiated in the stratosphere, there were later found many additional interesting events (in particular, by Tretyakova, Zhdanov (FIAN) and at several other laboratories in the Soviet Union).

At that time, the intra-nuclear cascade concept was widely accepted as reflecting reality. The view was that a primary hadron interacting with a nucleon of a nucleus produced secondary particles and transfer to other hadrons some of its energy. Each of these particles, in turn, produced within that nucleus new showers etc. This notion regarding the interaction of a very high-energy nucleon with an atomic nucleus was considered almost self-evident.

However, even in the fifties, the intra-nuclear cascade concept began to encounter difficulties. Calculations on the number of high-energy particles escaping from a target nucleus were in poor agreement with experimental data, in particular, those obtained in Alma-Ata (Takibayev, 1957).

Tashkent physicists came to similar conclusions on the basis of their experiments.

The installation of Uzbek physicists on the high-altitude station Kum-Bel began functioning at the end of the fifties. It consisted of a system of spark chambers, scintillators and Vavilov-Cherenkov radiation counters. The most interesting result obtained by means of this installation was the establishment of a very weak dependence of the partial coefficient of inelasticity K_γ for the transfer of energy by a nucleon to γ-quanta (π°-mesons) on the atomic number of the target atom nucleus. If the primary particle is a pion, then K_γ turns out to be somewhat larger, but also very weakly dependent on A (Azimov *et al.*, 1962, 1964).

This result is undoubtedly of important significance for understanding the properties of elementary particles immediately after interaction and the mechanism of intra-nucleus interaction.

At the Aragats Station, Armenian physicists constructed an installation consisting of many trays of hodoscopic counters located in a magnetic field. They used this installation to study the composition of cosmic rays at high-mountain altitudes.

In his work, Kocharyan *et al.* (1958) measured the energy spectrum of protons. In a joint undertaking by the Moscow University Scientific Research Institute of Nuclear Physics and the Erevan Physical Institute, there was performed a long series of studies into the pion production process by the controlled nuclear emulsion method developed by them earlier (Babayan *et al.*, 1964, 1966).

By the mid-fifties, there already arose the question that is still open, the most important question of cosmic-ray physics, namely, whether it was permissible to extrapolate data obtained by means of accelerators to the ultrahigh-energy region. In other words, at what energies is an asymptotic regime of interactions reached?

In effect, the answer to this question determines whether it is expedient to construct a new generation of accelerators, i.e., the overall approach to high-energy physics research.

One of the first indications that an asymptotic regime is not reached even at very high energies (incomparably higher than anticipated earlier) was the change in the dependence of the number of hadrons or the number of electrons in showers at energies of $10^{14}-10^{15}$ eV found on the Pamirs (Nikolsky *et al.*, 1956). Subsequently, more detailed investigations of various components of extensive air showers, performed by means of complex installations on the Pamirs and then more refined installations at the Tyan-Shan High Altitude Scientific Station and at other laboratories in the Soviet Union as well as abroad, confirmed the conclusion that there was not asymptotic regime. It turned out that in going over to primary energies of hadrons $> 10^{14}$ eV, there began to appear many phenomena that remained unnoticed (or do not occur at all) at lower energies. The proportions of the various components of extensive air showers and the energy spectra of these components changed, the degree of dissipation of primary particle energy increased and a noticeable number of events with anomalously large transverse momentum of secondary particles appeared. Perhaps the most effective is the appearance in the ionization calorimeter of so-called "long avalanches" for a total avalanche energy (the energy of the hadron producing such an avalanche) $> 10^{14}$ eV, which was studied in the works of Yakovlev and his co-workers (Aseikin *et al.*, 1974). While the range for absorbing the energy flux in an avalanche for an energy $< 10^{14}$ eV amounts to $\simeq 600$ g/cm^2 Pb, for an energy $> 10^{14}$ eV it approaches 1100 g/cm^2 Pb. Most likely, the explanation for this phenomenon lies in the multiple production of new particles with a relatively small interaction cross-section.

Returning to the results of cosmic-ray investigations performed in the Soviet Union up to the beginning of the fifties, one can say that in them there was revealed and studied the basic production mechanism of cosmic-ray secondary particles effective over a broad interval of energies. Electron-nuclear showers do not constitute separate events ("mixed showers" for Janossi (1940, 1942), Watagin (1940) and certain other investigators), but rather the phenomenon that characterizes and determines the passage of high- and ultrahigh-energy particles through matter.

Finally, it should be emphasized that the broad front of investigations into cosmic rays unfolded in the Soviet Union, particularly in the post-war years, played an important role in establishing the modern views on particle interaction processes at high and ultrahigh energies. A direct continuation of this work are the investigations of recent years, which in this essay are only mentioned in passing, showing quite clearly that at ultrahigh energies (10^{15} eV and higher) there appear new processes and regularities that are not noticed at energies reached by means of modern accelerators. This is the justification for the desire to study in more detail cosmic-ray particle interactions at higher and higher energies and to construct a new generation of accelerators (installations with beams of colliding particles) at equivalent energies of $10^{14}-10^{15}$ eV.

REFERENCES

Alikhanov, A.I., Alikhanyan, A.I., Nemenov, L. and Kocharyan, N.M.: 1944, *Phys. USSR* 8, 63.
Alikhanov, A.I. and Alikhanyan, A.I.: 1945, *JETF* 15, 145.
Anderson, C.D.: 1933, *Phys. Rev.* 43, 491.

Anderson, C.D., Millikan, R. and Pickering, W.: 1934, *Phys. Rev.* **45**, 352.
Aseikin, V.S. *et al.*: 1974, *Izv. AN SSSR, Ser. Phys.* **38**, 998.
Auger, P., Maze, R. and Robley, C.R.: 1938, *Compt. Rend.* **208**, 1641.
Azimov, S.A.: 1949, *Trudy FIAN* **IV**, 315.
Azimov, S.A., Abdullayev, A.M., Myalkovsky, V.M., Yuldashbayev, T.S. *et al.*: 1962, *Isv. AN SSSR, Ser. Phys.* **26**, 613; 1964, *Izv. AN SSSR, Ser. Phys.* **28**, 1773.
Babayan, Kh.P., Grigorov, N.L. *et al.*: 1964, *Izv. AN SSSR, Ser. Phys.* **28**, 1784.
Babayan, Kh.P., Grigorov, N.L., Sobinyakov, V.A. and Shestoperov, V.Ya.: 1966, Collected Vol. *Cosmic Rays*, No. 8, p.182, "Nauka" Publishers.
Baradzei, L.T., Rubtsov, V.I., Solovyov, M.V. and Smorodin, Yu.A.: 1955, *Izv. AN SSSR, Ser. Phys.* **19**, 4502.
Baradzei, L.T., Rubtsov, V.I., Solovyov, M.V. and Smorodin, Yu.A.: 1964, *Trudy FIAN,* **26**, 224.
Bhabha, H.I. and Heitler, W.: 1937, *Proc. Roy. Soc.* **159**, 432.
Birger, N.G., Veksler, V.I., Dobrotin, N.A., Zatsepin, G.T., Kurnosova, L.V., Lyubimov, A.A., Rozental, I.L. and Eidus, L.Kh.: 1949, *JETF* **19**, 826.
Blackett, P.M.S. and Occialini, G.: 1933, *Proc. Roy. Soc.* **A139**, 699.
Boos, E.G., Vinitsky, A.Kh., Takibayev, Zh.S. and Chasnikov, I.Ya.: 1958, *JETF* **34**, 3, 622.
Carlson, I.F. and Oppenheimer, J.R.: 1937, *Phys. Rev.* **51**, 220.
Ciok, P., Coghen, T., Gierula, J., Miesowicz, M. *et al.*: 1958, *Nuovo Cimento* **8**, 166.
Cocconi, G. and Festa, G.: 1946, *Nuovo Cimento* **3**, 293.
Cocconi, G.: 1958, *Phys. Rev.* **111**, 1699.
Dobrotin, N.A., Frank, I.M. and Cherenkov, P.A.: 1936, *Trudy Elbrusskoi Ekspeditsii AN SSSR and VIEM (1934 and 1935)*, p.33, Acad. of Sciences Publishers, Moscow-Leningrad, USSR.
Dobrotin, N.A., Ivanova, N.S. and Isayev, B.M.: 1938, *Izv. AN SSSR, Ser. Phys.* No. 5-6, 744.
Dobrotin, N.A., Guseva, V.V., Kotelnikov, K.A., Lebedev, A.M., Ryabikov, S.V., Slavatinaky, S.A. and Zelevinskaya, N.G.: 1962, *Nucl. Phys.* **35**, 152.
Dobrotin, N.A., Zelevinskaya, N.G., Kotelnikov, K.A., Maksimenko, V.M., Puchkov, V.S., Slavatinsky, S.A. and Smorodin, Yu.A.: 1965, *Izv. AN SSSR, Ser. Phys.* **29** 9, 1627.
Euler, : 1940, *Zs. Phys.* **116**, 73.
Grigorov, N.L., Murzin, V.S. and Rapoport, I.D.: 1958, *JETF* **34**, 2, 506.
Grigorov, N.L., Guseva, V.V., Dobrotin, N.A. *et al.*: 1960, *Proc. Internat. Conf. Cosmic Rays, Acad. Sci. Publ. USSR* **1**, 140.
Janossi, L. and Ingleby, P.: 1940, *Nature*, **145**, 511.
Janossi, L.: 1942, *Proc. Roy. Soc.* **A179**, 361.
Jentschke, W.: 1973, *Ann. Rep. CERN*, p.15.
Kocharyan, N.M., Saakyan, G.S. and Kirakosyan, Z.A.: 1958, *JETF* **35**, 1335.
Kolhörster, W., Mattes, O. and Weber, O.: 1938, *Naturwiss.* **26**, 576.
Kunze, P.: 1933a, *Zs. Phys.* **80**, 559; 1933b, *Zs. Phys.* **83**, 214.
Landau, L. and Rumer, G.: 1938, *Proc. Roy. Soc.* **166**, 213.
Lord, I., Fainberg, I. and Schein, M.: 1950, *Phys. Rev.* **80**, 970.
Mysovsky, L.V. and Tuvim, L.: 1925, *Zs. Phys.* **35**, 299.
Mysovsky, L.V. and Chizhov, P.: 1927, *Zs. Phys.* **44**, 408.
Mysovsky, L.V. and Tuvim, L.: 1928, *Zs. f. Phys.* **50**, 273.
Nikolsky, S.I., Vavilov, Yu.N. and Batov, V.V.: 1956, *DAN SSSR* **111**, 71.
Nikolsky, S.I.: 1970, *Trudy FIAN* **46**, 100.
Niu, K.: 1958, *Nuovo Cimento* **10**, 994.
Roinishvili, N.N.: 1961, *JETF* **41**, 919.
Rossi, B.: 1933, *Zs. Phys.* **82**, 151; 1930, *Nature* **125**, 636.
Skobeltzyn, D.: 1927, *Zs. Phys.* **43**, 354; 1929, *Zs. Phys.* **54**, 686.
Skobeltzyn, D.V.: 1942a, *C.R. USSR* **37**, 52; 1942b, **37**, 14.
Skobeltzyn, D.V. and Vernov, S.N.: 1940, *C.R. USSR* **26**, 33.
Slavatinsky, S.A., 1970, *Trudy FIAN* **46**, 40.
Steinberger, J.: 1948, *Phys. Rev.* **74**, 500.
Takibayev, Zh. S.: 1957, Doctor's Dissertation, Alma-Ata.
Tamm, Ig. and Belenky, S.: 1939, *J. Phys. USSR* **1**, 177.

Vavilov, O.N.: 1949, *Trudy FIAN* **IV**, 5.
Veksler, V.I. and Dobrotin, N.A.: 1941, *Phys. Rev.* **59**, 12, 1044.
Vernov, S.N.: 1934, *Phys. Rev.* **46**, 822.
Vernov, S.N.: 1936, *C. R. USSR* **14**, 263.
Watagin, G., de Souza Santos, M.D. and Pompeia, P.A.: 1940, *Phys. Rev.* **57**, 61, 339.
Zatsepin, G.T.: 1949, *DAN SSSR* **67**, 993.
Zhdanov, G.B. and Lubimov A.L.: 1947, *DAN SSSR* **55**, 119.
Zhdanov, G.B. and Khaidarov, A.A.: 1949, *DAN SSSR* **65**, 287.

AIR SHOWER EXPERIMENTS AT MIT

George W. CLARK[*]

The techniques of "density sampling" and "fast timing," which have been widely used in the study of extensive cosmic ray air showers, were developed by members of the MIT Cosmic Ray Group during the period 1947–55 and applied in a number of experiments aimed at determining the composition, spectrum and arrival directions of very high energy cosmic rays. In "density sampling" the core location and size of an extensive air shower incident on an array of proportional detectors are estimated by an analysis of the amplitudes of the signals generated in the detectors by the shower particles. In "fast timing" the arrival direction of a shower is determined from the differences between the arrival times of the shower front at the various detectors in a similar array. The development of large organic scintillation detectors and improved photomultipliers in the 1950's made it feasible to combine these techniques in very large arrays of detectors with which the properties of individual showers were measured in detail, and the energy spectrum and arrival directions of the primary particles were determined up to energies of more than one joule (10^{19} eV).

Robert Williams introduced the density sampling technique in an experiment, suggested by Bruno Rossi, and carried out on Mt. Evans, Colorado in 1947. Williams employed pulse ionization chambers that provided quantitative electron collection and microsecond time resolution. These chambers, built at MIT according to the design principles which had been worked out by Rossi and his associates at Los Alamos (Rossi and Staub, 1949), were being used as proportional detectors in a variety of cosmic ray experiments. In William's experiment the signals from four chambers were displayed and photographed for measurement on separate oscilloscopes with a common sweep triggered when coincident pulses above a certain size occurred in any two of the chambers. In principle, measurements of particle densities at four different positions over determine the core location and size of any shower with an axis near the vertical, provided the lateral distribution of the shower particles is known. Using a graphical method of analysis, Williams showed that the data on 47 well-measured showers were consistent with the assumption that all showers have the structure function derived theoretically by Moliere (1946). He determined the spectrum of shower sizes up to $N = 10^8$, and demonstrated the existence of primary particles with energies greater than 4×10^{16} eV. During portions of his runs, Williams measured the projected zenith angle of some showers by photographing the tracks of shower particles in a cloud chamber triggered by the coincidence signal. The data showed a strong concentration of axes near the vertical which justified the neglect of the spread in zenith angle in the graphical analysis of core location and size.

When I joined the Cosmic Ray Group as a student in 1950 the construction of pulse ion chambers and Geiger tubes was in full swing. Most of the electronic equipment was also built in the laboratory according to designs developed at Los

[*]Massachusetts Institute of Technology, Cambridge, U.S.A.

Y. Sekido and H. Elliot (eds.), Early History of Cosmic Ray Studies, 239–246.
© *1985 by D. Reidel Publishing Company.*

Alamos. These were used in a wide variety of investigations to determine the properties of the primary cosmic rays and to elucidate the phenomena of high energy interactions and secondary particles produced by cosmic rays in the atmosphere. My first assignment was to explore the use of sodium iodide scintillation detectors in cosmic ray measurements, and this led eventually to my thesis project in which I used sodium iodide scintillation detectors in a study of the nucleonic component at sea level and mountain altitudes during 1951 and 1952.

Pietro Bassi came to MIT in the fall of 1952 on leave of absence from the University of Padua. Professor Rossi suggested that Bassi and I look into the possibility of determining the arrival directions of air showers by fast timing measurements with scintillation detectors. To do this we needed to verify the idea that the particles in a shower, all traveling at nearly the speed of light, are concentrated in a fairly thin disc which is symmetric about the shower axis and propagates along that axis. If this is true, then the arrival direction of a shower can be readily calculated from measured differences in the arrival times of the particles at three or more noncolinear points on the ground. Three recent developments had made such measurements feasible. Ageno et al. (1949) had discovered that solutions of certain aromatic compounds in cheap aromatic solvents are efficient scintillators. Reynolds et al. (1950) had shown that the rise times of scintillation pulses in these liquids is less than 20 nanoseconds. And efficient, fast, end-window photomultipliers had become commercially available.

We constructed three detectors from five gallon drums half full of a solution of terphenyl in benzene viewed from above by RCA 5819 photomultipliers. The signals were amplified by fast distributed line amplifiers that had been developed recently by William Kraushaar for his measurement of the mean life of pi mesons produced by the MIT synchrotron. The amplified signals were delayed with respect to one another by fixed lengths of cable, added and displayed on a single fast oscilloscope for photographic recording. We operated the experiment with the detectors arranged in various configurations on the roof of the Physics Department at MIT for several months during the winter of 1952–53. We were delighted to find by a statistical analysis of the relative occurrence times of the pulses that shower discs are in fact nearly flat and only 1 to 2 meters thick. This confirmed the feasibility of using fast timing to measure the orientation of the disc and thereby the arrival direction of the shower (Bassi et al., 1953).

Several members of the Cosmic Ray Group, including in particular James Earl, William Kraushaar, John Linsley, Bruno Rossi and Frank Scherb and I, then jointed forces to develop a large array of large-area scintillation detectors with which we could measure both the density and arrival times of shower particles and thereby determine simultaneously the arrival direction, core location, structure function and size of

individual air showers that struck within the boundary of the array. Minoru Oda, on leave from Osaka City University, and Bassi participated in the planning and early stages of construction. Our primary purpose was to cast new light on the problem of the origin of cosmic rays by a determination of the primary energy spectrum and distribution of arrival directions of primaries with energies above 10^{14} eV. In the process we expected to gain new information about the structure and development of air showers in the atmosphere.

For the new experiment we placed fifteen 1 m² liquid scintillation detectors in concentric rings with one in the center. Each detector was a galvanized steel drum containing about 25 gallons of a liquid scintillator viewed by a 5″ Dumont 6364 photomultiplier. The amplified signals were displayed on a bank of fast oscilloscopes and recorded photographically. Their amplitudes and relative arrival times were measured by hand on the photographic records. The diameter of the outer ring of detectors was nearly 500 m. The difficult problem of finding a suitable site for so large an

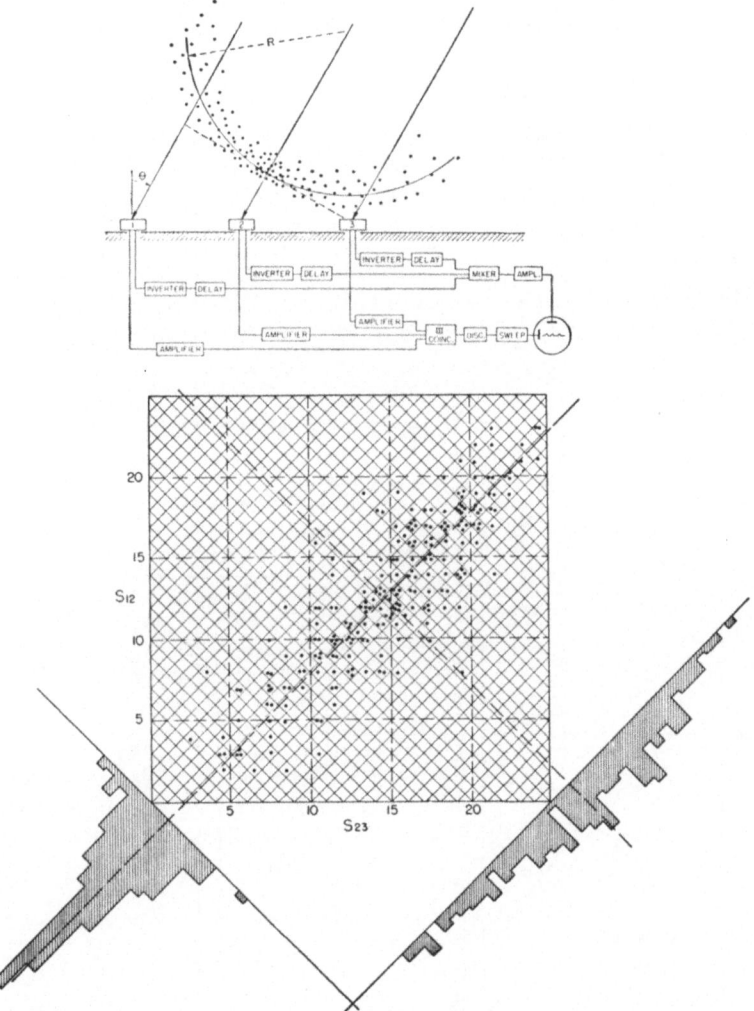

Fig. 1. Shematic diagram of experimental arrangement used by Bassi, Clark and Rossi (1953) to measure the distribution in arrival times of air shower particles, and graphical summary of the results. The feasibility of the fast timing method for determining arrival directions was demonstrated by the narrow spread of the data points about the diagonal line representing a linear correlation between the arrival time differences s_{12} and s_{23}.

apparatus near Boston was solved when one was made available at the Agassiz station of the Harvard College Observatory through the cooperation of the Observatory director, Professor Donald Menzel, and Harvard University. By July 1955 the detectors were in position, the cables were draped through the woods and connected, and measurements began with occasional interruptions to repair cables damaged by gnawing rabbits.

Meanwhile, two approaches to the formidable task of data analysis were tried. One employed an analogue device with strings and weights that displayed a plot of the logarithm of the shower density versus distance from the core when the core locator was moved to the correct position on a map of the array. The other was based on

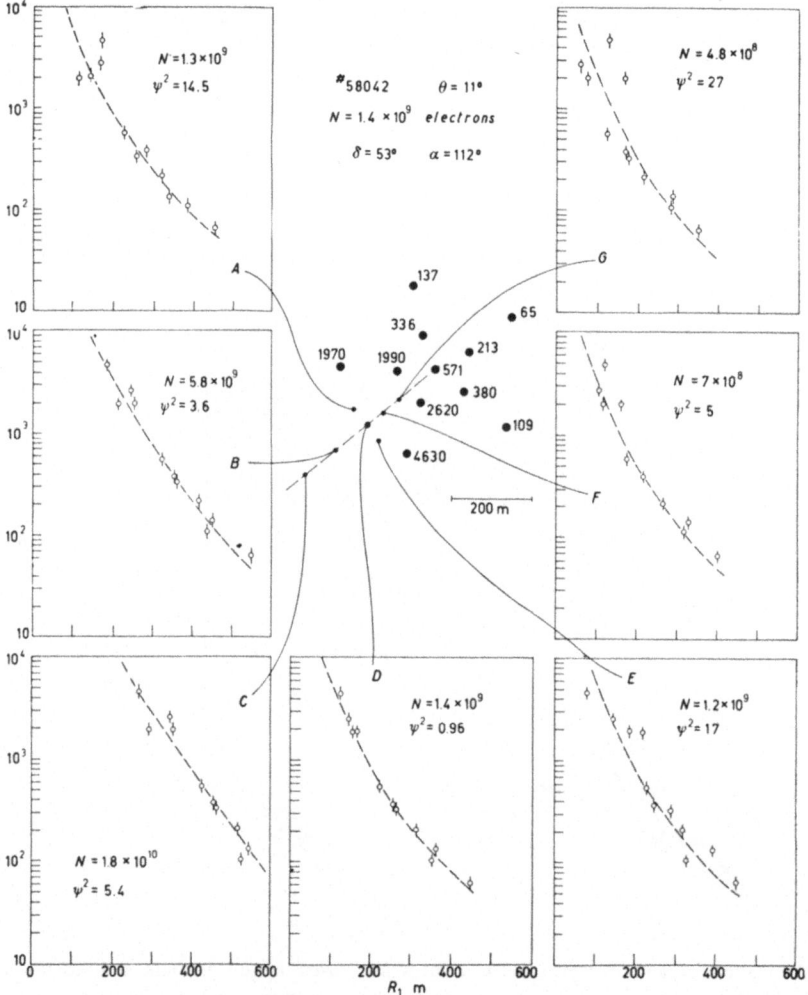

Fig. 2. Detailed analysis of the particle densities measured in the largest shower recorded in the Agassiz experiment. The apparent lateral distributions implied by various choices of core locations are shown in the plots of measured density versus distance from the shower axis. Core location D gave the best fit to a smooth curve.

digital computation which had only recent become feasible with the availability of electronic computers. The latter approach provided least squares solutions to the problem of fitting the five parameters specifying the arrival direction, core location and size to the set of measured arrival times and particle densities. The speed and flexibility of digital computation, implemented initially on MIT's Whirlwind computer and later on successively faster IBM computers, were soon apparent, and the analogue approach was abandoned.

Just as the experiment was getting well underway in the summer of 1955 and shortly after we had recorded our first really big shower one of the detectors burst into flames during a stormy night, an apparent victim of lightning. When the fire company of the town of Harvard arrived, they found a pillar of fire erupting from a mysterious metal contraption in the middle of the forest surrounding the Harvard College Observatory. The torch of burning toluene was eventually extinguished, the rain-damped woods did not catch fire, and the Observatory did not burn down. Six months later, after much negotiation between MIT and Harvard and extensive rebuilding of the detector housings with installation of clever fire prevention devices, the experiment operations were resumed.

Shortly after the fire Williams suggested that it might be practical to replace the flammable liquid scintillator with plastic scintillators which had been invented by Schorr and Torney (1950). At the time, however, plastic scintillators of the size we required were not yet commercially available, and the cost of even much smaller pieces was prohibitive. After a few test tube trials we built a pilot plant for making 16-inch diameter discs and a factory for 42-inch diameter discs (Clark et al., 1957). By the time the factory was dismantled in 1962 it had produced more than six tons of plastic scintillators that were used in the MIT air shower experiments at Agassiz, Kodaikanal Bolivia and Volcano Ranch and in the Cornell air shower experiment at Ithaca.

In the late spring of 1956 the liquid scintillators at Agassiz were replaced by the new plastic scintillators, and the experiment was resumed in June. The principal results were 1) accurate measurement of the lateral density distribution which was found to be in agreement with the theoretical distribution calculated by Nishimura and Kamata; 2) measurement of the distribution of shower sizes from which we deduced the primary spectrum in the energy range from 3×10^{15} to 10^{18} eV; 3) detection of one shower with a primary energy of 5×10^{18} eV ($\sim$ one joule!); 4) demonstration that the distribution of celestial arrival directions of several thousand ultra high energy cosmic rays showed no significant anisotropy (Clark et al., 1957, 1961).

During the preparation of the Agassiz experiment operations of the small detectors on the roof of MIT were continued with the addition of a fourth detector to form a square array with which we determined the celestrial arrival directions of 2660 showers in the size range corresponding to primary energies greater than 10^{14} eV. No evidence of anisotropy was found (Clark, 1957). We then decided to extend to southern declinations the search for evidence of an anisotropy which might provide a clue to the origin of high energy cosmic rays. For this we developed an improved fast timing apparatus utilizing four 16-inch plastic scintillators and the new 14-stage high gain photomultipliers which eliminated the need for fast amplifiers. The array was set up at Kodaikanal, India ($10°$ N latitude) in collaboration with Vikram Sarabhai and Eknath

Chitnis of the Physical Research Laboratory in Ahmedabad. Again no evidence of anisotropy was found in the celestial arrival directions of over 100,000 showers with an average size corresponding to a primary energy of approximately 10^{14} eV (Chitnis *et al.*, 1960).

In 1958 the Agassiz equipment was moved to a location at an altitude of 4200 m (13780 feet) on the Altiplano near the airport of La Paz, Bolivia in collaboration with Ismael Escobar and Juan Hersil of the Laboratorio de Fisica Cosmica, Universidad Mayor de San Andres, La Paz. The measurements made there demonstrated that the characteristics of showers at that high altitude differ markedly from those of showers at sea level. In particular, vertical showers with about 3×10^7 particles were shown to be near their points of maximum development, and to have lateral distributions which are much steeper than at sea level (Hersil *et al.*, 1961). Again, no evidence of anisotropy was found.

Satio Hayakawa (1952) had pointed out that interactions of cosmic rays with interstellar matter must be a source of cosmic gamma rays because such interactions produce neutral pions which decay into gamma rays. In 1958 William Kraushaar began the development of satellite instruments to search for cosmic gamma rays with energies near 100 MeV produced by this and other possible mechanisms, an effort that eventually succeeded with the detection of both a galactic and an extragalactic component of cosmic gamma rays in the MIT experiment on the OSO-3 satellite. Meanwhile, Minoru Oda proposed that the MIT, Tokyo and La Paz air shower research groups collaborate in an effort to detect air showers initiated by the very high-energy gamma rays predicted by Hayakawa. The idea was to search for rare showers with unusually few penetrating muons. The latter arise from the decay of charged pions produced in the nucleonic cascades generated by primary nuclei, and they constitute about 1 percent of the particles in most showers. In contrast a shower initiated by a primary high-energy photon should produce relatively few pions and therefore relatively few muons. The plan called for construction of a large heavily shielded detector in the middle of an air shower array where it would be used to measure the proportion of penetrating particles (muons) in individual air showers. The detection of a distinct group of "low-mu" showers would be evidence for a flux of primary high-energy photons. If, in addition, "low-mu" showers exhibited a significant anisotropy with higher intensities in the plane of the Milky Way where the rate of cosmic ray interactions is higher, then the gamma-ray nature of their primaries would be established. Estimates of the relative proportion of photon-initiated showers of a given size, based on Hayakawa's production mechanism, were about 10^{-3}.

The project was called the Bolivian Air Shower Joint Experiment and came to be known as "BASJE". To implement the plan Koichi Suga of the Institute for Nuclear Research, Tokyo, came to MIT in 1959 and developed a new 4 m^2 square detector module with commercial plastic scintillators viewed from below with a 14-inch diameter photomultiplier, together with a transistor amplifier with logarithmic response (Suga *et al.*, 1961). Fifteen of these units were shipped to Bolivia and installed at 5200 m (17060 feet) altitude under a massive concrete structure supporting 160 tons of native galena (Pbs) in the center of the old Agassiz array, which had been moved from El Alto to a site near the summit of Mt. Chacaltaya. Extensive measurements were made of the characteristics of air showers at this great altitude which is well above the altitude of maximum development of the showers that had previously been studied at sea level. In particular, the distribution in the ratio of the numbers of muons to the number of electrons was measured, and the presence of "low-mu" showers with less than one-tenth the average number of muons and a relative abundance of about 10^{-3} was demonstrated (Suga *et al.*, 1963). However, the low-mu showers did not form a clearly separate group in the ratio distribution, nor did they exhibit any significant anisotropy in the distribution of their arrival directions. Thus we could not exclude the possibility that the low-mu showers are produced by rare fluctuations in the initial interactions of primary protons in which nearly all of the incident energy is converted into neutral pions and thence into a nearly pure electromagnetic shower. The BASJE equipment was subsequently used in a variety of studies by the MIT and Tokyo groups, including measurements of the Cerenkov light pool and efforts to determine the composition of the primary particles in the energy range above 10^{14} eV.

During the later phase of the Agassiz experiment Linsley undertook the development of a new and much larger air shower array to explore the spectrum, composition and arrival directions of primaries with energies in the range above 10^{18} eV. Above 10^{19} eV the radii of curvature of the primary particles in the galactic magnetic field is comparable to or larger than the thickness of the galactic disc so that particles cannot be magnetically contained within the disc. Therefore, if they do arise in some remarkable galactic process, their distribution should be concentrated toward the Milky Way. On the other hand, isotropy at these energies would indicate an extragalactic origin, and this would present a new problem in view of the opacity of intergalactic space to nucleons of these energies due to photonuclear reactions with starlight and the microwave background radiation. In 1959 Linsley and Livio Scarsi completed construction of the Volcano Ranch air shower experiment located in the desert outside of Albuquerque. It had 20 scintillation detectors in an array whose maximum diameter for a period of time was 3.5 km. With it they detected showers with primary energies up to 10^{20} eV, thereby extending measurements of the primary spectrum well beyond the range of possible galactic confinement (Linsley, 1963).

It seems now that the various giant air shower arrays have exhausted the practical possibilities of the density sampling and fast timing techniques for extending knowledge of the high energy end of the primary cosmic ray spectrum. Radically new approaches are currently under development based on the idea of Kenneth Greisen to detect and track the "fire ball" of atmospheric scintillation light produced by the particles in a shower disc as it traverses the atmosphere. With a fireball detector, such as the complex Fly's Eye now in operation at the University of Utah, giant showers striking anywhere within a radius of many kilometers from the detector can be studied so that effective areas far exceeding those achievable with density sampling arrays can be achieved. The role of a large air shower array in this new kind of measurement will be to provide calibration data in the region of the primary spectrum near 10^{20} eV where the same showers can be measured simultaneously by the new and the old methods.

REFERENCES

Ageno, M., Chiozzotto, M. and Querzoli, R.: 1949, *Acad. Naz. Lincei* 6, 626.

Bassi, P., Clark, G. and Rossi, B.: 1953, *Phys. Rev.* 92, 441.

Chitnis, E. V., Sarabhai, V. A. and Clark, G.: 1960, *Phys. Rev.* 119, 1085

Clark, G.: 1957, *Phys. Rev.*, 108, 450.

Clark, G. W., Scherb, F. and Smith, W. B.: 1957, *Rev. Sci. Inst.* 28, 433.

Clark, G., Earl, J., Kraushaar, W., Linsley, J., Rossi, B. and Scherb, F.: 1957, *Nature* 180, 353.

Clark, G. W., Earl, J., Kraushaar, W. L., Linsley, J., Rossi, B. B., Scherb, F. and Scott, D.: 1961, *Phys. Rev.* 122, 637.

Hayakawa, S.: 1952, *Prog. Theor. Phys.* 8, 517.

Hersil, J., Escobar, I., Scott, D., Clark, G. and Olbert, S.: 1961, *Phys. Rev. Lett.* 6, 528.

Linsley, J.: 1963, *Phys. Rev. Lett.* 10, 521.

Moliere, G.: 1946, *Cosmic Radiation,* ed. by Heisenberg, W., Dover Publications, New York, Ch. 3.

Reynolds, G. T., Harrison, F. B. and Salvini, G.: 1950, *Phys. Rev.* 78, 488.

Rossi, B. and Staub, H.: 1949, *Ionization Chambers and Counters,* McGraw-Hill Book Co., Inc., New York.

Schorr, M. G. and Torney, F. L.: 1950, *Phys. Rev.* 80, 474.

Suga, K., Clark, G. and Escobar, I.: 1961, *Rev. Sci. Inst.* 32, 1187.

Suga, K., Escobar, I., Murakami, K., Domingo, V., Toyoda, Y., Clark, G. and LaPointe, M.: 1963, *Proc. Int. Conf. on Cos. Rays,* Vol. IV, 9.

Williams, R. W.: 1948, *Phys. Rev.* 74, 1689.

PART 6 PRIMARIES AND SECONDARY PRODUCTS

COMPOSITION OF EACH GENERATION

ACTIVITES SCIENTIFIQUE DE LOUIS LEPRINCE-RINGUET ET DE SON GROUPE DE TRAVAIL SUR LES RAYONS COSMIQUES DE 1933 A 1953

Louis LEPRINCE-RINGUET*

1.

Etude du rayonnement cosmique selon la latitude du lieu d'observation (Auger et Lerince-Ringuet, 1933. Un voyage en bateau de Hambourg à Buenos-Aires et retour à permis d'élucider, au moyen de compteurs d'électrons, l'effet de latitude qui venait d'être découvert par Clay avec une chambre d'ionisation. Cet effet est important parce qu'il permet de savoir si la composante primaire du rayonnement cosmique est électriquement chargée ou non. Son étude a fait l'objet depuis 1932 d'un grand nombre de travaux, en particulier une vaste expédition sur toute la surface du globe organisée par A.H. Compton. Les recherches que nous avons faites dès 1933 ont montré l'importance de cet effet au niveau de la mer sur les différentes composantes du rayonnement: elles ont permis aussi de vérifier l'effet azimutal (prédominance de rayons venus de l'ouest) qui s'interprète par une prédominance dans le rayonnement primaire, de particules chargées positivement.

2.

Au laboratoire international du Junfraujoch : La séparation des différentes composantes du rayonnement en groupes de propriétés bien définies, et la connaissance de ces propriétés nous amenèrent à expérimenter en altitude au laboratoire international du Jungfraujoch (3400 mètres au-dessus du niveau de la mer) où nous fîmes plusieurs séjours, définissant ainssi les deux principales composantes (composante pénétrante et composante molle) et les propriétés d'absorption de ces rayonnements en fonction du numéro atomique de l'élément traversé (Auger *et al.*, 1934, 1935).

3.

Etude des particules de grande énergie du rayonnement cosmique à l'aide d'une grande chambre de Wilson. Nous avons utilisé l'électro-aimant de l'Académie des Sciences à Bellevue pour la déviation des particules de grande énergie qui l'on observe dans le rayonnement cosmique : à partir de 1936, soit à Bellevue, soit plus tard dans des stations des Alpes, je m'efforcai de mettre au point de grandes chambres de Wilson commandées par compteurs d'électrons afin de déterminer le spectre d'énergie des rayons cosmiques et de définir leur nature. Ces recherches furent effectuées soit seul,

* Ecole Polytechnique, Paris, France.

Y. Sekido and H. Elliot (eds.), Early History of Cosmic Ray Studies, 249–253.
© *1985 by D. Reidel Publishing Company.*

soit avec J. Crussard, puis E. Nageotte, S. Gorodetzky, R. Richard Foy, M. Lheritier; j'ai pu obtenir en 1936, sur les rayons cosmiques, des mesures d'énergie jusqu'à vingt milliards d'électron-volts, cette limite étant d'ailleurs fréquemment dépassée par les plus durs.

Au cours des expériences effectuées à Bellevue, plusieurs propriètîs intéressantes ont été mises en évidence: tout d'abord la prédominance des particules positives dans l'ensemble du rayonnement pénétrant; d'autre part nous avons trouvé avec J. Crussard qu'il existait des particules des deux signes dont l'énergie n'était pas très grande (inférieure à 300 millions d'électron-volts) et capables cependant de traverser une grande épaisseur de plomb (14 cm). Ces particules ne pouvaient être des protons et il était extrémement improbable qu'elles pûssent être des électrons. Nous avons émis l'hypothèse que ce pouvait être des particules nouvelles, mais sans toutefois affirmer l'existence de ces rayons car l'on ne connaissait pas bien, à ce moment, les propriétés des électrons aux très grandes énergies. Néanmoins, il y avait là une indication en faveur de l'existence du méson qui a été reconnu par Anderson deux ans plus tard. Enfin, des résultats sur les secondaires des rayons cosmiques et sur certains effects nucléaires ont été aussi obtenus à Bellevue (Leprince-Ringuet and Crussard, 1935, 1937).

Sur un des clichés obtenus à Bellevue en 1939, nous avons pu (avec Gorodetzky, Nageotte, Richard-Foy) déterminer la masse d'un méson en appliquant les lois de la mécanique, avec naturellement les corrections relativistes, à une collision particulièrement avantageuse entre un méson et un électron du gaz de la chambre à détente. Cette mesure nous a donné la valeur de 240 ± 30 par rapport à la masse de l'électron prise pour unité (Laprince-Ringuet *et al.*, 1940).

4.

A partir de 1937, une grande chambre de Wilson a été installée dans les Alpes françaises à Largentière avec une bobine magnétique sans fer parcourue par 8000 ampères. Cette installation a fonctionné depuis 1938 et pendant l'occupation. Elle a permis aux physiciens restés en France d'effectuer des recherches, soit sur la masse des mésons, soit sur le spectre d'énergie ou l'excès positif de ces particules, soit sur leur vie moyenne, soit sur les effets nucléaires du rayonnement: nous avions monté à côté de Largentière, un petit centre expérimental au col du Lauraret (2000 mètres); et c'est à cette époque que nous avons entrepris, grâce à une subvention du Centre National de la Recherche Scientifique, la construction d'un laboratoire de haute montagne de 3600 mètres d'altitude à l'Aiguille du Midi près de Chamonix.

Sur un cliché, la collision, supposée élastique, d'une particule incidente de 600 MeV/c et d'un électron, permet de donner, pour la particule incidente, une masse de (990 ± 12%) me. Cette mesure peut être considérée comme la première indication favorable à l'existence d'un méson lourd (Leprince-Ringuet et Lheritier, 1944).

5.

Depuis 1947 nous avons étudié les phénomènes nucléaires produits par la rayonnement cosmique, grâce à la méthode des émulsions photographiques épaisses et concentrées mise au point à Bristol par Powell et Occhialini. Nous avons envoyé des plaques, soit à l'aiguille du Midi de Chamonix, soit au refuge Vallot, soit dans la haute atmosphère à 20000 mètres environ, grâce aux ballons-sonde de la Météorologie nationale.

Avec J. Heidmann, L. Jauneau, Hoang T.F., D. Morellet, nous avons examiné des milliers de désintégrations dont plus d'une centaine provoqués par mésons. En plus des plaques exposées aux rayons cosmiques, nous avons pu détecter, grâce à l'amabilité du Professeur Lawrence et de E. Gardner et G. Lattes, des plaques ayant subi une irradiation au grand cyclotron de Berkeley et contenant de nombreuses trajectoires de mésons artificiels: sur ces plaques nous avons pu observer plus de 500 étoiles nucléaires provoquées par mésons. L'ensemble de tous ces phénomènes nous a permis d'étudier d'une part l'excitation du noyau sous l'action des mésons légers ou lourds, d'autre part les grandes explosions nucléaires.

Compte-rendu Académie des Sciences du 10 décembre 1947, 24 mai 1948, 7 Juin 1948. Ces recherches se sont poursuivies jusqu'en 1953 avec J. Crussard, D. Morellet, G. Kayas, A. Orkin-Lecourtois, J. Tremblay, sur les masses des mésons lourds par ionisation, scattering et parcours (Crussard *et al.*, 1951, 1952, 1953a, b).

6.

Entre 1950 et 1955, l'équipe de l'Ecole Polytechnique avec Ch. Peyrou, B. Gregory, A. Lagarrigue, R. Armenteros, F. Muller, A. Astier, a installé au Pic du Midi (2950 mètres d'altitude) deux grandes chambres de Wilson superposées, de 200 litres chacune. Il semblait intéressant de faire une grande expérience qui combinerait les renseignements fournis sur l'énergie des particules par la courbure de leur trajectoire dans un champ magnétique avec l'observation de leur comportement à la traversée d'écrans de matière. La grande dimension des chambres était destinée à augmenter le rendement de l'expérience. Le champ magnétique était fourni par une bobine de Helmoltz de grande dimension fournissat un champ de 6000 gauss. Un ensemble de compteurs Geiger et d'appareils spéciaux d'électronique déclanchaient les chambres lorsqu'un phénomène nucléaire de grande énergie se produisait au dessus de l'appareil. Cette installation a fonctionné d'une façon très régulière pendant quatre ans et a fourni 100000 clichés. Plus d'une douzaine de publications ont été faites sur les résultats scientifiques obtenus: elles portent essentiellement sur l'identification et l'étude des propriétés des mésons lourds et des hypérons dont nous avons pu obtenir plusieurs milliers d'exemples de désintégration au cours de notre expérience. L'appareil a été construit effectivement par B. Gregory, A. Lagarrigue, Ch. Peyrou. Ont de plus participé aux expériences: A. Astier, R. Armenteros, puis, partiellement, W.B. Fretter, université de Californie, R.R. Rau, Université de Princeton, J. Tinlot, Université de Rochester, H. de Staebler, Massachusset Institut de Technology. Les principaux résultats ont été les suivants:

a) Mesure de la masse du méson K^+, identification de ses modes de désintégration. La masse est mesurée à partir de la courbure de la trajectoire combinée avec la mesure du parcours lorsque la particule s'arrête dans une chambre à écran. Dès le début des expériences, au moment où l'incertitude sur la masse des mésons lourds était très grande, nous avons publié une masse précise du méson K^+: 920 ± 30 me. Par suite nous avons pu préciser encore cette valeur et prouver les premiers que le mode principal de désintégration du méson K^+ était une désintégration en deux corps avec production d'un méson μ^+ et d'un neutrino.

b) Anomalies du comportement du méson K^-. Dès le début de l'expérience nous avons noté une absence presque complète d'une contrepartie négative du méson K^+. Ce fait semblait alors en contradiction avec la symétrie de charge. L'explication actuelle de ces observations est que le méson K^- est l'anti-particule du méson K^+ et qu'il faut une énergie plus considérable pour le fournir: on dit que le méson K^- a une étrangeté opposée à celle du méson K^+. Nous avons aussi publié en 1953 un évènement remarquable montrant la production d'un hypéron Λ°, par une particule négative d'énergie insuffisante pour créer une paire de particules étranges.

c) Etude des désintégrations du type V chargé. Un grand nombre de photographies comportaient des désintégrations dans la chambre supérieure (à champ magnétique) du type V chargé, c'est-à-dire se présentant sous la forme d'un angle vif sur une trace rapide. Le sommet de cet angle est le point où la particule primaire s'est désintégrée en vol pour donner un secondaire chargé visible dans la chambre et un secondaire neutre que nous avons pu, dans certains cas, détecter par ces intéractions dans les écrans de la chambre du bas. En particulier nous avons montré qu'une proportion importante de ces désintégrations correspondant à la mort d'hypérons Σ dont la masse est plus élevée que celle des protons. Cette identification a pu être faite notamment par l'observation des intéractions, dans la chambre à écrans, du neutron émis dans la désintégration. Nous avons pu également identifier un certain nombre de modes de désintégration de particules K chargées. Enfin, dans une photo remarquable, nous avons pu identifier la nature du secondaire chargé d'une désintégration rare à l'époque, celle d'une particule dite Ξ^- dont le mode de désintégration est le suivant:

$$\Xi^- \to \Lambda^\circ + \pi^-.$$

Le Λ° étant observé dans la chambre par sa désintégration en un proton et un π^-. Nous avons pu, dans l'ensemble de ce travail, donner des renseignements sur les vies moyennes de ces différentes particules par l'observation de la distribution, dans la chambre, des sommets de désintégrations (voir thèse de doctorat de F. Muller). A. Hendel a également rédigé une thèse de doctorat sur la désintégration des mésons τ dans la chambre supérieure.

d) Etude des désintégrations neutres (hypérons Λ° et mésons θ°). Notre dispositif expérimental s'adaptait spécialement bien à l'étude des désintégrations de particules neutres. Elles se présentent sous la forme d'un V renversé. On observe ainsi deux particules de signes opposés produites par la désintégration de la particule neutre. On peut mesurer l'angle au sommet et les courbures des trajectoires. Les deux sedondaires traversent la chambre à plaques et donnent ainsi des informations sur leur nature. Vers 1955, le lot de désintégrations neutres que nous avions ainsi était bien supérieur à l'ensemble des évènements de même type observés dans tous les laboratoires du monde. En particulier nous avons pu donner une meilleure valeur de la masse du Λ°, résultat important car il corrigeait certaines anomalies d'interprétation sur les énergies de liaison des noyaux excités (ou hyperfragments). Nous avons pu donner également une meilleure valeur à la masse du θ° et établir de façon définitive que ce méson θ° se désintégrait en deux mésons π par l'observation dans la chambre à écrans des intéractions de ces deux secondaires.

La plupart des résultats ont été exposés au congrès international sur le rayonnement cosmique de Bigorre de 1953.

A partir de 1955 tout le laboratoire de l'Ecole Polytechnique s'est orienté vers les chambres à bulles et l'accélérateur de particules du Cern en abandonnant le travail sur le rayonnement cosmique.

Appendix

Voici les principales publications:

Production d'un V° par un primaire négatif de 1 BeV, Congrès international sur le Rayonnement Cosmique de Bagnères de Bigorre (1953).

Quel ques résultats sur les V chargés, Congrès International sur la Rayonnement Cosmique de Bagnères de Bigorre (1953).

Mesures de masse de partioules S par Moment-Parcours, Congrès International sur le Rayonnement Cosmique de Bagnéres de Bigorre (1953).

Mass measurements of primaries of S events by a momentum range method, *Phys. Rev.* **92**, 1583, (1953).

Etude des mésons K chargés au moyen de deux chambres de Wilson superposées, *Nuovo Cimento* **11**, 292, (1954).

Résultats sur les particules V chargées, *Supplément au Nuovo Cimento* **12**, 324 (1954)

Sur le signe de la particule $K\mu$, *Supplément au Nuovo Cimento* **12**, 324 (1954).

Further discussion of the $K\mu$ decay mode, *Nuovo Cimento* **1**, 915 (1955).

A V° decay with an electron secondary, *Nuovo Cimento* **4**, 917 (1956).

Mésons in the momentum cloud chamber of the Ecole Polytechnique, *Supplément au Nuovo Cimento*4

Cimento **4**, 217 (1956).

V^- events, *Supplément au Nuovo Cimento* **4**, 541 (1956).

V^+ events, *Supplément au Nuovo Cimento*, 529 (1956).

Results of the study of S-events in the double cloud-chamber of the Ecole Polytechnique, *Supplément au Nuovo Cimento* **4**, 520 (1956).

On the Q-values of the Λ° and the θ° and on the anomalous V° decays, *Nuovo Cimento* **6**, 1135 (1957).

REFERENCES

Ref. 1. Auger, P., et Leprince-Ringuet, L.: Compte-rendu Académie des Sciences, 13 novembre 1933: Etude de la variation du rayonnement cosmique entre les latitudes 45 degrés Nord et 38 degrés Sud.

Ref. 2 Ref. 3 Auger, P., Leprince-Ringuet, L. et Ehrenfest, P.: Compte-rendu Académie des Sciences, 22 octobre 1934: analyse du rayonnement cosmique en haute altitude; et 29 avril 1935: absorption de la fraction molle du rayonnement cospusculaire cosmique.

Ref. 4 Crussard, J., Worellet, D., Kayas, G., Orkin-Lecourtois, A. et Tremblay, J.: Compte-rendus Académie des Science 234,84 (1951); 234,1359 (1952); 236,64 (1953a); 236,872 (1953b).

Ref. 5 Ref. 6 Leprince-Ringuet, L. et Crussard, J.: Compte-rendu Académie des Sciences, 9 décembre 1935 et janvier 1937: étude des particules de grande énergie dans le champ magnétique de l'électro-aimant de Bellevue.

Ref. 7 Leprince-Ringuet, L. et Lheritier, M.: Compte-rendu Académie des Sciences, 13 décembre 1944: existence probable d'une particule de 990 me dans le rayonnement cosmique.

Ref. 8 Leprince-Ringuet, L., Gorodetzky, S., Nageotte, E. et Richard Foy, R.: Compte-rendu Académie des Sciences, 4 novembre 1940: mesure de la masse d'un mésoton par choc élastique.

COSMIC RAY NEUTRONS, 1934–59

Serge A. KORFF[*]

This is a largely personal account of the development of the study of the neutrons associated with the cosmic radiation during the first twenty five years. We shall refer to the work done since 1960 only briefly, where necessary to complete the logic. Post-1960 work has been reviewed by others elsewhere. The convenient stopping-date of this account comes just about at the termination of the International Geophysical Year 1957-58, which led to a great expansion of our knowledge in this field and to the establishment of the international chain of neutron monitors. The references in this work are cited as pertinent to the text. Many other persons have worked in this field, and the author hopes that those not cited, for reasons of economy of space, will not think that the author is not appreciative of their contributions.

1. Early Work

Few historical accounts of scientific problems have clearly identifiable starting points. Almost invariably, several people were working and thinking independently along the lines of the later-acclaimed discovery. However, the history of the neutrons associated with the cosmic radiation appears to be an exception.

In this case, to review a partly personal account, we will start with the basic phenomenon of the day, a major economic depression. This unfortunate event had made jobs very hard to find, and those that did exist carried extremely low salary or fellowship stipends. This had one important result, namely that no one went into physics unless he really liked it and really wanted to do so. Those of us who did, used to talk with our colleagues about the subject a great deal. In 1933, having just finished a National Research Fellowship at the Mt. Wilson Observatory and the California Institute of Technology, working mostly on a topic in spectroscopy, and having come on to the CalTech payroll at a microscopic stipend, I was looking for "gainful employment", as were many of my colleagues. When an opportunity rose to take two of the CalTech ion chambers to Peru, to make ionization measurements at various altitudes, I went. I also took with me a Geiger counter telescope to measure the East-West asymmetry of the radiation.

Returning to the U.S., I visited the Department of Terrestrial Magnetism of the Carnegie Institution of Washington, where Merle Tuve, Lawrence Hafstad and their colleagues had just built a Van deGraaff accelerator. With this they made some of the classical proton-proton scattering experiments, which clearly pointed to the existence of a strong short-range force in nuclei.

At about the same time, Chadwick and his colleagues, in 1932, had just published his experiments pointing to the existence of neutrons. We discussed these at great

*New York University (emeritus), New York, U.S.A.

Y. Sekido and H. Elliot (eds.), Early History of Cosmic Ray Studies, 255–262.
© *1985 by D. Reidel Publishing Company.*

length, and tried to understand all the implications of this work. At the same time, and also in Washington, L.F. Curtiss and A.V. Astin (the latter to become Director of the National Bureau of Standards some years later) were interested in developing a technology of radiosonding, which seemed to me to be ideally suited for the problem which had been intriguing me for some time, that of measuring the cosmic radiation at high altitudes. Together, we made several high altitude balloon flights from Washington, and learned much about what the techniques needed for perfection while also finding out how much the radiation increased at the higher altitudes which we reached. At this time an old friend, Dr. Clyde Fisher, then Curator of the Hayden Planetarium, invited me to joint the Hayden Planetarium-Grace Eclipse Expedition to Peru, to observe the June 1937 eclipse. This event was one of the series of long eclipses, totality lasting of the order of seven minutes. While our primary mission was obtain good high resolution phtographs of the corona, I was also interested in knowing if any "glitches" in the cosmic radiation were associated with the event. We found none.

Again returning to the United States, I accepted an invitation to join the staff at the Bartol Research Foundation of the Franklin Institute, then located on the grounds of Swarthmore College in Pennsylvania. There, I had the opportunity to work both on Geiger and proportional Counters and on Cosmic ray radiosonding.

Soon after the discovery of neutrons by Chadwick in 1932, several persons became interested in the question as to whether there were neutrons associated with the cosmic radiation. Since the primary radiation was known to have very high energies, it seemed reasonable to suppose that its impact on the nuclei of atoms or molecules in the upper atmosphere might generate various nuclear component particles. We discussed these possibilities with various people including Prof. Gilbert Lewis who visited the Bartol Laboratory at that time and who encouraged my first efforts in this direction. Soon I had a neutron detection system attached to a radiosonde and started a flight program (Korff, 1939).

Another visitor to our laboratory was Werner Heisenberg, who came to the U.S. for a visit at that time. It was a particular pleasure to talk to him, since nothing ever needed to be said twice, and he instantly understood all the implications. Indeed he was usually ahead of the speaker, anticipating and asking very pertinent questions. This remarkable man was torn emotionally as he strongly disapproved of the such excesses of the Nazi government as he knew about, yet he was a patriotic German. Among other things we discussed the mechanisms by which the neutrons were produced. He was much interested in nuclear evaporations, and for example we considered the fact that the neutrons increased with elevation faster than did the cosmic radiation (mostly mesons) that had survived to reach sealevel, and at approximately the rate of the "soft component" in the higher levels. This was years before the concept of the nucelonic cascade had developed, and we were feeling around the leading edges of this phenomenon.

As a member of the Scientific Advisory Committee of the U.S. Antarctic Service, I was also interested in considering what cosmic ray studies could usefully be made in that continent and on the various trips through a wide range of latitudes and longitudes. I was able to recommend my colleagues, first Dr. Eric T. Clarke and later Dr. Dana K. Bailey to these expeditions. Dr. Clarke made a good latitude effect study, from which the temperature coefficient could be determined, and from which the meson lifetime could be calculated, but our neutron-detector had not been sufficiently developed or stabilized to permit the latitude effect in neutrons to be determined.

The first successful experiment appears to have been carried out by Rumbaugh and Locher (1936), who made use of the National Geographic balloon flight. On November 11, 1935 this balloon reached an altitude of 72400 feet (about 23 kms), which remained for years a record for man-carrying flights. They sent aloft some nuclear emulsions, which when developed showed tracks of protons, starting and terminating in the emulsion. They interpreted these as recoils, generated by "fast" neutrons.* Today it appears that their interpretation was indeed correct, but at the time, few people took them seriously. The present author several years later, after several of his own neutron-measuring balloon flights, recalls hearing a distinguished physicist, who had a Nobel Prize, say: "There is absolutely no evidence whatsoever that there are any neutrons associated with the cosmic radiation." Thus he learned that the ownership of a Nobel Prize did not automatically endow the owner with infallibility.

Experiments followed in rapid succession. Fünfer (1938) operated a boron lined and argon-filled ionization chamber at the summit of the Zugspitze (8600 feet) in Austria. He also found a counting rate which he attributed to neutrons. His detector responded preferentially to "slow" neutrons.* The present author flew boron-trifluoride-filled proportional counters (Korff, 1939) to balloon altitudes in a series of flights. C.G. Montgomery and his wife D.D. Montgomery (Montgomery, 1939) made BF3 counter experiments. Von Halban, Kowarski and Magat (1939) took a bottle of ethyl bromide up to about 30000 feet (9 kms) above Paris in an airplane and measured the neutron-induced activity in the bromine, using isotope separation. Several other investigators (Schopper, 1937) made studies with photographic emulsions and other devices (see Locher et al., 1933). All concluded that there were indeed neutrons, and some also measured the rates of increase with altitude. Counters were used, either filled with BF_3 gas, or boron-lined, in which ejections from the walls of the counter of alphas or recoil nuclei were observed, using standard proportional counter procedures.

At about this point, World War II intervened. For a while almost all cosmic ray neutron work ceased. Yet the war brought several important developments, which included improved electronic instrumentation, isotopically enriched BF_3 gas, knowledge about neutrons in general, slowing-down mechanisms, and much better absorption and scattering cross section data. Further, better balloons became available, as did aircraft with higher altitude ranges, and new radiosonding equipment.

BF_3 neutron counters, first built and used simultaneously and independently by Korff and Libby, operate by making use of the $^{10}B(n, \alpha)$ reaction, the alpha particle being easily detectible. When operated as proportional counters, such devices give much larger pulses from the neutron-capture reactions than from the omni-present background of betas, gammas, and ordinary cosmic-ray mesons. Since the alpha reaction does not take place in the ^{11}B nucleus, and since ordinary commercially available boron is an isotopic mixture of only 20% ^{10}B/80% ^{11}B, it is clear that isotopic enrichment can greatly improve BF_3 counter efficiency. Such enriched gas was early available from the Oak Ridge Laboratory.

* Here we employ the common loose terminology, in which a "fast" neutron usually is considered to have an energy in the range of perhaps a few hundred kev to a few MeV, and a "slow" neutron may be thermal or just above. A fast neutron and a slow proton can have the same energy.

Fig. 1. Large balloon being inflated, Churchill, Manitoba, Cosmic Ray flight. NASA Photo.

An interesting sidelight is that such a proportional counter is also sensitive to any other ionizing event producing approximately as much ionization as the alpha emerging from the ^{10}B capture event. Such ionizing events occur in the atmosphere, and increase rapidly with elevation, due to nuclear disintegrations or "stars" or slow protons or heavy primaries. The result is that a background is generated which looks to the electronic circuit much like the alpha capture reaction. This background must be determined and subtracted from the counting rate observed. To determine the background, the present author developed a differential measurement system, in which two otherwise identical counters differing only in the amounts of isotopic enrichment were used. This work was later substantially aided when BF$_3$ isotopically depleted in ^{10}B became available, also thanks to the Oak Ridge Laboratory. The amount of correspondence engaged in by the present author with that Laboratory, and with others, to encourage the start of production of the depleted ^{10}B was considerable, since the depleted fraction had previously been discarded after the enrichment process. It was thus possible to design and operate a slow-neutron detection system (see Korff, 1955) which had a good detection efficiency for the desired neutrons, and a very low background due to all other events.

A second important point was the need to pressurize all detectors to be used in the upper atmosphere, for neutron counters generally employ fairly high operating voltages, up to several thousand volts, which at high elevations and reduced ambient pressure tends to generate sporadic corona discharge, which looks to the electronics much like the counter-response, and therefore generates spurious counts. After several flights yielding too-high counting rates, the problem was identified and eliminated through pressurization.

Fig. 2. High Altitude Cosmic Ray Balloon flight taking off. The balloon fills as it rises, becoming approximately spherical at top altitude. Below balloon is, first, flight termination mechanism, then parachute for instruments, and at bottom the cosmic ray instrumentation package.

Whereas the BF_3 detector measures preferentially slow neutrons, fast neutrons can be detected using counters in which recoiling nuclei are generated by the fast neutrons. Such detectors were also flown (Korff, 1941) and proportional counters filled for example with methane (CH_4) were used often in the early work.

It is also important to realize that a slow-neutron counter measures the density, or number of neutrons per cc, while a fast neutron detector measures the flux, or number of neutrons crossing unit area in unit time. Thus the slow and fast neutron measurements are not immediately and directly comparable. This property of the slow neutron counter comes about due to the fact that the slow neutron capture cross section varies inversely with the velocity of the neutrons. Because the velocity term cancels out in the equations, the flux cannot be directly determined by our slow-neutron detecting system.

The entire problem of the neutron energy spectrum, or velocity spectrum, soon emerged. It turned out that its evaluation could be accomplished by observing the counting rate of a detector when surrounded by shields whose absorption cross sections for various velocities was known. In a balloon flight this could be done by arranging a motor-driven set of shields, one of which covered the counter for a specific time, then was withdrawn and replaced by a different shield. The present author (Korff and Hamermesh, 1946) made ballon flights with such a set of three shields in 1945.

2. First Theoretical Analysis

The first analysis of the production and equilibrium of neutrons in the atmosphere was undertaken in 1939 by Bethe, Korff and Placzek (BKP). In this analysis these authors considered that the neutrons were generated in the atmosphere mostly with energies below 30 MeV, with a few surviving from still higher energies, from various nuclear disruption processes, such as the "stars" and evaporations seen in photographic emulsions. Since nitrogen nuclei contain 7 neutrons, complete evaporation could generate up to 7, and partial evaporation any smaller number. Slightly larger numbers could come from oxygen, and still larger ones from argon nuclei. The originating entity might be a charged particle or a photon. Any entity with more than the nuclear binding energy would suffice to generate neutrons.

The neutrons thus generated would be slowed down by collisions with atmospheric constituents. The collisions would be inelastic for fast neutrons, which would very rapidly degrade the energy down to the lowest level in the target nuclei, which is perhaps 6 MeV for oxygen and 2.8 MeV for the more abundant nitrogen. Below this level, scattering would be elastic and can be treated by classical billiard-ball mechanics. Since nitrogen has a substantial $(1/v)$ cross section for low energy neutrons, these would be captured, and there would not be any appreciable accumulation of thermal neutrons.

Modern cross section data, obtained after BKP analysis, has changed this history only in detail. It remains true that more than half the neutrons will be captured by nitrogen to form radio-carbon. Small contributions by other processes will also take place. (See summary by Newkirk, 1963)

Another isotope of geophysical interest, produced in this process is tritium, also from nitrogen. This isotope has been identified by v. Grosse and his associates (v. Grosse et al., 1951), and is also of importance in geophysics. It is useful, for example, in oceanography and glaciology. Another important isotope is ^{10}Be, but this is mostly produced as a spallation product from heavier nuclei. It is clearly important in the long run, because of its long half-life (about 2.5 million years) in sedimentation studies and other geological processes.

The BKP analysis considered the energy spectrum of the neutrons in the atmosphere, and also the boundary effects. Near the earth's surface, the energy spectrum is determined by the chemical composition of the ground. Only close to the surface of water and at the same time far from ground which in general has a mixed chemical composition, can it be calculated with any approximation to reality. Experimental tests of this effect were made by Swetnick (1954), and in the free atmosphere by Korff and Hamermesh (1946) as mentioned earlier.

Since neutrons are generated by the incoming primaries and high energy secondaries, they show a "Pfotzer maximum" or transition effect in the atmosphere, the maximum being at between 60000 feet and 100000 feet (roughly 20 to 30 km), depending on the latitude (see Haymes, 1960). At the top of the atmosphere some neutrons will diffuse out into space. Neutron observations were carried out in a rocket by Reidy, Haymes, and Korff in 1962 up to an altitude of 200 kms., and have since then been carried further out into space by others. Eventually they are lost in the background of the neutrons generated in the body of the carrying vehicle and detector.

It has been suggested that an "absolute altimeter" might be built, which would signal to a spacecraft an approach to a planet or large meteorite, or one that might indicate approach to the surface of the earth by a variation in the ratio of fast to slow neutrons.

One of the more important features of the neutrons is their use in the study of the fluctuations induced by solar and geomagnetic variations. There are two very different types of such fluctuations, the decreases associated with changes in the geomagnetic field, such as the well-known "Forbush" decreases, and second the increases which result from large fluxes of solar particles arriving at the earth. While such variations also can be detected in the ionizing component of the radiation, the neutrons show an amplification in the amplitudes, since neutrons are often produced in multiples, or more than one per ionizing primary (for example, Korff *et al.*, 1946).

At the time of the IGY (1957—58), it was decided to set up a world-wide grid of neutron monitors which could make a long-term set of measurements, to help in analyzing the solar effects. At the same time ionospheric, geomagnetic and other parameters were measured and time-correlated. The entire set to studies was most successful. We learned for example of the "solar wind," which "blows" past the earth, and of the magnetic field associated with it. This field modulates the cosmic radiation reaching the earth from beyond the solar system. Some large fluxes of solar particles have been observed, in which the observed neutron rates suddenly increase by large factors over the normal values (see for example, Meyer and Simpson, 1956).

In considering the effects which the large increases may have on the long-term radiocarbon levels, on which in turn much archaeological dating depends, one must realize that the long lifetime (some 5700 years) of radiocarbon tends to smooth out the spectacular increases. On the other hand slow but long-term variations, such as the Maunder minimum when for close to a century there were very few sunspots, can and do produce observable changes in the radiocarbon accumulation and shows up in the comparison of radiocarbon dates with dendrochronological dates (see for example, Korff and Mendell, 1980).

Thus neutrons have given us an important tool for archaeological research, and have given us new timing determinations in pleistocene geology, both quite unanticipated in pre-1940 studies. Further they have also given us an important new tool for analyzing solar-terrestrial relationships and effects, as well as a new means of analyzing the changes in the geomagnetic field, the field far from earth, the solar wind, its possible effects on manned space flights, and a much expanded understanding of solar physics.

REFERENCES

Bethe, H. A., Korff, S. A. and Placzek, G.: 1940. *Phys. Rev.* 57, 573—587.
Chadwick, J.: 1932, *Proc. Roy. Soc. (London)* A116, 602.
Fünfer, E.: 1938, *Zeits. f. Phys.* 111, 351.
von Grosse, A., Johnston, W. M., Wolfgang, R. L. and Libby, W. F.: 1951, *Science* 113, 1.
von Halban, H., Kowarski, L. and Magat, M.: 1939, *Comptes Rendus* 208, 572.

Haymes, R. C. and Korff, S. A.: 1960, *Phys. Rev.* **120**, 1460. Soberman, R., Beiser, A. and Korff, S. A.: 1955, *Phys. Rev.* **100**, 859. Staker, W. P., Pavalow, M. and Korff, S. A.: 1951, *Phys Rev.* **81**, 889.

Korff, S. A. and Danforth, W. E.: 1939, *Phys. Rev.* **55**, 980; Korff, S. A.: 1939, *Phys. Rev.* **56**, 1241.

Korff, S. A.: 1955, *Electron and Nuclear Counters*, Page 81, 2nd edition, The D. Van Norstrand Co., New York.

Korff, S. A.: 1939, *Phys. Rev.* **56**, 1241; 1941, **59**, 214.

Korff, S. A. and Hamermesh, B.: 1946, *Phys. Rev.* **69**, 155.

Korff, S. A. and Hamermesh, B.: 1946, *Phys. Rev.* **70**, 429; 1947, **71**, 842; 1937, *J. Franklin. Inst.* **241**, 335.

Korff, S. A. and Mendell, R.: 1980, *Radiocarbon.* **22**, 159–165.

Locher, G. L.: 1933, *Phys. Rev.* **44**, 774; 1934, **45**, 296; 1936, **50**, 394; 1937, *J. Franklin. Inst.* **224**, 555; Korff, S. A.: 1939, *Phys. Rev.* **56**, 210; Froman, D. K. and Strearns, J. C.: 1938, *Phys. Rev.* **54**, 969.

Meyer, P., Parker, E. N. and Simpson, J. A.: 1956, *Phys. Rev.* **104**, 771.

Montgomery, C. G. and Montgomery, D. D.: 1939, *Phys. Rev.* **56**, 10.

Newkirk, L. L.: 1963, *J. Geophys. Res.* **68**, 1825; Lingenfelter, R. E.: 1963, *J. Geophys. Res.* **68**, 5633.

Reidy, W. P., Haymes, R. C. and Korff, S. A.: 1962, *J. Geophys. Res.* **67**, 459.

Rumbaugh, L. H. and Locher, G. L.: 1936, *Phys. Rev.* **44**, 855.

Schopper, E.: 1937, *Naturwiss.* **25**, 557; Schopper, E. and L.: 1939, *Phys. Z.* **40**, 22; Heitler, W., Powell, and Fertel: 1939, *Nature*, **144**, 283.

Swetnick, M.: 1954, *Phys. Rev.* **95**, 793.

REMINISCENCES ON COSMIC RAY RESEARCH
AT THE
UNIVERSITY OF MINNESOTA

Edward P. NEY*

Cosmic ray research at Minnesota had an unusual beginning. The famous balloonist Jean Piccard and his colleagues in Aeronautical Engineering became interested in the use of plastics like cellophane for the construction of lightweight high altitude balloons. Jean's primary motivation was manned flight to record breaking altitudes. He constructed a plastic cell and flew it from Memorial Stadium. The Aeronautical Division of General Mills soon became involved and Otto Winzen (later to found Winzen Industries) led the successful effort to manufacture and launch polyethylene balloons which could carry payloads of 60 pounds to altitudes of 90000 feet. Piccard's dream of launching a cluster of balloons with a manned gondola never materialized, but John T. Tate at the University realized the potential for high altitude research, especially in cosmic rays, and the Department attempted to recruit a well known cosmic ray physicist. When he turned them down they decided to take a chance on several young people with interest and enthusiasm, but with little or no experience in either cosmic rays or ballooning. In the summer of 1946 Ed Lofgren, Frank Oppenheimer and I, all recent employees of the Manhattan project, found ourselves designing cloud chambers and balloon gondolas and tracking equipment to have a look at primary cosmic rays. At the same time Charlie Critchfield and Joe Wienberg came to Minnesota as theorists and both took an active interest in the research program. Phyllis Freier was the first and for some time the only graduate student. Phyllis was immediately accepted as an essential member of the team and did an effective job of keeping us on the right track. The early cloud chambers were tempermental, the Ilford C-2 emulsions would not show the tracks of relativistic charge one particles, and the balloons did not consistently float at ceiling altitude. But we were lucky and on a flight made in April, 1948, we discovered (in collaboration with Helmut Bradt and Bernard Peters of the University of Rochester) that primary cosmic rays contained the nuclei of elements heavier than hydrogen, including helium and probably as heavy as iron (Freier *et al.*, 1948).

This first paper must have set some kind of record for fast publication. The balloon flight was in April, the paper was submitted in June, and it was published in Phys. Rev. in July. We spent the month of May convincing our colleagues and ourselves that we had indeed seen the tracks of stripped energetic nuclei of atoms. There were at least three crucial questions: 1) Were they primary, coming from above the gondola?, 2) Did they undergo nuclear interactions with nuclei of the equipment?, and 3) Did they capture electrons to become neutral atoms as they slowed down in matter? The first and third questions were answered in the May flight. Fortunately we had emulsions above and below the one inch of lead in the cloud chamber and the cosmic rays were in the top emulsions, but not in those below the lead. We soon found that

*University of Minnesota, Minneapolis, Minnesota, U.S.A.

Y. Sekido and H. Elliot (eds.), Early History of Cosmic Ray Studies, 263–275.
© *1985 by D. Reidel Publishing Company.*

Fig. 1. Photograph of some of the participants at the Symposium in June 1948 honoring R. A. Millikan's eightieth birthday. The ones I can identify are Robert Oppenheimer, W. Heitler, Pierre Auger, Robert Millikan, W. F. G. Swan, H. P. Robertson, J. Clay, Bruno Rossi, Lee DuBridge, John Wheeler, Scott Forbush, I. I. Rabi, George Valley, Manuel Vallarta, Marcel Schein, Bob Serber, Bill Fretter, Bob Brode, Bob Sard, Helmut Bradt, Jesse DuMond, George Reynolds, Guissepe Cocconi, Willy Fowler, Serge Korf, L. I. Schiff, W. Pickering, Willie Lauritsen, Carl Anderson, G. Rochester, Urner Liddel, Ed Ney, Bob Christy, Seth Neddermeyer, R. J. Finkelstein, and Victor Neher.

we could follow the heavy nuclei from plate to plate, but there were large gaps because the early emulsions had glass backing. I can still remember Phyllis Freier's war whoop when she successfully followed the first "heavy." Following tracks showed that as they passed through matter their ionization first increased due to the z^2/β^2 in the energy loss expression and then decreased at the end of the range as they captured electrons in atomic orbits finally becoming neutral atoms. Finding nuclear interactions took longer and about three mean free paths were required before the first interaction was observed (never trust a two Sigma answer!). By the end of 1948 many of the fundamental questions were answered and described in two papers (Freier *et al.,* 1948a; Bradt and Peters, 1948).

The early days of ballooning were exciting. The flights were made at dawn from a military base, Camp Ripley, about 100 miles north of Minneapolis. We would be up all night getting the equipment ready and then would chase after the balloon in National Guard airplanes and Frank Oppenheimer's open car. Finally when the gondola was returned to the lab we had to see whether the cloud chamber worked so we developed the film. We followed all this by giving some pretty sleepy lectures to our physics classes. Tearing after a balloon with Frank was a thrill in itself. He had an old Chrysler touring car with wooden doors tied shut with ropes. With the top down Frank would scan the sky while driving at top speed on whatever road seemed to go in the right direction.

Our exciting collaboration had a short life. At the end of 1948, Ed Lofgren returned to Berkeley. Ernest Lawrence had persuaded him to supervise the construction of the Bevatron. In the summer of 1949, Frank Oppenheimer fell victim to the House Unamerican Activities Committee and was forced to resign from the University. For many years his talents were wasted on his ranch at Pagosa Springs, Colorado. However, he returned briefly to academic science at the University of Colorado and subsequently created one of the best science museums in the world, the Exploratorium in San Francisco.

Before Frank left I had a chance to join him in a balloon flying expedition from the aircraft carrier Saipan off the coast of Cuba. Phyllis and Frank and I wanted to study the effect of geomagnetic cutoffs on the heavy nuclei and we needed to expose a stack at low latitude. Our research was supported by the Office of Naval Research and our sponsors arranged to have balloon flights from the deck of the carrier. Jack Winckler had already shown this to be feasible with his classic counter telescope flights from the Norton Sound in which he established the geomagnetic effects on primary particles. The deck of a carrier is an ideal launching platform because the ship may be maneuvered so that there is no wind across the deck and the balloon can be vertically inflated and launched. Our plan was to launch from the carrier and to have a destroyer at the "down point" to pick up the gondola before salt water could leak in and damage the equipment. However, our most successful flight did not land in water, but on Mount Torquino in Cuba, the part of Cuba where Castro's revolution began. The captain of the carrier did not want to send military personnel to recover the load so a helicopter landed Frank and me as close as possible. We climbed the mountain and as we progressed we accumulated a troop of interested Cubans who climbed with us and finally numbered at least thirty souls. Frank's enthusiasm and excellent Spanish had recruited our own army. When we arrived at the parachute it was hanging high in a tree and with the Cubans politely watching, Frank proceeded to climb the tree with climbing irons loaned us by the Navy. He had never used them before and he was exhausted, and kept slipping down the tree, to the great amusement of the assembled multitude. Finally one of the group volunteered and scooted up the tree like a monkey to the cheers of the crowd. When we returned to the helicopter rendezvous point the Cubans had a banquet for us—goat meat fried in sweet oil and Frank gave each of them a dollar for a present. It doesn't sound like much money now, but it meant that we ate one meal a day in Miami while we waited for a MATS flight back to Minneapolis.

In the summer of 1948, Cal Tech held a cosmic ray symposium to honor R.A. Millikan's eightieth birthday. The papers were printed in the Reviews of Modern Physics. Frank and Helmut Bradt and I were there to tell about the heavy nuclei. The participants were like a Who's Who of Cosmic Rays and Theoretical Physics. Figure 1 shows a photograph of most of the group. In the first row are Robert Oppenheimer, W. Heitler, Pierre Auger, Robert Millikan, W.F.G. Swan, H.P. Robertson, J. Clay, and Bruno Rossi.

Millikan gave the first paper "Present Status of the Atom Annihilation Hypothesis" (1949). He still believed that cosmic rays were produced by the annihilation of atoms of hydrogen, helium, carbon, etc. He didn't pay much attention to the presence of these elements at high energy in the primary cosmic rays so recently discovered. However, at a later meeting of the Physical Society when one of us gave a paper, K.K. Darrow, the raspy-voiced secretary of the Physical Society, pronounced "We used to think that cosmic rays were the death cries of the elements, but now we know that they are the elements themselves!"

Another exciting conference was the Echo Lake, Colorado Conference on Cosmic Rays in the summer of 1949. One of the amusing events was Edward Teller's composition of a poem at the bar of the Teller House Hotel in Central City. I repeat it because it indicates where we were in understanding the nuclear physics and high energy aspect of cosmic rays thirty years ago.

The Meson Song
Second Generation Edition

Edward Teller

Recited by Dr. Teller at the Banquet in the Teller House, Central City, Colorado, on Saturday, June 25, 1949.

> There are mesons pi, and there are mesons mu.
> The former ones serve as nuclear glue.
> There are mesons tau—or so we suspect—
> And many more mesons which we cannot detect.
> Can't you see them at all?
> Well, hardly at all,
> For their ranges are short and their lifetimes are small.
> The mass may be small, and the mass may be large.
> There may be a positive or negative charge.
> And some mesons will never show on a plate

For their charge is zero, though their mass is quite great.
    What, no charge at all?
    No, no charge at all.
    Or, if there's some charge, it's exceedingly small.
There are mesons lambda at the end of our list
Which are hard to detect but are easily missed.
In cosmic ray showers they live and they die,
But you can't get a picture—they are camera shy.
    Well, do they exist?
    Or don't they exist?
    They are on our list, but are easily missed.
From mesons all manner of forces you get;
The infinite part you may simply forget.
The divergence is large, the divergence is small:
In meson field quanta there is no sense at all.
    What, no sense at all?
    No, no sense at all.
    Or, if there's some sense, it's exceedingly small.

At the Echo Lake Conference Frank and I presented a paper entitled, "Wide Angle Sprays of Minimum Ionization Particles," (Oppenheimer and Ney, 1949). These were electron showers seen in the lead plates of the cloud chambers flown at high altitude. They probably were produced by primary electrons, but we preferred to think that they were something "mysterious." It was many years before the primary electrons were discovered and understood.

In 1949 Jack Winckler who had just completed his high altitude study of geomagnetic effects came to Minnesota to join the group and to contribute his considerable expertise in cosmic ray detectors and electronics. One of his many clever ideas was the use of the Cerenkov effect to distinguish the direction of motion of the particles at the top of the atmosphere and thereby to measure the albedo of the earth for charged particles.

Phyllis Freier got her Ph.D. in 1950 and her thesis "The Heavy Component of the Primary Cosmic Radiation" is a remarkable contribution. Figure 2 from her thesis is one of my favorite illustrations. It shows the increase in ionization of a heavy nucleus as it slows down in the emulsion. This is a nucleus probably as heavy as Iron. Following Phyllis' example, a number of excellent Ph.D. students contributed to the success and vitality of the program in the next decade. Among these were the following (the dates are the dates on which they obtained their Ph.D. degrees): Phyllis Freier (1950), Sophie Oleksa (1951), John Linsley (1952), John Naugle (1953), George Anderson (1953), Nahmin Horwitz (1954), Nelson Mitchell (1955), Kinsey Anderson (1955), Frank McDonald (1955), Bill Erickson (1956), Bob Danielson (1958), Larry Peterson (1960), Robert Hoffman (1962), Roger Arnoldy (1962), Wayne Stein (1964), and Dave Hoffman (1966).

By 1950 it was clear that the majority of cosmic rays were positively charged nuclei of the elements with abundances resembling cosmic composition, but with heavier elements relatively more abundant than those derived by astrophysicists for the sun. The big mystery was "where are the electrons?" Hulsizer and Rossi (1949) flew a lead

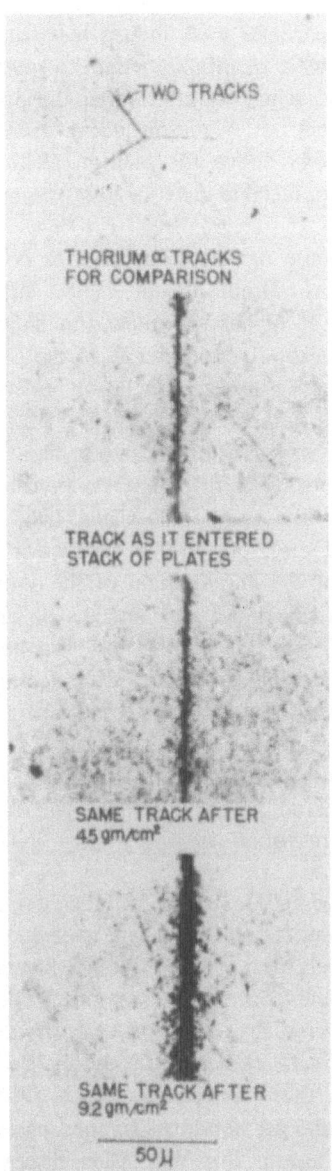

Fig. 2. From P. S. Freier's Ph.D. Thesis. The track of a heavy nucleus in Ilford C2 emulsion, showing the increase in ionization as the particle slows down.

shielded ionization chamber and concluded that the burst producing radiation (electrons) could not constitute four percent of the primary flux.

We had, of course, observed electron showers in the lead plates of our cloud chambers and Charlie Critchfield, Sophie Oleksa and I carried out a series of experiments with vertical and horizontal lead plates to try to understand the soft component at balloon altitudes (Critchfield et al., 1950, 1952). We showed that the sprays of minimum ionizing particles were indeed electron showers and that if all the soft radiation at 18 millibars were of primary origin, it would constitute 0.6 percent of the cosmic ray flux. However, when we considered the production of soft component by the gamma rays from neutral π meson decay from π's produced by nuclear interactions of primary protons above the balloon altitude, we concluded that, "it seems most likely that no appreciable part of the primary radiation is composed of electrons or gamma rays."

The true discovery of cosmic ray electrons did not come for more than a decade when Jim Earl who came to Minnesota in 1958, improved the cloud chamber technique and flew balloons at higher altitudes. Jim showed that there were indeed primary electrons constituting about one percent of the cosmic radiation (Earl, 1961). Peter Meyer at Chicago reached a similar conclusion independently (Meyer and Vogt, 1961). But in 1961 when the primary electrons were identified they had been observed in the universe in another way. Iosif Shklovsky had worked out the theory of synchrotron radiation for electrons in the magnetic fields of the filaments of the Crab Nebula and had shown that the radio and visual light from this object led to a knowledge of the density of electrons in the supernova remnant. He concluded "a factor of great significance for the development of the above theory is the question of the electron component of the primary cosmic rays observed at the earth. Electrons have not yet been detected in primary cosmic rays. Furthermore, radio astronomy has shown that if we could increase the sensitivity of the techniques for recording primary cosmic electrons by only one order of magnitude, then electrons could in fact be detected in the primary component," Shklovsky, 1960). I consider this prediction of Shklovsky's to be an outstanding triumph of astrophysical theory. Our own attempt to pawn the electrons off as meson progeny allowed the discovery of electrons in the Crab Nebula before they were found to surely exist in the cosmic ray beam at the earth.

The Guanajuato Conference in Mexico in 1955 was organized by Manuel Vallarta. Vallarta had made great theoretical contributions to the theory of geomagnetic cutoff phenomena refining the models first discussed by Störmer. The current state of our knowledge of phenomenological cosmic rays was much discussed at this meeting. We were used to thinking of cosmic ray trajectories which connected with infinity and then using Liouville's theorem to calculate intensities. S. Fred Singer gave a notable paper in which he considered those trajectories which did not connect to infinity. His discussion was the closest to the prediction of trapped particles in the earth's magnetic field of any of which I am aware. Jim Van Allen discovered the trapped radiation three years later with the spacecraft Explorer I.

By the time of the International Geophysical Year (summer of 1957 to the end of 1958) a number of observations had suggested that the solar cycle had pronounced effects on cosmic rays. Scott Forbush had shown in solar cycle 18 with a minimum in 1944 and maximum in 1947 that there was an inverse correlation between the counting rate of ion chambers at sea level and the sunspot numbers. (Forbush, 1954). He also had shown that magnetic storms were accompanied by decreases in cosmic ray intensity (Forbush decreases). The sun was modulating galactic cosmic rays in the sense of higher solar activity being associated with fewer galactic cosmic rays. The balloon flights of Neher showed that the changes of intensity at high altitudes were much larger than those at sea level (Neher and Forbush, 1958).

The International Geophysical Year was chosen to span the maximum of solar cycle 19 which in late 1957 reached the highest recorded level of sunspot numbers. The counting of sunspots has been carried out as a crude measure of solar activity since 1749 and of all the cycles the one which reached its maximum in 1957—58 appears to be the most intense. Principally through the efforts of Jack Winckler, the Minnesota group mounted a continuous monitoring program involving counters, ion chambers and photographic emulsions. The goal was to have high altitude balloons in the stratosphere as nearly continuously as possible.

On the first day of the I.G.Y. Winckler and Larry Peterson were rewarded by the discovery that their detectors flying at a latitude of 45° showed strong bursts of X-rays coincident with the appearance in the zenith of a visible aurora. Observed many times during IGY, we now know that the auroral X-rays are produced by the release of electrons trapped in the Van Allen radiation belts. The auroral X-rays were discovered almost a year before the radiation belts.

The previous sunspot minimum was in 1954 and it was the lowest level of solar activity of recent cycles. The balloon measurements of this period compared with the maximum of 1957 showed that low energy galactic cosmic rays were modulated by a factor of three in total ionization at the top of the atmosphere. This large change has effects on the C^{14} age dating (Elsasser et al., 1956) and perhaps on terrestrial weather (Ney, 1959).

Monitoring of cosmic rays at high altitude during IGY led to a number of discoveries. On March 20, 1958 Larry Peterson and Jack Winckler detected a burst of 200 to 500 KeV gamma rays coincident with a solar flare and showed that the radiation could be interpreted as Bremsstrahlung from high energy electrons in the flare region. The sun was showing us it could accelerate high energy particles.

Perhaps one of the most important discoveries of IGY was the "solar protons" later to become "Solar Cosmic Rays." Jack Winckler had made a series of counter telescope flights over the years and on one of these on February 23, 1956, following a great solar flare he detected a fourfold enhanced flux of particles at the top of the atmosphere (Winckler, 1956). In this same event Simpson's sea level neutron monitor network recorded increases from the equator to high geomagnetic latitudes and Simpson and his collaborators concluded that these were due to an increase in the nucleonic component at the top of the atmosphere (Meyer et al., 1956). Kinsey Anderson was the first to demonstrate that high energy protons were accelerated in solar flare events. Using single counters, counter telescopes and ionization chambers, he showed that high energy protons arrived at the earth on August 22, 1958, and that their flux was about 10 times the flux of galactic cosmic rays (Anderson, 1958).

The first emulsion detection was in a smaller event which happened on March 26, 1958, but which had not been analyzed until after Kinsey's discovery. In the March event the flux was only three times cosmic rays and Phyllis Freier had to understand the proton background in the plates before we could be sure the extra particles were real (Freier et al., 1959). But larger events were in store. On May 12, 1959 protons were detected with a flux of 1000 times cosmic rays. Figure 3 is a photomicrograph of

TOP OF EMULSION

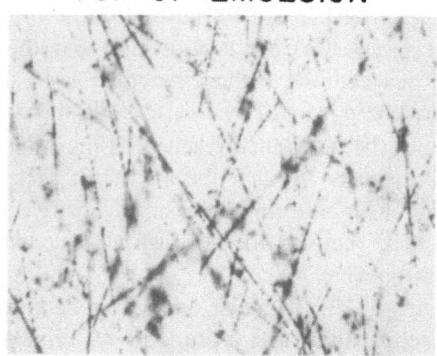

Fig. 3. "Proton Solaire." This photomicrograph is a 2000X enlargement of a photographic plate flown on May 12, 1959 in an intense beam of solar cosmic rays. It was exposed in a balloon flight at 100,000 feet. A normal cosmic ray exposure would have had approximately 1/10 of a track in this area.

nuclear emulsion in which the flux of solar particles is so high that the same emulsion would have shown no cosmic ray tracks in the same area (Ney *et al.*, 1959). This photograph shows a number of ending protons. It seemed incredible that at the end of each of these tracks is a nucleus of hydrogen which was on the sun only a day before it came to rest in the emulsion.

The largest solar cosmic ray events were recorded in November 1960 when a series of flares produced plasma clouds and accelerated cosmic rays. The intensity at the earth reached 10000 times galactic cosmic rays and an astronaut in space would have been subjected to about 2.5 roentgens/hour and a total dose of 60 roentgens, about one tenth the lethal dose. The photographic emulsions exposed in this event were completely blackened. We developed half of the stack and the plates were so dense with proton tracks that the exposure was too great to get light through for microscopic study. We used a medical microtome to slice the unprocessed 600 micron emulsions to 25, 50 and 100 microns (Ney and Stein, 1962). These thinner sections could be studied in the usual way. An important aspect of the solar proton events is their production of polar cap absorption of cosmic radio noise. Leinbach and Reid (1959) found that they could monitor the arrival of proton streams with a device called a riometer ("relative ionospheric opacity meter") and invariably their polar cap absorptions were accompanied by increases observed at ballon altitudes, when balloons were in the air (Leinbach and Reid, 1959).

Tommy Gold produced a model for the propagation of solar produced cosmic rays in which magnetic loops can guide the solar particles and produce a magnetic connection to the earth (Gold, 1959). Since the same magnetized clouds also produce the Forbush decreases, the model of Gold gave a good morphological picture of the role of the solar ejecta.

In the history of cosmic rays the high latitude knee of the cosmic ray intensity curve was a familiar subject. It was first thought to be due to a solar dipole magnetic field. This was the predominant view at the time of the discovery of heavy nuclei. A collaborative effort between Jake Waddington and Peter Fowler at Bristol and the group at Minnesota helped to clarify this subject (Fowler et al., 1957; Waddington, 1956; Fowler et al., 1958). In 1956–57 Peter Fowler spent a year at Minneapolis and we made balloon flights at such unlikely places as Waukon, Iowa and Kirksville, Missouri. The net result of these efforts, which involved the study of primary alpha particles, was the realization that the primary spectrum itself was deficient in low energy particles, and that there was no sharp magnetic cutoff as a dipole solar field would have produced. The combined British and American results also showed that geomagnetic cutoffs could not be calculated from an earth centered dipole magnetic field. Pam Rothwell suggested that the cosmic ray results were in good agreement with the geomagnetic dip coordinates which better represent the real earth field (Rothwell and Quenby, 1958).

Although the alpha particle collaboration is the only one I have mentioned, the ties between Minnesota and Bristol were very close. Two British photographic plate scanners, Pamela Curry and Janet Mitchell, spent several years at Minnesota and in addition to Peter Fowler, Jake Waddington joined us for a year. Later Jake came to Minnesota in 1961 to join the faculty.

During the IGY period Paul Kellogg came to Minnesota to do theoretical nuclear physics. However, he was soon tempted by the problems in solar plasma physics and cosmic rays. Among other things, he worked out the neutron decay albedo theory for injecting particles into the inner radiation belt (Kellogg, 1959) and predicted the bow shock wave from the interaction of the earth's magnetic field with the solar wind (Kellogg, 1962).

Paul and Jack Winckler and I also had an interesting and sobering experience with classification and the bureaucracy. In 1958 shortly after the discovery of the radiation belts we became interested in the dumping of the belts as a mechanism for the production of the aurorae. We realized that the energy stored in the magnetic field of the earth was equivalent to the energy of a 200 megaton atomic weapon and made some calculations about the effect of the detonation of a nuclear explosion in the magnetosphere. Our sponsor was the Office of Naval Research and we sent them a letter outlining our ideas. Things happened in a hurry. I had a call from a friend who was an Atomic Energy Commission commissioner and who told me "you have a classified document on your desk." I was incredulous when I was told that a government employee had told me it was classified. When I denied it, he said, "A government employee is telling you now." In several days Herb York, who was the new Director of ARPA (The Advanced Research Projects Agency of DOD) called to ask us to come and discuss our ideas with him. By this time we had a preliminary paper prepared for Nature. Paul and I spent several sweaty hours in Herb's Pentagon office, I now realize, persuading him that we did not have and did not want to acquire any classified knowledge. He asked us to delay our publication for six months, which we did. In the meantime, we became aware of a military detonation of a nuclear weapon at Johnston Island at an altitude in excess of 70 miles. This explosion funneled particles down the lines of force to a magnetic observatory in Apia and produced the first recorded aurora at this station on August 1, 1958. We were unaware of the classified Argus experiment to study the effects of a small nuclear explosion high in the magnetosphere. It had apparently been suggested much earlier by

Christofilos. We published in Nature "Geophysical Effects Associated with High Altitude Explosions" in February (Kellogg *et al.*, 1959) and heard about the real thing at a symposium of the National Academy in the spring. The recently declassified results of the Argus experiment were the topic. I now believe that nuclear weapons are not experimental toys for scientific experiments and am sorry that any were detonated in the magnetosphere. The probability of nuclear war is the greatest threat to civilization in the history of mankind.

Where are they now? My own last active participation in cosmic rays was at the Pontifical Academy study on "The Problems of the Cosmic Rays in Interplanetary Space" in Rome in 1962. The cosmic ray program at Minnesota thrives under the leadership of Jake Waddington and Phyllis Freier. Jack Winckler and Paul Kellogg study the plasma physics of the magnetosphere by injecting intense beams of electrons from rocket payloads (much better than nuclear explosions). John Linsley pursues the highest energy cosmic rays with extensive air showers. John Naugle played a central role in determining the course of NASA's scientific missions for two decades, finally retiring as chief scientist. Frank McDonald leads the high energy astrophysics group at Goddard Space Flight Center. Bob Danielson went to Princeton and worked with Martin Schwarzschild on Project Stratoscope until Bob's untimely death in 1974. Larry Peterson has led the group at La Jolla in X- and γ-ray astronomy, and Kinsey Anderson had exploited his clever ideas (such as anchored IMP around the moon) at Berkeley. He was elected to the National Academy in the spring of 1980. Probably the most important result from the cosmic ray program at Minnesota has been the stimulation which helped to motivate outstanding students. I have always thought, and still do think, that cosmic rays are one of the most elegant topics in physics, and that the study of them gives us an important part of our vision of the astrophysical universe.

REFERENCES

Anderson, K. A.: 1958, *Phys. Rev. L.* **1**, 335.
Bradt, H and Peters, B.: 1948, *Phys. Rev.* **74**, 1828.
Critchfield, C. L., Ney, E. P. and Oleksa, S.: 1950, *Phys. Rev.* **79**, 402.
Critchfield, C. L., Ney, E. P. and Oleksa, S.: *Phys. Rev.* **85**, 461.
Earl, J. A.: 1961, *Phys. Rev. L.* **6**, 125.
Elsasser, W., Winckler, J. R. and Ney, E. P.: 1956, *Nature* **178**, 1226.
Forbush, S. E.: 1954, *J. Geophys. Res.* **59**, 525.
Fowler, P., Freier, P. S. and Ney, E. P.: 1958, *The Primary Alpha Particle Spectrum over North America and Geomagnetic Cutoff Energies,* Nuovo Cimento Supplemento, Vol. 8, Series X, 492.
Fowler, P., Waddington, C. J., Freier, P. S., Naugle, J. and Ney, E. P.: 1957, *Phil. Mag.* **2**, 157.
Freier, P., Lofgren, E. J., Ney, E. P., Oppenheimer, F.: 1948, *Phys. Rev.* **74**, 1818.
Freier, P. Lofgren, E. J., Ney, E. P., Oppenheimer, F., Bradt, H. L. and Peters, B.: 1948, *Phys. Rev.* **74**, 213.
Freier, P., Ney, E. P. and Winckler, J. R.: 1959, *J. Geophys. Res.* **64**, 685.
Gold, T.: 1959, *J. Geophys. Res.* **64**, 1665.
Hulsizer, R.: 1949, *Phys. Rev.* **76**, 164; *Proceedings of the Echo Lake Cosmic Ray Symposium,* p.218.
Kellogg, P. J.: 1959, *Nuovo Cimento* **11**, 48.
Kellogg, P. J.: 1962, *J. Geophys. Res.* **67**, 3805.
Kellogg, P. J., Ney, E. P. and Winckler, J. R.: 1959, *Nature* **183**, 358.

Leinbach, H. and Reid, G. C.: 1959, *Phys. Rev. L.* **2**, 61.

Meyer, P., Parker, F. N. and Simpson, J. A.: 1956, *Phys. Rev.* **104**, 768.

Meyer, P. and Vogt, R.: 1961, *Phys. Rev. L.* **6**, 193.

Millikan, R. A.: 1949, *Mod. Phys.* **21**, 1.

Neher, H. V. and Forbush, S. E.: 1958, *Phys. Rev. L.* **1**, 173.

Ney, E. P.: 1959, *Nature* **183**, 451.

Ney, E. P., Winckler, J. R. and Freier, P.: 1959, *Phys. Rev. L.* **3**, 183.

Ney, E. P. and Stein, W. A.: 1962, *J. Geophys. Res.* **67**, 2087.

Oppenheimer, F. and Ney, E. P.: 1949, *Phys. Rev.* **76**, 1418.

Rothwell, P. and Quenby, J.: 1958, *Cosmic Rays in the Earth's Magnetic Field,* Nuovo Cimento Suppl., Vol. 8, Series X, p. 249.

Shklovsky, I. S.: 1960, *Cosmic Radio Waves,* Harvard U Press, Cambridge, Mass., p. 197.

Waddington, C. J.: 1956, *Nuovo Cimento* **3**, 930.

Winckler, J. R.: 1956, *Phys. Rev.* **104**, 220.

REMINISCENCES OF SOME EARLY COSMIC RAY WORK IN INDIA

Devendra LAL*, Yash PAL** and Bernard PETERS***

The Bombay monsoon was heavy in 1955. From plastic lined tarpaulins on the leaky roof of a deserted army barrack, the rain water rushed gurgling into carefully cleaned large glass carboys. Inside the hut, in the only dry corner, we used home made ion exchange columns to recover less than a million atoms of Be^7 which cosmic rays had produced in their passage through the atmosphere and which had been washed out by the condensing moisture.

In another scenario, at the Khilanmarg plateau (at 13000 ft. altitude) in the Himalayan range of Kashmir, the longer lived isotope of beryllium, Be^{10} (1.5 million years) was being recovered from a few thousand tons of glacier snow melt. The location was majestic and the wind, which, one cold night, blew away our tents saw to it that the experiment had a dramatic flavour.

Simple calculations had convinced us that although spallation reactions of cosmic rays with air nuclei and especially with atmospheric argon were rare, nevertheless enough radioactive fragments of various types with half-lives shorter than 5 years were constantly produced to be detectable in a few hundred litres of rain water.

With little experience but complete faith we went about the task of developing microchemical techniques, some of them rather sophisticated even by present day standards. Distinguished radiochemists in the atomic energy establishment were somewhat discouraging about our enterprise, which involved preparation of chemically pure samples containing just a few thousand atoms of a radioisotope per milligram of carrier. Their scepticism was justified because we had to learn to overcome many unsuspected difficulties on the way. But we managed and in the process also learned to make beta counters which could measure radioactivity down to a few disintegrations per hour.

There is no dearth of rain water during a Bombay monsoon. But even collection of water samples became a bit of an adventure because we wished to tag individual rain showers to check on the possibility of using radioisotope concentrations as indicators of air-mass histories. So, day or night, any time the deluge started, D.L. and his wife donned their custom-made, supposedly waterproof, suits and raced ten miles on a motor cycle to the leaky barracks at the southern end of Bombay – which was then our laboratory.

Within a span of three years, about a dozen radioisotopes, with half-lives ranging from about an hour (Cl^{39}, 55 min) to several years (Na^{22}, 2.6 yr), were detected and

*   Physical Research Laboratory, Ahmedabad, India.
** Space Applications Center, Ahmedabad, India.
*** Danish Space Research Institute (Emeritus), Lundtoftevej, Lyngby, Denmark.

277

Y. Sekido and H. Elliot (eds.), Early History of Cosmic Ray Studies, 277–282.
© *1985 by D. Reidel Publishing Company.*

studied to understand gross features of atmospheric circulation; the isotope ratios permit one to estimate the air mass trajectory, time of descent from the stratosphere, and the time elapsed before being washed out from the air mass. The three long-lived isotopes, Be^{10}, Al^{26}, Si^{32}, with half-lives in the range of $10^2 - 10^6$ yrs proved useful later for studying chronology of glacier deposits and marine sediments (Lal and Peters, 1962).

In the late forties and early fifties cosmic ray science was still young and one was free to follow one's natural curiosity with self-chosen periods of intense involvement in a large number of seemingly unconnected subjects.

We were intrigued at that time by the prospect of using "tracers" (produced in nuclear transformations in matter by this perennial beam of particles) for dating archaeological samples and meteorites, and for the study of meteorological and oceanographic processes. We were equally intrigued by the newly opening channel of information from the universe; here was a possibility to learn about the nature of cosmic ray sources, cosmic acceleration mechanisms and fields and matter in the intervening space, through a study of the energy spectrum and chemical and isotopic composition of the primary particles.

The most revolutionary impact of cosmic rays at that time was however on theories on the nature of matter. Because of their enormous kinetic energies they had begun to open up a whole new world of sub-nuclear particles (and thereby subsequently provide the motivation for building the giant particle accelerators of today).

The following reminiscences are connected with this last and perhaps most exciting aspect of physics research in the early fifties. Our research involved the identification of several types of strange particles (including the very first examples of negative kaons). We measured their masses, decay modes, interaction properties, and observed the first examples of "associated production".

At that time nuclear emulsions became the primary tool for detecting energetic charged particles. They had been developed in the late forties and used by Cecil Powell in the important discoveries of the pi-meson and its generic relation with the muon. This same technique was used for studying the chemical composition and energy spectrum of primary cosmic rays, (Lal *et al.*, 1952a), the multiple production of mesons (Lal *et al.*, 1952b, c) and the nature of strange particles (Lal *et al.*, 1953). The total number of events of heavy unstable particles was at that time no more than a dozen or two and those were discussed in such detail that the picture of each event was known to all the workers in the field. Cosmic rays and subatomic physics had made a close connection. There was expectancy in the air and a general feeling of being on the verge of penetrating a deep mystery.

On the technical side our major concern at the Tata Institute of Fundamental Research in Bombay was to use nuclear emulsions in the form of solid blocks (composed of carefully aligned thin sheets) so as to contain most of the created particles within the detector. These blocks were then exposed to primary cosmic rays in the stratosphere using long trains of meteorological balloons filled with hydrogen. A major problem at that time was to demolish old psychological barriers by convincing young and raw graduates as some of us were at that time that even for people educated in India who had never been abroad in the famous centres of physics, and who disposed of only moderate means, there was no natural law against their turning

Fig. 1. One of the early rubber balloon launches of emulsion stacks in which the first nuclear
interactions of strange particles and improved mass determinations of
K-mesons were obtained. 1950, Bangalore.

into reasonably productive experimental scientists able to improve existing techniques or even invent new ones. The success attained in these efforst were accompanied by many rich and colourful adventures.

Balloon flights were always big events. They offered adventure, produced anxiety and gave a lot of room for innovation. In our earlier flights, hydrogen was prepared on location using heavy cast iron high pressure generators. Hydrogen generation for a single flight took a crew of five people several days of continuous operation. Later, when we had gotten hold of a compressor we tried to fly from open grounds located near factories where hydrogen under low pressure was available in plenty. Hunting for such location across the country was an enterprise in itself. Our earlier flights were made with a large number of natural rubber or Latex balloons. The Latex balloons were supposed to behave better if they were immersed in boiling water before inflation. Since some of our flights required as many as sixty balloons, the process of inflation would begin as early as midnight for a balloon launch around 6 A.M. The process would go somewhat as follows: Take a balloon out of its box, immerse it for several minutes in boiling water, fill it with hydrogen while measuring its free lift against a weight suspended from its neck, tie the neck and hand it over to a colleague, who walks with a flash-light over a network of pipes, often without tripping, then over the paved factory road, out the factory gate, across the city road (traffic was not much to speak of), sometimes followed by barking dogs, ducking in the dark under the low strung telephone and electric wires, into open ground, to finally hand over the balloon to another colleague holding on to one of the several cluster points along the load line, laid along the length of an often undulating field.

While this was repeated about sixty times in five hours, and the parachute, the payload and the barographs were tied down, someone rode a bicycle to the nearest telephone to inform the airport (if there happened to be one near that town) about the impending flight. The load had to be placed downwind at the end of the balloon line, to ensure that the balloon train would drift overhead when released. Somehow the wind always shifted at the last minute, often 180°, requiring a complete rotation of the balloon line. A good launch was always followed by resounding cheers from the crew and the crowd, which usually grew to several hundred by early morning.

The payload recovery depended on the goodwill of the finder in sending a telegram, claiming the reward of Rs.25 to Rs.100 ($6 to $24 in those days) in return for the payload and the parachute. Of course we did have a fair idea about the target area because the balloon was tracked by optical theodolite stations, set up at several "bench mark" locations down-range in the wind direction. Theodolite teams carried synchronized chronometers (meaning accurate handwound clocks with a second hand) and, on occasions, were in contact with the launch station or each other through V.H.F. sets manned and operated by the army signal corps personnel deputed to work with us. This was the only reliable method to obtain position and altitude profile of the balloon, though later some of our flights did carry barographs, which on occasions, were mercury manometers, imaged on to a photographic film or a paper roll, driven by a clock mechanism.

And then, if and after we got a telegram, there was a new stir in the tired team and we set out in a rented car to the end of a road and from there on bicycle or oxcart into the jungle to recover the payload. This, for many of us, was a "discovery of India" trip in a true sense of the word. Each recovery trip was different. Across swollen rivers, through deserts, in deep forests, on high mountains. Sometimes not one of us understood the local language, though nearly all of us belonged to the same country. But every time we arrived in the neighbourhood of the area, excited crowds would gather around, talking of strange things coming down the evening sky and hundreds might accompany us to the house of the headman or the school teacher who was usually elected as the custodian of the precious, celestial object. There were long sessions to explain the objectives of our crazy enterprise, feasting and invitations to stay and enjoy their hospitality. Once in the middle of a forest, we were invited to join a panther hunt. With two muzzle loading guns which "took only two minutes to recharge", a scythe and a rifle, we waited at the end of a narrow pathway in the jungle for the panther to come into view out of the bushes only a few yards away while it ran from the racket made by a hundred "beaters", supposedly converging on it from the sides. It was lucky for us that the panther decided that day to feast on a goat in a neighbouring village. We do not know what would have happened if he had not changed his routine for the day, because when the beaters were closing in on us, two of us were still arguing whether the safety catch on the muzzle loading gun (which we had learned to fire the same morning) should be removed after we see the panther or before!

Perhaps one story about the recovery of a very valuable emulsion block (200 emulsion pellicles, the largest such detector of that period anywhere) would be worth sharing. It was the winter of 1953-54, when we launched the balloon from an open yard near a textile mill in Delhi. The flight went up at six in the morning and one of us rushed to the airport where an airforce DC 3 was standing by for take off. We carried a V.H.F. set to talk to several theodolite teams which had started for their down range stations a day or two earlier. We also carried an air force parachutist who was supposed to jump after the payload when released from the balloon by a clock mechanism after four hours of flight at altitude. The airplane had a transparent dome and Y.P. stood on a platform and shouted instructions to the pilot to keep circling under the balloon. We soon learned that some theodolite teams had been unable to cross a river during the night and were left way behind. But everything was fine because the balloon was high and we were drifting right under it. A few minutes before the expected cut-off, the parachutist was briefed and ready to jump. The zero hour arrived and nothing happened. We kept under the balloon for two hours longer and still nothing happened. Finally the pilot informed us that much as he would like to continue the chase, we were approaching the alien airspace of Nepal and, in any case, he had barely enough fuel left to make it back to Delhi. We wistfully said good-bye to the balloon, majestically floating at 80000 ft. over the snow covered mountains and turned back. After that we moved to Agra, and then further south to Ahmedabad and made some good flights from an open space next to the Calico Mills. But the story of our Nepal flights goes a bit further.

A month after we had launched from Delhi, we received a letter from Nepal on selfmade ricepaper sent by messenger to the Indian border and on by mail. Someone in the village in the neighbourhood of Pokhra had found our payload and asked us to come and take it. He also gave instructions about getting there: two days bus trip from the Indian border, a day or two on ponies and just three days walk. We set out to Delhi to get the Nepalese visas. We were informed that we could get the visas to Kathmandu and would have to travel to the interior only out of Kathmandu, after getting proper permission from the foreign ministry. So we flew to Kathmandu. We arrived at three o'clock in the afternoon to be informed by the Indian mission that there was once a week flight to Pokhra and if we wanted to make the next day's plane we must rush and get the necessary permission before the foreign office closed at 4 P.M. that afternoon. So we hurried to the Secretariat in a taxi, with a loud horn, hardly any brakes and a dauntless driver, and requested to see the Foreign Minister of Nepal. The Foreign Minister being in a cabinet meeting, we agreed to meet the Deputy Foreign Minister. In spite of the letter we carried, the gentleman concerned was properly sceptical about our story of a balloon, launched from Delhi, drifting into Nepal to drop an instrument in the snows. He would have to get the approval of the Foreign Minister himself. So call him, we said. And he did. While he talked in Nepalese over a hand cranked telephone, he became more and more excited. When he finished we learnt that he had disturbed the cabinet meeting, and the Prime Minister, instead of getting upset, suddenly became interested and remarked that the Commander-in-Chief had just come back from the neighbourhood of Pokhra after investigating a widespread scare caused by objects floating down the evening skies. Indeed he had found something, which looked somewhat like a bomb, and that this object was lying downstairs in the verandah and we could see whether this was what we had come to look for. And so we trooped down in the company of several ministers, and the Commander-in-Chief himself and sure enough there was our payload, intact and in good shape, soldiers standing guard around it. Needless to say that when the Commander-in-Chief invited us to stay as State Guests for a couple of days or longer, we readily agreed. This stack later turned out to be an extremely valuable payload for much of our work during that period. In particular, because of availability of very long pathlengths in emulsion, the difference between interaction behaviour of K^+ and K^- mesons was clearly brought out.

We have wondered, along with many cosmic-ray physicists of that era, "what was so special about the time" and why we still consider it a privilege to have been a part of that largely forgotten scientific period. While being conscious of the fact that years of one's youth always acquire added lustre with passage of time, one cannot help feeling that there was a special spirit pervading this effort. There was an atmosphere of total involvement, not only intellectual but also emotional and physical. Perhaps the very nature of this work demanded this total immersion, very often including members of the family and friends. Practically all segments of the enterprise had to be put together from raw ingredients and the overall effort itself had value — made significant through creative labour, intellectual and physical — and not only through the ultimate results published in scientific journals. An almost total absence of "black-boxes" allowed vast areas for playful improvisation. And the subject itself, connected at once with the cosmic and the minute and intimate aspects of the universe, compelled urgent attention. The field was ripe, effort came easily and discoveries were both anticipated and surprising. The young amongst us grew minute by minute, in an atmosphere that was bouyant, charged with excitement and full of learning.

REFERENCES

1. Lal, D. and Peters, B.: 1962, *Elementary Particle and Cosmic Ray Physics* Vol. VI, North Holland Publ. Co.; 1955, *Proc. Ind. Ac. Sci.* **41**, 67. Lal. D. and Peters, B., Encyclopedia of Physics Vol. XLVI/2, Springer Verlag: 1962.
2. Lal, D., Pal, Y. and Peters, B.: 1953, *Phys. Rev.* **92**, 438.
3. Lal, D., Pal, Y., Peters, B. and Kaplon, M.F.: 1952a, *Phys. Rev.* **86**, 569.
4. Lal, D., Pal, Y., Peters, B. and Swami, M.E.: 1952b, *Proc. Indian Ac. Sci.* **36**, 75; 1952c, *Phys. Rev.* **87**, 545.

PART 7   FURTHER IN PARTICLE PHYSICS

THEORETICAL STUDY OF COSMIC RAYS IN JAPAN

Mituo TAKETANI*

No one would dispute the contributions of Japanese theoretical achievements to the development of cosmic ray research. However, the understanding of Japanese contributions is mostly limited to a general feeling, and many of the important details are overlooked and/or incorrectly advocated. This is due partly to the language barrier for foreigners, but also due greatly to talks and publications based on incorrect memory and unconscious distortions possibly affected by complex psychology. In order not to repeat such a mistake, I have written the present paper referring to my article "Formation of the Elementary Particle Theory Group" (Yukawa et al., 1951, 1965) written in 1950 when my memory was fresh and I was able to interview those who also had fresh memories.

In writing that article I referred to as many as possible of the documents available; Section 1 of the present paper is mainly based on such documents which told me of what I did not experience by myself. Other parts, Section 2 and Section 3, may be regarded as my personal recollections. In order to make these parts as objective as possible, I have asked Satio Hayakawa to review my article in comparison with his work (Hayakawa, 1972, 1981) on similar subjects.** I therefore hope that the present paper gives an objective story, even though it is primarily based on my own views.

1. Activities before Yukawa's Meson Theory

One might be under the impression that the discovery of the meson theory by Hideki Yukawa was a sudden blooming in a desert. This is however misleading; a highly developed culture does never bloom without matured basis. Theoretical physics in Japan had been matured during the first quater of the twentieth century. I remark two towering achievements, the Saturn-like atomic model by Hantaro Nagaoka in 1903 and the quantization condition in the generalized coordinates by Jun Ishihara in 1915. Following these two great men, Yoshio Nishina who stayed in Copenhagen contributed to the relativistic theory of Compton scattering, well-known as the Klein-Nishina formula (Klein and Nishina, 1929; Nishina, 1929).

The theory was worked out immediately after Dirac's equation of the electron was published. The Klein-Nishina formula was the first to demonstrate the advantage of Dirac's equation over the non-relativistic equation in the relativistic energy region. This encouraged Japanese physicists to devote their energy at the very frontier of modern physics. After coming back to Japan, Nishina opened a laboratory at the Institute of Physico-Chemical Research (abreviated in Japanese as Riken) and stimulated Sin-itiro Tomonaga and Hideki Yukawa by his lectures at Kyoto University.

* 1-16-26 Shakujiidai, Nerima-ku, Tokyo, Japan.

** I express my thanks to Professor S. Hayakawa for his help in preparing this paper.

Y. Sekido and H. Elliot (eds.), Early History of Cosmic Ray Studies, 285–294.

© 1985 by D. Reidel Publishing Company.

Tomonaga joined Nishina's laboratory in 1932. This was the time when the neutron was discovered. Because of its high penetrability, Nishina wondered if cosmic rays consisted of neutrons. Then they started calculations on the electromagnetic interactions of neutrons with electrons, the result being given by Y. Nishina, Kwai Umeda and S. Tomonaga at the Semi-Annual Meeting of Riken in November 1932.

In 1933 the positron was discovered. Nishina and Tomonaga (1933) immediately pointed out that a pair of electron and positron are created by a gamma-ray in the nuclear Coulomb field. They extended this idea, in collaboration with Shoichi Sakata, to perform a quantitative calculation of the cross section (Nishina *et al.*, 1934). The result was read at the Semi-Annual Meeting of Riken in May 1934. This may be the first paper of Japanese elementary particle theory group, which received international recognition.

2. Interpretation of Cosmic Ray Phenomena in Terms of Meson Theory

Immediately after I graduated from Kyoto University in 1934, Fermi's theory of β-decays came to our notice. Tamm and Iwanenko showed that the exchange of electron and neutrino between nucleons gave too weak a nuclear force. Having been stimulated by these two works, Yukawa (1935) introduced a field which mediates the nuclear force and takes part in nuclear β-decays. The quantization of this field yields a particle of about 200 electron masses.

Unfortunately, however, Yukawa's theory was hardly admitted by the international community of physics. One reason for this is regarded as due to the lack of circulation of Japanese journals. The main reason may be the difficulty in admitting the theory based on an as yet unobserved particle. Even inside Japan very few physicists besides Nishina and Tomonaga understood Yukawa's theory. However, Sakata and I highly appreciated this theory and were prepared to collaborate with Yukawa, since the theory could be considered to be worked out with the substantialistic method.

Fig. 1. S. Tomonaga at the party of congratulating on his Nobel Prize (1965).

Fig. 2. H. Yukawa (left) receiving Nobel Prize from Prince of Sweden (1949). Courtesy of Prof. Yukawa.

In 1936 Yukawa and Sakata made efforts to obtain relativistic equations of the U-field, the field associated with Yukawa's particle, extending the general spinor equation of Dirac. I was interested in cosmic rays, particularly in showers.

In those days the shower was considered as being due to the creation of many electrons and positrons by a single collision. However, the quantum theory of radiation predicted the probability of the n-th order process to be proportional to α^n, where $\alpha = 1/137$. Moreover, the theory predicted a rather low penetrability of electrons and photons which were regarded as the main constituents of cosmic rays. There existed a general feeling that quantum electrodynamics might have a limit of validity and most cosmic ray phenomena were located beyond this limit.

Such a feeling was further emphasized by Shankland's experiment (Shankland, 1936). In this experiment he measured the coincidence of scattered photons and recoil electrons in Compton scattering of gamma-rays from RaC. As he failed to obtain significant coincidence events, he and Dirac (Dirac, 1936) were inclined to give up the laws of conservation of energy and momentum and to revive the statistical conservation advocated by Bohr, Kramers and Slater (1924). I was unable to accept such an irrational interpretation and tried to interpret the experimental result in terms of multiple processes (Taketani, 1936). In fact, the angular dependence of the coincidence rate was found to agree with that expected if three particles are produced, for example, by double Compton scattering. However, the absolute value of the probability is much too small becasue of the small factor, $\alpha = 1/137$. I therefore pointed out that the high probability of multiple processes would be due to the same origin as the shower, so that the experimental result could be understood without giving up the conservation laws, if we understood the shower.

Shortly thereafter, the shower was explained by the cascade theory as the successive multiplication of electrons and photons through pair creation and bremsstrahlung (Carlson and Oppenheimer, 1936; Bhabha and Heither, 1936). This demonstrated the fact that quantum electrodynamics is still valid at very high energies. However, the problem of the high penetrability of cosmic rays remained to be solved.

In the spring of 1937 Niels Bohr visited Japan and delivered a series of lectures. He stimulated Japanese physicists in many respects, but did not show any interest in Yukawa's meson, because this was regarded as imaginary. Ironically enough, however, Neddermeyer and Anderson (1937) announced the discovery of the meson on his way back home. They had obtained a cloud chamber track of the meson even earlier (Anderson and Neddermeyer, 1936), but no one but Yukawa took this result seriously. Independently of them, Street and Stevenson (1937) obtained a track of the meson. At the Nishina laboratory, Takeuchi and others examined cloud chamber pictures they had taken and found a good track of the meson (Nishina et al., 1937).

The existence of mesons made clear the behaviour of cosmic rays. The soft component consists of electrons and photons, while the hard component of high penetrability is attributed to the meson. Difficulties in understanding cosmic ray phenomena were found irrelevant to a modification of quantum electrodynamics, though many eminent physicists had suspected to be so. The identification of this new particle with that predicted by Yukawa was remarked by Yukawa (1937) himself and also by Oppenheimer and Serber (1937). We were very excited by the experimental discovery and convinced ourselves to go forward immediately, as we had been prepared for further development of Yukawa's theory. Then Yukawa, Sakata and I started to work together to formulate the scalar and vector theories of the meson and published the second (Yukawa and Sakata, 1937) and the third papers (Yukawa et al., 1938a).

I shall not describe the whole story here since this can be found elsewhere, for example, in Hayakawa's paper (1972, 1981) but only mention an episode on the spontaneous decay of the meson.

When we had almost completed the vector theory, several short papers appeared in a Nature issue published in January 1938. They were similar to ours, except for the spontaneous decay pointed out by Bhabha (1938). One day when I walked into Yukawa's office, I found Yukawa and Sakata discussing the paper by Bhabha. They thought that the β-decay of the meson would be possible only as a virtual process, since otherwise the energy-momentum conservation would be violated. Then I pointed out that the spontaneous decay could occur if an electron and a neutrino have the same momentum of different signs. However, we noted the importance of the absolute value of its lifetime, since mesons would not survive to the ground level if its lifetime were too short. Sakata and I estimated the lifetime on a train going back together to Sakata's home and found it long enough to explain the hard component of cosmic rays. The third paper of meson theory was thus written, and its manuscript was submitted for publication on 15 March 1938 (Yukawa *et al.*, 1938a).

In April 1938 I was appointed to be an unpaid assistant of Osaka University. Yukawa's group became still more active, as Minoru Kobayasi joined us after spending a few years with Tomonaga. Kobayasi took part in problems related to cosmic rays and calculated the production and absorption of cosmic-ray mesons in collaboration with a new postgraduate, Daisuke Okayama. Okayama was an ingeneous student and had read a paper on cosmic ray showers at a meeting of the Physico-Mathematical Society of Japan.

Meanwhile Tomonaga then at Leipzing sent us a letter, telling us that Heisenberg appreciated the third paper and that Euler obtained the lifetime of two microseconds to explain cosmic ray phenomena. We were very much pleased by this information, since the lifetime he obtained was in approximate agreement with ours. Shortly after, however, we found the theoretical value to be smaller than the cosmic-ray value by an order of magnitude, adopting the correct value of a β-decay constant. Discrepancy was still greater for the Konopinski-Uhlenbeck interaction which was considered to be better for explaining nuclear β-decays.

After finishing the fourth paper (Yukawa *et al.*, 1938b), I was working to solve this lifetime problem with Sakata. The difficulty seemed to disappear if two neutrinos, a fermion and boson, were emitted in the β-decay. However, my scientific activity was forced to be interrupted on 13 September 1938, since the thought police arrested me for the reason that I joined a group publishing a liberal journal called "World Culture".

Since February 1939, I was allowed to spend some hours at a police office to study physics. Sakata sent me copies of new papers, and I carried them into jail concealed under a blanket, since it was forbidden to take anything into jail. Among them I was particularly interested in a long paper by Euler and Heisenberg (1938). This paper gave me the impression that all about cosmic rays were solved.

Through reading this paper, I again became interested in the lifetime problem and sent a letter to Sakata, in which I suggested separating the meson decay from the nuclear β-decay. This model was worked out successfully by Sakata (1940, 1941), that is, the meson dacay should take place through a virtual pair of nucleons, the latter decaying into an electron and a neutrino by the Fermi interaction. I also remarked that a better value of the meson lifetime could be obtained with a proper choice of the coupling constant by reference to the latest information on the nuclear force, as calculated at the police office. About a week later, I received a copy of a paper by Yukawa and Sakata (1939a,b) who had worked along the same line.

I was released by the police on 22 April 1939. Yukawa arranged an assistantship for me under Seishi Kikuchi to work in experimental nuclear physics. Kikuchi was so sympathetic that he gave me financial support. All the members, in particular Yuzuru Watase and Seitaro Yamaguchi, were sympathetic, and I enjoyed working with them. In addition to experimental work, I kept interest in theory and did some theoretical calculations. Because of overwork, I became sick in the summer of 1940 and had to stay in bed until early 1941.

3. Breaking Through Difficulties in Meson Theory

As the meson was admitted by a wider and wider circle of physicists as well as by the

general society, difficulties in incorporating the meson theory with cosmic-ray phenomena became more and more disclosed. In order to reveal how the meson should
behave in cosmic rays, various processes which were supposed to take place in
cosmic-ray phenomena were calculated.

In June 1940 Sakata and Yasutaka Tanikawa reported at the Semi-Annual Meeting
of Riken that the neutral meson should decay into two or three γ rays with very short
lifetimes, depending on spin 0 or spin 1 (Sakata and Tanikawa, 1940). This provided a
basis of my later contribution to the origin of the soft component.

Theoretical studies of cosmic rays were carried out mainly by Kobayasi and his
associates in Osaka and Kyoto as well as by Tamaki and his associates at Riken in
Tokyo. A problem of the greatest concern was the high penetrability of the hard
component. Although the meson was found penetrable because of its small bremsstrahlung cross section, the penetrability would not seem high enough to account for the
behaviour of high energy cosmic rays.

Since the meson should interact strongly with the nucleon, the mean free path for
nuclear interactions would be about 100 g cm^{-2} if the cross section were energy
independent. This mean free path was known to be too short. Still worse results were
expected for interactions with derivatives, since the cross section would increase with
energy. In order to get rid of this difficulty, the damping theory was proposed to
suppress the divergence of the cross section at high energies.

Electromagnetic interactions of the vector meson were also subject to similar
difficulties because of an interaction with a derivative. Although Tomonaga (1940)
showed unappreciable spin dependences of the meson-electron collision cross sections
with a semi-classical method, bremsstrahlung appeared to give the energy loss rate
increasing with energy. As it was considered difficult to obtain a particle picture of the
vector meson, Sakata and I (Sakata and Taketani, 1940; Taketani and Sakata, 1940)
developed a theory of representing the vector field by 6 × 6 matrices, extending
Kemmer's method. Further detail was worked out by Tamaki (1942).

In the summer of 1940 when we were deeply concerned with the vector meson,
Shuichi Kusaka who had been a student of Oppenheimer, visited Japan. He was already

Fig. 3. M. Taketani (left) and S. Sakata (1941). Photo by Ken Domon.

known to us through his work on the quadrupole moment of deuteron that the vector meson gave the wrong sign. He was also working on the bremsstrahlung of vector meson with Kemmer's method. We enjoyed fruitful discussions with him. He then completed his work in collaboration with Christy (Christy and Kusaka, 1941), showing that the vector meson would give too high a frequency of bursts (Schein and Gill, 1939).

After returning home, Kusaka tried to arrange a fellowship for me to work with Oppenheimer's group, but this was unsuccessful because of political tension between Japan and U.S.A. Instead, I was awarded a fellowship by Iwanami Shoten, so that I was supported for the first time in my life for research in theoretical physics. Thanks to the fellowship I was able to continue research with Tomonaga at Nishina's laboratory since April 1941. There was a bright, young physicist, Tatuoki Miyazima (1941), who tried to introduce the spin 1/2 meson which was found by Christy and Kusaka to give the burst frequency in agreement with experiment. As for nuclear forces, he adopted the pair theory similar to that proposed by Marshak (1940).

In those days it became more promising that the primary component of cosmic rays consists mainly of protons (Schein et al., 1941), but not of positrons, as Euler and Heisenberg (1938) assumed. This should result in that mesons are produced by nuclear interactions with a large cross section, whereas the mesons once produced interact only weakly with nuclei. This forced us to rewrite the picture advocated by Euler and Heisenberg. This had already been noted by Yukawa after his trans-world trip in 1939, and effort had been made by Yukawa, Sakata and Tanikawa, as given at the semi-Annual Meeting of Riken in November 1939. I was concerned with the question as to how the interaction cross sections of the meson could be reduced.

As mentioned before, the perturbation theory gives cross sections increasing with energy. While I was sick, I thought that this might be analogous to the resonance radiation in which the higher order processes otherwise diverging could converge by taking into account the term of radiation damping. Tomonaga was also interested in this problem and tried to expand the interaction term with respect to the inverse power of coupling constant. As this method had been worked out by Wentzel (1940), he invented an intermediate coupling theory (Tomonaga, 1941). This resulted in the scattering cross section similar to resonance scattering, containing both the damping and inertia terms. This succeeded in reducing the cross section to some extent.

As soon as I moved to Tokyo, I felt that a more frequent exchange of ideas between the east and the west of Japan would be more fruitful. I proposed Tomonaga if we could have meetings to discuss problems of common interest in an informal atomosphere. Tomonaga seconded my proposal and convened the so-called meson meetings twice a year. At these meetings reports on ideas in the premature stage, on research programs, and on preliminary theoretical and experimental results were encouraged, in contrast to reports given at formal meetings. At the first meeting Tomonaga's theory of intermediate coupling was one of the main topics. Satoshi Watanabe, who proposed a five-dimensional field theory, and I were the talkative members arguing against each other. It was so noisy that discussions were known as the performance of Satoshi Watanabe and his music group.

In those days I was mostly concerned with the experiment by Schein, Jesse and Wollan (1941). According to them, the cosmic ray flux near the top of the atmosphere did not change as the thickness of lead to penetrate increased from 4 cm to 18 cm. Neither were showers produced by the lead. If cosmic rays impinging on their apparatus are protons, they should produce positive, negative and neutral mesons. Among them neutral mesons would have to decay into γ-rays, as predicted by Sakata and Tanikawa (1940), and in turn to produce showers. At first I thought this as evidence against the neutral meson and was led to favor the pair theory which did not require the neutral meson to explain the nuclear force between like nucleons. In a paper given at the Semi-Annual Meeting of Riken in June 1943, however, I referred to this experiment to determine the lower limit of the lifetime. Namely, the lifetime of the neutral meson is long compared with the flight time over the experimental apparatus but short enough to dacay in the upper atmosphere. People at Nishina's laboratory proposed fluffy mesons which were produced with low energies and contributed to the soft component.

My prediction was compared with the careful analysis of cosmic ray phenomena by Tamaki. He showed that the observed intensity-altitude curve of the soft component is higher by a factor of about two than that expected from the decay of the charged mesons into an electron and a neutrino. He also estimated the neutrino loss by comparison between the incident energy and the energy lost by ionization in the atmosphere. The contribution of neutral mesons was found just enough to explain the soft component, as I later published in an English version (Taketani, 1948), whereas this resulted in too small a value of neutrino loss. I therefore suspected that some part of the neutral meson would decay into a pair of neutrinos. Nowadays, however, the theoretical neutrino loss is found to be just as large as that observed, according to the two-meson theory.

At the same meeting a remarkable paper was given under the co-authorship of Sakata, Tanikawa, Seitaro Nakamura and Takesi Inoue (The English version of their work was published by Sakata and Inoue, 1946; Tanikawa, 1947). In this paper they assumed that the meson responsible for nuclear forces was different from the meson observed in cosmic rays, thus the theory being called the two-meson theory. The theory attempted to fill the gaps for the nuclear scattering cross section and the decay probability by a substantialistic method. Numerically these values observed in cosmic rays were found smaller by a factor of about one hundred than those predicted by the Yukawa theory. Although I was in favor of their general attitude, I felt that they were going too far, since the gap of 1/100 appeared too small to claim the existence of a new substance and this might be got rid of by damping and inertia, as in Tomonaga's theory. Another difficulty was found in a very short lifetime of the nuclear force meson (10^{-21} s) they adopted, so that neutrinos produced in this rapid decay might give observable effects.

According to later developments, both the original proposal by Sakata et al. and my criticism are correct. The points I criticized have disappeared in the following way.

Conversi, Pancini and Piccioni (1947) performed an experiment to observe the decay of mesons stopping in solid materials. According to Tomonaga and Araki (1940), positive mesons should decay, while negative ones would have to be captured by nuclei. They separated positive and negative mesons by a magnetic lens and observed decay electrons in the respective cases. The result agreed with the theoretical prediction for heavy stopping materials such as iron, but negative mesons were also found to produce decay electrons in carbon with an appreciable probability. This implies that the mesons they observed interact very weakly with nuclei, weaker by a factor of 10^{-12} than predicted for the nuclear force meson. This would readily lead one to assume that the cosmic-ray meson and the nuclear force meson are different. In fact, Marshak and Bethe (1947) proposed a two-meson theory along this line.

Towards the end of 1947 I saw a paper in Nature by the Bristol group (Lattes *et al.,* 1947). They observed that a heavy meson (π) stops to emit a light meson (μ) and that a meson (σ) produced by a nuclear disintegration stops and produces another nuclear disintegration. On the other hand, a negative μ-muon on stopping did not produce a nuclear disintegration. The π- and σ-mesons were identified with the nuclear force mesons, while the μ meson was regarded as the cosmic-ray meson. A nuclear disintegration does not occur for the pair theory which Marshak and Bethe adopted, since a substantial part of energy is carried away by a counter particle of the pair. The capture of the μ-meson does not produce a nuclear disintegration, since energy is taken away by a neutral particle, in disagreement with the model adopted by Marshak and Bethe. All the experimental information is in essential agreement with Sakata's model. In addition, the mass of π-meson heavier than we had thought called for a revision of nuclear forces. I discussed these problems in collaboration with Nakamura, Ken-ichi Ono, and Muneo Sasaki (Taketani *et al.,* 1949).

Turning back to 1943, I report on a symposium of the Meson Club in September 1943. As we were afraid of the continuation of research because of the War, we planned to organize a symposium to review the status of meson theory. Tomonaga proposed to invite review papers according to my three-stage strategy. In the phenomenological stage, Tamaki reviewed experimental results of cosmic rays and their phenomenological analysis. In the substantialistic stage, Sakata and Gentaro Araki discussed models, and I discussed the necessity of introducing the neutral meson with regard to the analysis of experimental results. Finally in the essential stage, Tomonaga talked about methematical treatments of meson theory. Their manuscripts were distributed among participants who were expected to present further contributions. Indeed, several more papers were presented at the symposium. This symposium was said to be epoch-making in the sense that all the important results by that time were summarized and that all papers presented were appreciated after the War. The proceedings were printed immediately thereafter and published in 1949 by Iwanami Shoten under the title "Research of Elementary Particle Theory I."

In conclusion, I would like to emphasize that the development of cosmic ray theory has been made by constant discussions with experimental physicists, particularly at colloquia held once a week at Nishina's laboratory. Such a close collaboration between theory and experiment has been rarely seen in other fields of science in Japan. This tradition still survives in the cosmic ray group in Japan, but I am afraid it is getting less and less active.

REFERENCES

Anderson, C.D. and Neddermeyer, S.H.: 1936, *Phys. Rev.* **50**, 263.
Bhabha, H.J.: 1938, Nature **141**, 117.
Bhabha, H.J. and Heitler, W.: 1936, *Proc. Roy. Soc.* **159**, 432.
Bohr, N., Kramers, H.A. and Slater, J.S.: 1924, *Phil. Mag.* **47**, 785.
Carlson, J.F. and Oppenheimer, J.R.: 1936, *Phys. Rev.* **51**, 220.

Christy, R.F. and Kusaka, S.: 1941, *Phys. Rev.* 59, 405, 414.

Conversi, M., Pancini, E. and Piccioni, O.: 1947, *Phys. Rev.* 71, 209.

Dirac, P.A.M.: 1936, *Nature* 137, 296.

Euler, H. and Heisenberg, W.: 1938, *Erg. Exakt. Naturwiss.* 17, 1.

Hayakawa, S.: 1972, *Cosmic Rays* (Chikuma Shobo, Tokyo); 1981, in *Proc. International Symposium on the History of Particle Physics,* Cambridge University Press, Cambridge.

Kemmer, N.: 1939, *Proc. Roy. Soc.* 173, 91.

Klein, O. and Nishina, Y.: 1929, *Z. Phys.* 52, 853.

Lattes, C.G.M., Muirhead, H., Occhialini, G.P.S. and Powell, C.F.: 1947, *Nature* 159, 659.

Marshak, R.E.: 1940, *Phys. Rev.* 57, 1101.

Marshak, R.E. and Bethe, H.A.: 1947, *Phys. Rev.* 72, 506.

Miyazima, T.: 1941, *Sci. Pap. Inst. Phys. Chem. Res.* 39, 28.

Neddermeyer, S.H. and Anderson, C.D.: 1937, *Phys. Rev.* 51, 884.

Nishina, Y.: 1929, *Z. Phys.* 52, 869.

Nishina, Y., Takeuchi, M. and Ichimiya, T.: 1937, *Phys. Rev.* 52, 1198.

Nishina, Y. and Tomonaga, S.: 1933, *Proc. Phys. Math. Soc. Japan* 15, 248.

Nishina, Y., Tomonaga, S. and Sakata, S.: 1934, *Sci. Pap. Inst. Phys. Chem. Res.* 24, Supp. 17, 1.

Oppenheimer, J.R. and Serber, R.: 1937, *Phys. Rev.* 51, 113.

Sakata, S.: 1940, *Phys. Rev.* 58, 576; 1941, *Proc. Phys. Math. Soc. Japan* 23, 283, 291.

Sakata, S. and Inoue, T.: 1946, *Prog. Theor. Phys.* 1, 143.

Sakata, S. and Taketani, M.: 1940, *Proc. Phys. Math. Soc. Japan* 22, 757.

Sakata, S. and Tanikawa, Y.: 1940, *Phys. Rev.* 57, 548.

Schein, M. and Gill, P.S.: 1939, *Rev. Mod. Phys.* 11, 267.

Schein, M., Jesse, W.P. and Wollan, E.O.: 1941, *Phys. Rev.* 59, 615.

Shankland, R.S.: 1936, *Phys. Rev.* 49, 8.

Street, J.C. and Stevenson, E.C.: 1937, *Phys. Rev.* 51, 1005.

Taketani, M.: 1936, *Kagaku* 6, 234.

Taketani, M.: 1948, *Prog. Theor. Phys.* 3, 319.

Taketani, M. and Sakata, S.: 1940, *Sci. Pap. Inst. Phys. Chem. Res.* 38, 1.

Taketani, M., Nakamura, S., Ono, K. and Sasaki, M.: 1949, *Phys. Rev.* 76, 60.

Tamaki, H.: 1942, *Sci. Pap. Inst. Phys. Chem. Res.* 40, 11.

Tanikawa, Y.: 1947, *Prog. Theo. Phys.* 2, 220.

Tomonaga, S.: 1940, *Sci. Pap. Inst. Phys. Chem. Res.* 37, 399.

Tomonaga, S.: 1941, *Sci. Pap. Inst. Phys. Chem. Res.* 39, 247.

Tomonaga, S. and Araki, G.: 1940, *Phys. Rev.* 58, 90.

Wentzel, G.: 1940, *Helv. Phys. Acta* 13, 269.

Yukawa, H.: 1935, *Proc. Phys. Math. Soc. Japan* 17, 48.

Yukawa, H.: 1937, *Proc. Phys. Math. Soc. Japan.* 19, 712.

Yukawa, H. and Sakata, S.: 1937, *Proc. Phys. Math. Soc. Japan.* 19, 1085.

Yukawa, H. and Sakata, S.: 1939a, *Proc. Phys. Math. Soc. Japan* 21, 138: 1939b, *Nature* 143, 761.

Yukawa, H., Sakata, S. and Taketani, M.: 1938a, *Proc. Phys. Math. Soc. Japan* 20, 319.

Yukawa, H., Sakata, S. and Taketani, M.: 1951, *Standing in the Field of Truth* (Mainichi Press, Tokyo); 1965, reprinted in *Exploration of Elementary Particles* (Keiso Shobo, Tokyo).

Yukawa, H., Sakata, S., Kobayashi, M. and Taketani, M.: 1938b, *Proc. Phys. Math. Soc. Japan* 20, 720.

THE FIRST INTERNATIONAL COSMIC RAY CONFERENCE

Marian MIESOWICZ[*]

The first of a series of International Cosmic Ray Conferences organised by the IUPAP Cosmic Ray Commission took place in Cracow (Poland) more than 30 years ago, from October 6 to 11, 1947. Professor J. Clay (Fig. 1), the famous discoverer of the geomagnetic effects, was the chairman of the Cosmic Ray Commission and P. Auger (Fig. 2) was the secretary. One may say that the Conference took place in the historical period when, in cosmic ray physics, a new fundamental branch of science, Elementary Particle Physics in the present meaning first came into existence. The participants of that Conference will never forget the historical report of Cecil Frank Powell entitled in the Conference schedule "Evidence for the Existence of Mesons of Different Mass". It was one of the first reports on the discovery of π-mesons (Fig. 3). However, also at that Conference, Professor Leprince-Ringuet (Fig. 1) in his report "Estimations et measures des masses de mésons" presented examples of particles with masses about 1000 times that of an electron. So, K-mesons were also presented at the Cracow-Conference.

Fig. 1. Sitting; P. Fleury, M. Fórró, J. Clay, P. Blackett. Standing; first row; W. Heitler, J.A. Wheeler sec. row; J. Weyssennoff, L. Janossy, L. Leprince-Ringuet.

* Institute of Nuclear Physics, Kraków, Poland.

Y. Sekido and H. Elliot (eds.), Early History of Cosmic Ray Studies, 295–298.
© 1985 by D. Reidel Publishing Company.

Fig. 2. Prof. P. Auger, Prof. P. Fleury, Prof. Cosyns.

Fig. 3. C. F. Powell.

Fig. 4. J. Barnothy, M. Fórró.

Fig. 5. P. Blackett, J. Blaton, J. A. Wheeler, W. Heitler.

About 20 prominent physicists from Brasil, Czechoslovakia, England, France, Holland, Hungary, Ireland, Italy and USA arrived in Cracow and 16 lectures were given. Besides the physicists shown in Figs. 1–7, G. Bernardini, A. Duperier, A. Fréon, B. Gross, R. Maze and J.A. Montgomery were also contributors. Very important and perspective problems were discussed. These were concerned especially with what we now call "High Energy Interactions".

Many social events were organized, which contributed very much to the extremely friendly atmosphere of the Conference. During the excursion at a very old salt mine in Wieliczka, which was in a big cave more than 100 m underground, one of the Conference lectures was given by Professor (Mrs) Barnothy (Fig. 4) from Budapest. The Conference itself took place in the building of the old Cracow-University (Jagiellonian University is now 617 years old). This University was the host of the Conference in Cracow of which I was secretary of the local Committee. Our University is proud not only of its old age, but also because Copernicus had been one of our students. Most of the participants of the Conference did not know this. But this was found out during a sightseeing tour at the University. A ceremony honouring Copernicus was spontaneously organized by the participants of the Conference under the leadership of Wheeler (Fig. 5). The ceremony took place in front of the Copernicus monument and Prof. Wheeler gave a very beautiful speech about the philosophy of astronomy on the basis of Copernicus's fundamental works. Blackett (Fig. 5) and Auger placed a beautiful wreath of flowers on the monument.

The Cosmic Ray Commission organized some official meetings during the Conference. One of them was devoted to the discussion of names for elementary particles. The resolution and also some reminiscences on the Cracow Conference, about which I gave a talk at Plovdiv in 1977, are included in the Proceedings of the Plovdiv Conference.

Fig. 6. W. Heitler.

Fig. 7. L. Janossy.

The historical photos included in this paper were kindly taken by Prof. A. Hrynkiewicz during the time of the first cosmic ray conference.

THE EARLY HISTORY OF THE STRANGE PARTICLES

George D. ROCHESTER*

1. Introduction

The history to be related in this article is concerned with the discovery of the principal strange particles in cosmic rays starting with the work on the so-called 'penetrating showers' in 1938 and ending with the advent of the great accelerators. It is only part of the wider history of what may with some justice be termed the second vintage period of cosmic ray particle physics covering roughly the years 1947-54 when the whole field of sub-atomic particle physics was opened up by a series of remarkable discoveries at the Powell School of Cosmic Ray Physics at Bristol University and the Blackett School at Manchester University.

2. Pre-War Manchester Cosmic Ray Research

The Manchester School began with the appointment in the autumn of 1937 of P.M.S. Blackett to the Langworthy Chair of Physics and the transfer of most of the Birkbeck College cosmic ray group to Manchester in 1938. Prominent among those transferred were John G. Wilson, a cloud-chamber specialist, Lajos Jánossy, an outstanding young Hungarian physicist, and Arthur H. Chapman, Blackett's personal technical assistant. The major piece of equipment brought to Manchester was a cloud chamber and an 11-ton magnet, a magnet destined to play an important role in the discovery of the new particles. Bernard Lovell and the writer were already in post in Manchester and quickly made the transition to cosmic rays. At the time the writer had an offer of a permanent post in Molecular Spectrosocopy at another University, but an interview with Blackett and Hartree in which the case for cosmic rays was put in an unanswerable form by Blackett, settled the issue once and for all!

By the end of 1938 the main lines of research were clearly delineated. Blackett and Wilson went on with their magnet cloud chamber work, and Blackett pursued his brilliant analysis of the temperature and anomalous absorption effects. Lovell took up cloud chamber work and in collaboration with Jánossy and Wilson investigated extensive air showers. Bound and the writer built a cloud-chamber and measured the intensity of slow protons and Jánossy, who had a hand in most of the experiments in the laboratory, collaborated with Rossi and others on the non-ionising component. Jánossy also began to study with Ingleby showers of penetrating particles (pp).

The laboratory was a stimulating place, a consequence of Blackett's inspiring leadership and dynamism, and the large number of distinguished visitors he attracted, among whom may be mentioned Auger, Bhabha, Carmichael, Cosyns, Dymond, Occhialini, Duperier, Heisenberg, Heitler, Rossi, Wataghin and Williams. The shadow of war was slowly spreading across Europe, however, and in the summer of 1939 most

* University of Durham (emeritus), Durham, U. K.

299

Y. Sekido and H. Elliot (eds.), Early History of Cosmic Ray Studies, 299–321.
© *1985 by D. Reidel Publishing Company.*

U.K. members of the Blackett School were sent to Scarborough for work on a secret Radar Station at Staxton Wold. At that time the Station was manned by R.A.F. personnel and it was the intention to replace them by the physicists. After six weeks it was realized that the Station did not require highly trained physicists who were then assigned other war duties. The writer was therefore sent back to Manchester with J.M. Nuttall to help physicists still there run the Honours School. It was this unexpected circumstance which led to the writer becoming a close colleague and collaborator of Jánossy, to the study of meson showers and ultimately the strange particles.

Many members of the group eventually returned to Manchester but Peter Ingleby who was posted to St. Athan Aerodrome to work with Lovell was tragically killed in the crash of a Hudson aircraft early in 1940.

3. Penetrating Showers

The war work at Manchester additional to the teaching of honours students and cadets consisted in running a University Fire Brigade Unit and helping with Civil Defence. This involved long periods of duty at the University and because of the infrequency of air-raids much free time during daytime. Because of this, and as an aid to the honours teaching, Blackett was asked if he would permit research to continue on a modest scale. He agreed, stipulating that only the resources available in the Laboratory be used. The Jánossy-Ingleby experiment on penetrating showers (ps) met these conditions and was, in fact, in working condition. It also seemed the right experiment in principle for it offered the possibility of studying interactions in which cosmic ray mesons were created and thus of throwing light on one of the key problems of cosmic rays. An alternative approach in the same period was being made by Schein and his co-workers in Chicago who were sending complex counter arrangements to great altitudes in balloons by which the interactions of the primary cosmic rays could be studied directly.

Given that the Jánossy-Ingleby interpretation of ps was correct the next questions to be answered were the following:

1) Could it be shown by a cloud chamber experiment that ps contain charged pp?
2) What was the nature of the radiation which produced ps?

To answer the first question a deep cloud chamber controlled by a counter system which would select ps was built by Jánossy, McCusker[1] and the writer. The counter selection was sixfold and the total thickness of lead absorber was 30 cm. A plate of lead 2.3 cm thick was placed across the chamber. The lighting system was by incandescent lamps and the chamber was slow but nonetheless some quite good pictures were obtained, one of which was published by Jánossy et al. (1941) which showed three pp apparently coming from a common origin. This photograph was not unique but was typical of many published in the years 1940-42 in which cloud chambers were triggered by counter-sets sensitive to ps or were exposed at high altitudes to cosmic ray primaries which induced high-energy interactions in metal plates. Notable examples were published by Powell (1940, 1941), Herzog (1941), Wollan (1941) and Bostick (1942). Thus the answer to the first question was in the

---

[1] McCusker worked in Civil Defence on a large commercial building in Manchester at nights and with Jánossy and the writer during day!

affirmative, that is, ps did contain pp but it was not known what these were: indeed, it was commonly assumed that they were cosmic ray mesons.

The answer to the second question was given by Schein and his co-workers in 1941 who found that the intensity of mesons increased steadily to the highest altitude reached by balloons and did not, as had been found previously by the same workers, go through a maximum. Taken with the discovery of the E-W effect by Johnson this showed that pp could not be produced by electrons and photons but were probably produced by protons. Indeed, Rossi *et al.* (1940) had already shown that not more than 1% of the sea-level meson flux could be produced by the photon component. These results, whether at altitude or sea-level, referred only to the penetrating component of cosmic rays and still left the source of the soft component uncertain. Cocconi (1941) and others showed by cogent argument that there must be a mechanism for producing electrons in addition to meson decay and knock-on processes.

Notwithstanding the definitive experiment of Schein and his colleagues, Jánossy and the writer decided to investigate experimentally the primaries producing ps, in particular production by neutral as well as charged particles. In consequence, after some preliminary experimentation, the elaborate counter-set shown in Fig. 1 was constructed. This consisted of a sevenfold ps set B_{123} C_{12} D_{12} in a mass of 15 tons of lead with a producing layer, T, of variable thickness, surrounded by an anticoincidence shield A and an absorber Σ which could be varied from 0 to 35 cm lead. Great care was taken to ensure that each of the counters was almost 100% efficient. The result,

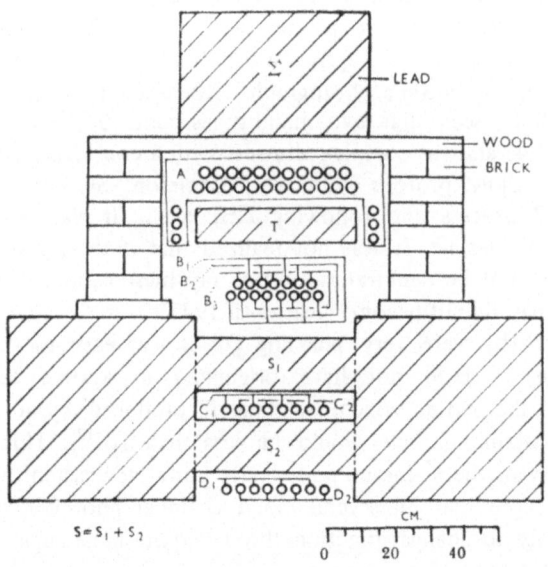

Fig. 1. A complex counter set built by Jánossy and Rochester in 1942 to investigate the production of ps. The penetrating showers were selected by a sevenfold counter set designed rigidly to exclude electron showers and events such as triple knock-ons, which might simulate ps. T was the lead producing layer. A was an anticoincidence shield to indicate whether the primaries were charged or neutral. Absorber Σ of variable thickness, gave further information about the primaries.

reported by Jánossy and Rochester (1943) showed that ps were produced almost equally by charged and neutral primaries, presumably protons and neutrons. Little theoretical advice was available and indeed our only contact with theoreticians was with Heitler in Dublin who in 1943 put forward a new version of the two meson theory of meson production of Møller and Rosenfeld. According to Hamilton *et al.* (1943) nucleons in collision produced two types of Yukawa particle, termed pseudo-scalar and vector mesons, of lifetimes 10^{-6} and 10^{-8} s. and spins 0 and 1 respectively, via the following reactions.

$$n + p \rightarrow p + p + y^{+} \text{ and } p + p \rightarrow n + p + y^{-}$$

The pseudoscalar mesons were identified arbitrarily with the ordinary penetrating cosmic ray particles, decaying to electrons; the vector mesons were assumed to decay rapidly to electrons at the top of the atmosphere and after cascade multiplication to produce the soft component. Single production of mesons was predominant so that showers of mesons could only be produced by a cascade process. Cloud chamber photographs, often showed pp starting at a point, a fact in apparent contradiction with the HHP theory. However, Jánossy (1943), a keen advocate of the theory, showed that a cascade could be built up within a heavy nucleus like lead in such a manner as to give the appearance of production at a point and many cloud chambers had producing layers of lead. Nonetheless the frequency of explosive-type interactions in thin plates in cloud chambers led to a number of American physicists, notably, Wollan and Hazen, to cast doubt on HHP. Schein *et al.* (1943) also found clear evidence of multiple production in paraffin, a light material.

The writer's main contribution at this time was to build in 1944 a deep cloud chamber counter-controlled by a ps set and to obtain a representative set of 'snap-shots' of ps. Similar photographs were published by Hazen (1944) and Powell (1946). Examples of the writer's photographs, Rochester (1946), are shown in Plate 1 from which it will be seen that two main types could be identified, (a) showers of several pp (prints 1-4) and (b) complex showers with a considerable electronic element, (prints 5 and 6). Slow protons were very common. Showers (a) were taken to represent the HHP process and following HHP again, it was assumed the pp were ordinary cosmic ray mesons. It was not realized that if the pp were indeed Yukawa particles they could not be penetrating cosmic ray mesons, i.e., they must be strongly interactive, a fact that did not become clear until 1947.

Showers of type (b) were very puzzling for it was easy to show that the large electronic component could not have originated in known decay or knock-on processes. Indeed, the writer concluded that if produced by a decay process the lifetime of the unstable particle would be less than 10^{-10}s. This was several years before the discovery of the π°-meson and clearly quite inconsistent with HHP theory.

It is worthy of note that on several cloud chamber photographs charged particles were seen apparently to change direction through quite large angles in the gas (see for example, Daudin (1944) and Jánossy *et al.* (1945)). These were thought to be decay events but because there was no magnetic field their nature could not be established. They could, in fact, have been the familiar $\mu \rightarrow e$ decays but in the light of subsequent discoveries they were more probably $\pi \rightarrow \mu$ decays.

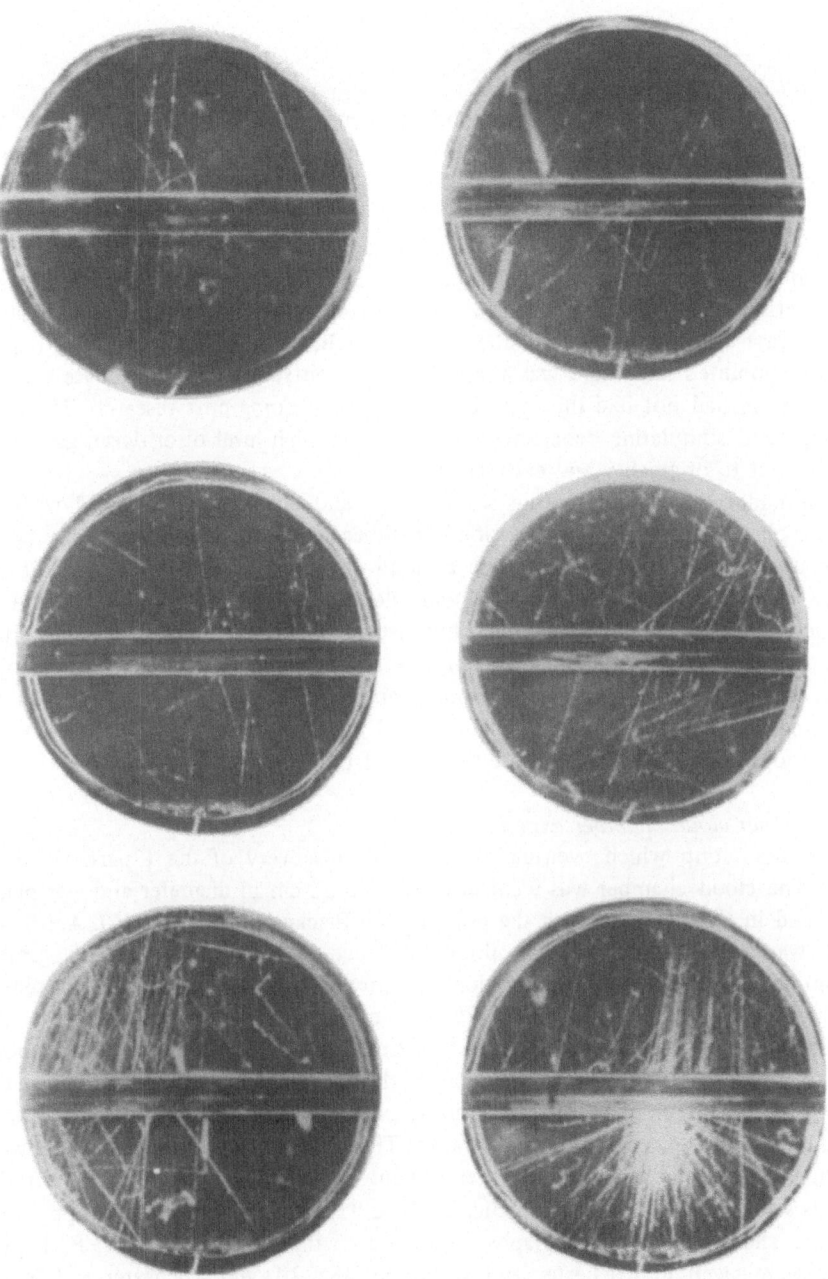

Plate 1.   Typical early photographs of penetrating showers (1944–45).

4. The V-Particles

4.1 General

With the end of World War II Blackett and other members of the pre-war Cosmic Ray School, notably Braddick, Lovell and Wilson, returned to Manchester and initiated a period of intense activity and great fruitfulness. They were joined by an eminent theoretician, Professor L. Rosenfeld, who built up a powerful research group on the theory of elementary particles. Blackett, who had lost none of his dynamism, threw himself into a wide range of researches in cosmic rays, and started up new lines in paleomagnetism and radioastronomy, the latter being under the control and direction of Lovell, first in Manchester, and then at Jodrell Bank. Moreover, new members of staff were appointed and there was a large influx of outstanding postgraduate students many of whom had not had the opportunity before of doing pure research. The result was a large and stimulating department in which the high level of criticism gave every encouragement to originality and excellence.

Blackett had followed closely the work on ps and was much impressed with the interesting phenomena shown by the cloud chamber photographs and he readily agreed that the urgent next step was to place a cloud chamber controlled by a ps set in a magnetic field so that the momenta and electric charges of particles could be determined. In 1945 a bright young physicist with considerable skill in technical matters, C.C. Butler of Reading University, was appointed to the staff and although originally asked to work with Lovell soon transferred to the writer to help in the modernizing of the Blackett magnet chamber and the building of a ps set round it. For a short period S.K. Runcorn was also a member of the group.

4.2 The magnet cloud chamber experiment

Details of the set-up which eventually led to the discovery of the V-particles are as follows: The cloud chamber was 9 cm in depth and 30 cm in diameter and was placed between and in close contact with the poles of the Blackett magnet (1936). Across the chamber was a lead plate 3 cm in thickness faced above and below with 1.8 mm chromium-plated brass plates. Illumination was provided by an entirely new system of two Siemens S.F.4. Flash Tubes[2] each 25 cm long. The pulse was delayed 0.05 s after the chamber expansion in order to let the droplets grow sufficiently to be photographed. Stereoscopic photographs were taken through one pole of the magnet at F/4 on Kodak R55 film in fields of 3500 and 7200 gauss. The clearing field was always switched off before photographing the tracks. The usual precautions for cleanliness of the chamber and temperature control were taken. The chamber was counter-controlled by two ps sets, F and P, as shown in Fig. 2: both were fivefold sets but two pp would set off the chamber. 745 photographs were taken with F and 4664 with P. Typical examples were given in Rochester and Butler (1948a, b), and Rochester and Wilson (1952).

4.3 Penetrating shower particles

Since the original purpose of the experiment was to study pp in ps the following

---

[2] F.A. Vick from his wartime experience kindly put us in touch with this development.

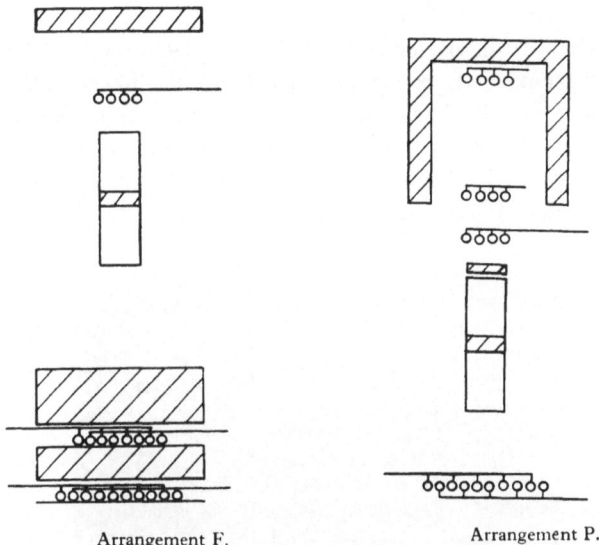

Arrangement F.   Arrangement P.

Fig. 2.   The ps sets used by Rochester and Butler in their early V-particle experiments.

measurements were made on the tracks which appeared in the showers: the momenta and sign of charge: the ionization, determined visually by a number of observers against the ionization of fast particles: the scattering in the lead plate. The main results were the following:

1) Slow particles Some 30 slow particles were plotted in an ionization v momentum diagram theoretical lines being drawn for masses of $200m$, $320\ m$ and $1837\ m$ where m is the mass of the electron. The mass $320\ m$ was chosen because it was the value of the pion mass in current use at Bristol. The particles fell into two mass groups, one roughly on the proton line and the other the $200\ m$ line. The conclusion, later known to be erroneous, was that the light particles were mainly ordinary cosmic ray mesons and hence were being created in the lead. By 1949-50 the emulsion photographs of high energy interactions and the creation of mesons at the Berkeley synchrocyclotron showed beyond doubt that the profusely created light meson was the pion. The occasional event which seemed to show the creation of the cosmic ray meson (the muon) was later found to be the rare β-decay of the pion.

The end of the ps saga was the convincing experimental proof by Piccioni (1950) that slow mesons in ps were strongly interacting mesons, (pions), and not muons.

2) Fast particles By 'fast' is meant a particle of $v\sim c$. These particles were positive and negative in charge but in the small sample measured a large fraction was positive. A striking feature was the frequent occurrence of large-angle scattering in the 3 cm lead plate. This work was extended and correlated with the Bristol emulsion work, Fowler (1950) and Camerini *et al.* (1950) in the years 1949-50, by Butler *et al.* (1950), and Barker and Butler (1951) and the fast particles were shown to be mainly pions and protons.

4.4 The first V-particles

The first V^0-particle was discovered on 15th October 1946, on the photograph of a ps taken with arrangement F. It is the inverted V on the lower R.H.S. of the cloud chamber picture in Plate 2a. Each track of the V was at minimum ionization which meant that the velocities of the secondaries were of the order of c; their momenta were approximately equal to 350 MeV/c, with an uncertainty (i.e. SD) of $\sim$ 30% for the positive particle and 50% for the negative particle. Accepting the correctness of the

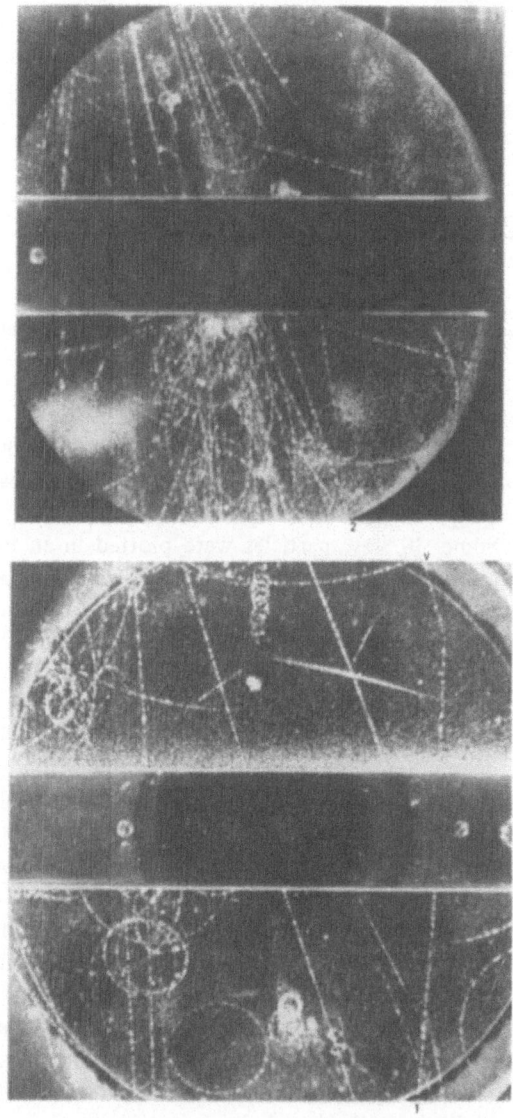

Plate 2.   The original V-particle decays (1947).
(a) V^0-decay.
(b) V^+-decay.

momenta led immediately to the conclusion that neither particle could be as heavy as a proton since the ionization would then have been five times minimum. Stereoscopic projection showed that the tracks were copunctal and lay entirely within the gas of the chamber, well away from the plate and the walls in a well-illuminated region with good condensation conditions as shown by a thin fog which pervaded the entire chamber. The data are summarised in Table 1.

Table 1

Photograph	H (Gauss)	α (deg)	Track label	p (eV/c)	Δp (eV/c)	Sign	Probable type (Modern nomenclature)
2a	3500	66.6	1	3.4×10^8	1.0×10^8	+	$K^0_{\pi 2}$
			2	3.5×10^8	1.5×10^8	−	
2b	7200	161.1	V	6×10^8	3.0×10^8	+	$K^+_{\mu 2}$
			1	7.7×10^8	1.0×10^8	+	

The fact that the two tracks were accurately copunctual showed that the V was not due to the accidental crossing of two separate tracks. All collision processes e.g. nuclear disintegrations, nuclear scattering, pair production etc. were excluded because very few similar events were seen to originate in the lead across the chamber although the plate contained at least 800 times as much matter as the gas. This general point emerged as a result of protracted discussion with Blackett and other colleagues during the winter of 1946-47 and led to the important deduction that the event did not depend on the amount of matter traversed but on the linear distance travelled, i.e., it was a decay process.

Detailed arguments supported this conclusion. Thus the elastic collision of a charged particle coming up the cloud chamber would have produced an easily recognisable recoil since the momentum transferred would be of the order of 580 MeV/c. If the recoil were a hydrogen atom the recoil would have a range of tens of metres and if the heaviest atom present in the gas, argon, 3mm (Blackett and Lees, 1932). Again the V could not be electron-positron pair production because the opening angle would have been no larger than 0.1° ($mc/p = 0.511 \times 10^6 \times 60/3.5 \times 10^8$) instead of the observed angle of 67°. Bremsstrahling loss was also excluded.

Attention was then directed to possible schemes and possible mass values. Blackett took a keen interest in this and in the winter of 1946-47 when travelling across Europe for various conferences, obtained some possible masses. He thought the most probable decay was to two identical particles and found that if the secondary particles had masses of 200 m the mass of the primary particle would be about 800 m. At the same time Butler and the writer were working at the decay schemes rather more generally using the Einstein relativistic energy formula and the usual conservation laws. For the special case of symmetrical decay into two equal particles of mass m_0 it was easily shown that

$$\frac{M}{m} = \frac{2m_0}{m} \left(1 + \frac{p_1^2 c^2}{m_0^2 c^4} \sin^2 \theta_1\right)^{\frac{1}{2}}$$

where M is the mass of the primary particle, θ_1, the angle between the track of the primary V° and particle 1, that is , half of the observed angle, and p_1, the momentum or this particle. Inserting the values of p_1, and θ_1, the incident mass was $(850 \pm 200)m$ for $m_0 = 200\ m$ and $(1100 \pm 150)m$ for $m_0 = 400\ m$. Clearly then the V^0-particle was an entirely new unstable particle in a different mass range from the lighter mesons already known. A number of the assumptions made still needed verification by experiment but none affected this general conclusion.

Although the significance of the V^0 was realized in early 1947 the results were not published immediately in the hope that other V-particles would appear. Patience was rewarded for on 23rd May, 1947, the event shown in Plate 2b appeared. This was a very convincing example because the secondary particle of the decay was a long track for which the momentum could be determined with reasonably high accuracy. The M.D.M. was 8 BeV/c on a track of length 6 cm. and a simple application of the Einstein relation using the transverse momentum p_t showed that the minimum possible mass was $(976 \pm 150)m$. Thus from Table 1 and Fig. 5.

$$\frac{M_{\min}}{m} = \frac{2p_t}{mc} = \frac{2 \times p_1 \times \sin \theta_1}{mc} = \frac{2 \times 7.7 \times 10^8 \times \sin 18.9}{0.51 \times 10^6} = 976 \pm 150.$$

On the assumption of two-body decay to particles of mass 200 m the mass of the V^+ became $(1080 \pm 100)m$. This result together with the analysis of the V^0-particle was published in Rochester and Butler (1947) where a fuller treatment of the decays was given. Claims[3] had been made previously for the existence of *charged* particles of mass about 1000 m, but the V^0-particle provided the first clear evidence of an *unstable neutral* particle of this mass or greater.

The first public announcement of the V-particles outside Manchester was made at the Institute of Advanced Studies in Dublin in December, 1947, with Mr. de Valera in the audience. It is of interest that it was also in Dublin in July 1947 that Powell and his group arrived at the essential elements of the correct form of the two meson hypothesis (Lattes *et al.* 1947). Again it was at another conference in 1947, The Shelter Island Conference in New York, in June, that Marshak and Bethe (1947) put forward the same two meson hypothesis quite independently of Bristol[4]. If to the discovery of the new unstable particles be added the highly important results on the decay and absorption of cosmic-ray mesons by Conversi, Pancini and Piccioni in 1947 this year is seen to be a high point in the history of the elementary particles.

5. The Tau Meson ($K_{\pi 3}$)

No further V-particles were discovered for some years and the next important advance was the discovery of the τ-meson by Brown *et al.* (1949) on one of the first Kodak electron-sensitive emulsions (NT4) exposed to cosmic rays. The event consisted of a long-range, heavily-ionizing particle which stopped in the emulsion producing a slow π^--meson, a fast π^+-meson and an unidentified meson. The direction of the τ-meson was known and its mass was determined by two methods, namely, g* v R (gap density v

---

[3] Notably, Leprince-Ringuet, L. and Lhéritier, M, : 1944, *Compt. Rend. Acad. Sci.*, **219**, 618.

[4] Unknown to most Western physicists the same hypothesis had been first put forward by the Japanese physicists (Sakata and Inoue, 1946).See Mashak (1980).

range) and $\bar{\alpha} \cdot v$ R (scattering v range) leading to the values $(1080 \pm 160)m$ and $(990 \pm 270)m$. The three secondaries were co-planar suggesting that there were no other products of the decay. Application of the conservation laws to find the momentum of the third particle led to a mass value from the secondaries of $(946 \pm 16)m$, close to the directly measured values. Strong arguments led to the conclusion that the event represented the decay of a heavy meson into three π-mesons. With a Q-value of 65 MeV and pion masses of 286 m the mass of the τ-meson was found to be 985 m. This was the first strange particle for which the mass had been determined with reasonably high precision.

The discovery of other τ-mesons proved to be as slow as V-particles and it was not until 1950 that further examples were found (Harding, 1950). These confirmed the decay scheme of Brown et al. and led to similar mass values.

6. V-Particles 1948-50

The two years following 1947 were tantalising and embarrassing for the Manchester group because no more V-particles were found. It was therefore decided to move the equipment to a mountain station where the rate would be much greater. For a typical mountain station at, say, 3000 m in height the rate would be expected to increase by more than an order of magnitude. At the suggestion of Occhialini, Blackett selected the Pic du Midi, a mountain 2867 m high in the French Pyrenees. Dr. Rösch, the Director of the Observatory, was exceedingly helpful and provided a building and power for the magnet. A group consisting of Barker, Chapman and Green working under the direction of Butler carried through the installation of the Blackett magnet late in 1949. The cloud chamber was put in later. The group had many adventures. Access to the mountain was difficult in winter and often involved long ski climbs on one of which Green sadly collapsed and died.

In the meantime, however, because of the publicity which the V-particles had received, developments were taking place in many parts of the world. A major event in 1949 was the receipt of a letter to Blackett from Carl Anderson, dated 28th November, containing the following exciting news:

"Rochester and Butler may be glad to hear that we have about 30 cases of forked tracks similar to those they described in their article in Nature about two years ago, and so far as we can see now their interpretation of these events as caused by new unstable particles seems to be borne out in our experiments".

The cloud chamber used in the Pasadena experiments was the freely falling one used by Leighton et al. (1949) to determine the momentum spectrum of the decay electrons from muons. A sixfold coincidence ps set was used as a trigger. Two trays of counters were above the chamber and one below and a trigger required three particles in the top tray, two in the second and one in the third. Above the top tray was a transition layer of 20 cm lead and above the second tray there was a block of 5 cm of lead. Early in 1950 Anderson sent the MS of a paper on the V-particles by Seriff et al. (1950), in which it was reported that 6 V^0 particles had been observed in Pasadena, and 24 V^0 and 4 $V^{\pm}$-particles at White Mountain 3200 m above SL. He stated that there was good evidence for two-body decays since in 12 cases the plane of the V contained the point of production. Again, although distortion prevented accurate momentum measurements above 200 MeV/c it was clear that there were secondaries in the mass range $(200 - 300)m$ and that some secondaries were of protonic mass. The mean lifetime of the V^0 was estimated at 3×10^{-10} sec. An example of an Anderson V^0-particle is given in Plate 3a.

The Pic du Midi ps set is shown in Fig. 3(1) and it will be seen that compared with the arrangements F & P the whole set was more compact and the producing layer of lead much closer to the chamber. This led to a considerable improvement in the geometrical efficiency of detection for particles of short lifetime. Two combinations of counters could trigger the chamber, namely a fourfold $A_2 B_2$ or a fivefold coincidence

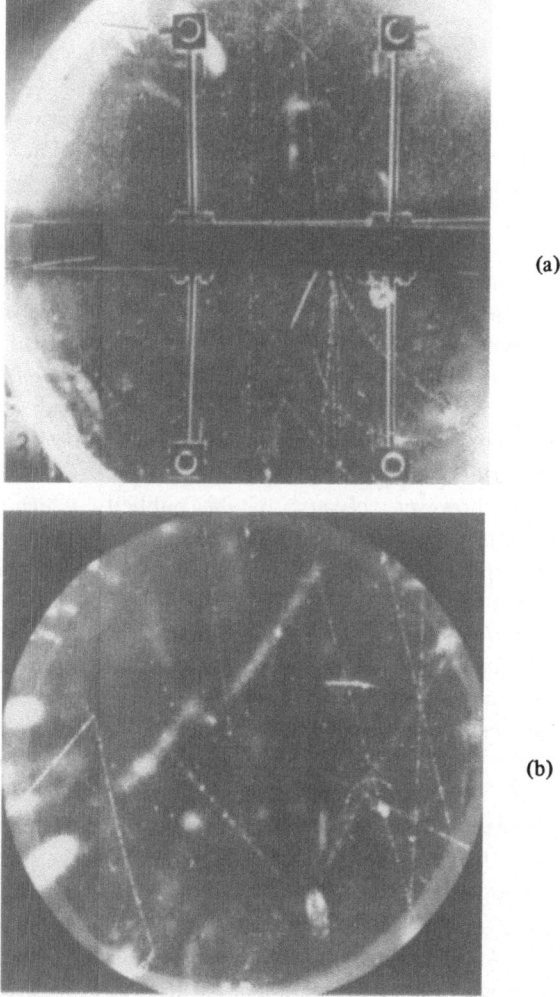

(a)

(b)

Plate 3. (a) One of the earliest photographs of a V^0-decay, taken by the Pasadena group in 1950; (b) A V_1^0-decay showing a proton secondary, taken by the Manchester group on the Pic du Midi in 1951.

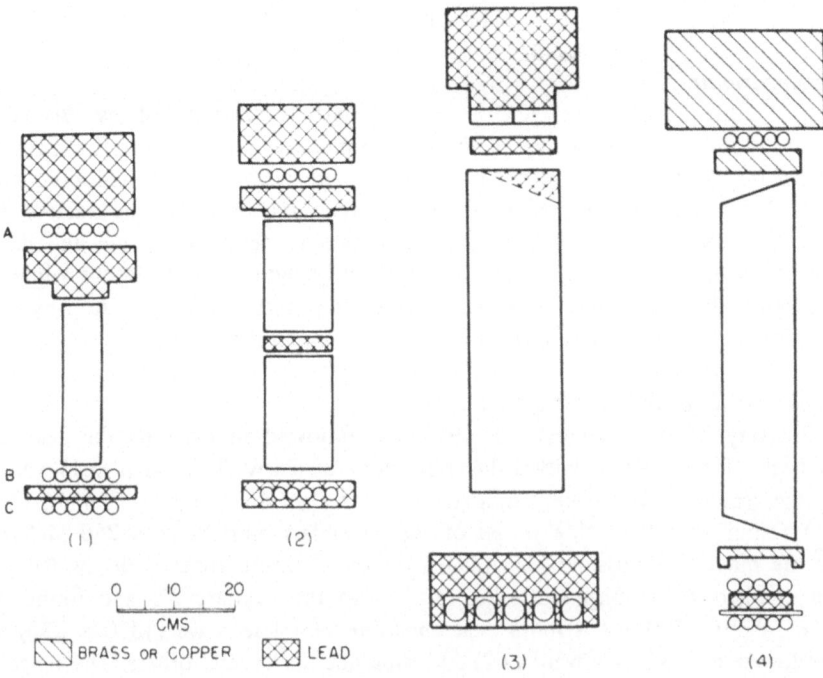

Fig. 3. Some typical cloud chambers and counter selection systems used in the study of the strange particles in the period 1950–54, (after Bridge, 1956). (1) The cloud chamber set used on the Pic du Midi by the Manchester group. (2) The double cloud chamber used by the Pasadena group. (3) The Manchester Jungfraujoch set-up. (4) The Indiana chamber and counter set.

$A_1 B_2 C_2$; (i.e. one particle at tray A_1 and two at each of the trays B and C).

The first six months running, starting in July 1950, produced 43 V-particles of which 36 were V^0 and 7 $V^{\pm}$. Again, many of the secondaries were found to be light particles of mass about the pion mass but protons also occurred. The M.D.M. was the same as for Photograph 2b. Clear evidence was found for two types of V^0 particle, namely,

$$V_1^0 \rightarrow p^+ + \pi^- \quad \text{(later term } \Lambda^\circ)$$

$$V_2^0 \rightarrow \pi^+ + \pi^- \quad \text{(later term } \theta^\circ, \text{ then } K^\circ)$$

(The original V^0 was of the V_2^0 type). A fine example of a V_1^0 is shown in Plate 3b.

There was no proof at this time that the light particles were pions. The Manchester work was published by Armenterous *et al.* (1951) but in the same year, quite independently, similar conclusions were reached by the Indiana group, Thompson *et al.* (1951).

Apart from one possible observation of a V^0-particle in the photographic emulsion by Hopper and Biswas (1950) neutral V-particles were only found in cloud chambers for many years.

7. K-Mesons 1951 — 1954

7.1 Nomenclature

So far in this chapter the generic term V-particle, introduced by Blackett and Anderson, has been used but in this section the modern nomenclature first proposed at the Bagnéres Conference[5] in 1953 by Amaldi *et al.* (1954), will be adopted wherever possible. Thus, particles whose masses lie between the pion and the nucleon will be termed K-mesons, (symbol K), and those between the nucleon and the deuteron, hyperons, (symbol Y). Small Greek letters identify light (or L-) mesons while capital Greek letters distinguish hyperons. An S-event is a generic term introduced by the MIT Group, for the decay or capture of a K-meson or hyperon at rest.

7.2 The first K-particles from S-events

The discovery of the $K_{\pi3}$ meson in 1949 was followed in 1951 by the discovery by O'Ceallaigh of a K-particle with a different mode of decay. This event ($K1$) was picked up in the search for μ-e decays in a study of the electron decay spectrum. O'Ceallaigh (1951) found that the value of $p\beta$ of the secondary particle was 250 MeV/c, well above the maximum possible (55 MeV/c) for the electron from μ-e decay. Rochat and Menon applied their range-scattering relation to the primary $K1$ and found a mass value of about 1000 m. A more exact measurement later gave (1320 ± 250)m. This was followed by a second event ($K2$) and subsequently several others. The range in the values of $p\beta$ indicated either an admixture of decay modes or decay to three particles. With the discovery of $K5$ the secondary was definitely identified as a muon. It is now thought that the $K1$ event represented the decay of a particle, $K_{\mu2}$ to two particles and the second $K2$, a particle, $K_{\mu3}$, which decayed to a muon and two neutral particles. Later, other modes of decay of the K-meson in S-events were found both in emulsions and in cloud chambers. The unravelling of the various modes of decay proved to be unexpectedly arduous, and became possible only with the introduction of new techniques.

7.3 Technical developments
7.3.1 Introduction

The emulsion and cloud chamber techniques played vital and in many respects, complementary roles in the study of the new particles. By 1951 emulsions with a wide range of properties capable of recording all types of charged particles were available and well understood. In general they had the property of showing the whole event for charged particles and of giving limited information about neutral particles within a volume of linear dimensions of the order of centimetres. Precision in range and association with origin for very short times was quite unique and unequalled by the cloud chamber. The emulsion was not, however, suitable for the detection of neutral particles for association with the origin was usually not possible because of the lack of time coincidence and confusion with background events. Thus it was not an accident that all neutral heavy unstable particles were discovered in the cloud chamber. Again,

---

[5] Perhaps the best of all the excellent particle Conferences held in the period 1947–1954. Entirely devoted to the strange particles and in situation and timing unique.

cloud chambers could be placed in magnetic fields and hence the electric charges of particles and their momenta could be measured. Time association made it relatively easy to pick up gamma rays and distingusish electrons from other particles. Combinations of various types of cloud chambers proved of great value.

7.3.2 Emulsions

The major development in the period 1951-54 was the introduction of the technique of stripped emulsions by which a large block of emulsion could be exposed and then separated into sheets for scanning and study. The size and form of the block was such that particles from interactions could be followed for great distances and over a solid angle of 4π. This enormously increased the versatility of the emulsion technique. Stripped emulsions were introduced in 1952-53 (Powell, 1953). In this technique the emulsions of a batch of plates, produced in the usual way, were stripped off and packed together to form a solid block. Registration of one emulsion against its neighbours was made by fine X-ray marks. After exposure the emulsion sheets were separated, attached again to glass plates, and then processed in the usual way.

A typical large stack was the one flown in 1954 by the universities of Bristol, Milan and Padua. It had dimensions of $36 \times 25 \times 15 \text{ cm}^3$ and weighed 63 kg. When broken up parts of the block were distributed to many scientists in many laboratories leading to a collaborative effort on a far larger scale than ever before achieved in emulsion particle physics. This stack and similar ones made notable contributions to our konwledge of the types and relative proportions of K-mesons in the cosmic radiation. The technique was introduced later with notable success at the great accelerators.

7.3.3 Cloud chambers

The major developments in cloud chamber techniques are best illustrated by a selection of some typical set-ups shown in Figs. 3 and 4, and taken mainly from Bridge (1956). All diagrams are side views: illuminated volumes are shown and the disposition of counters and absorbers is indicated. All the cloud chambers were triggered by ps sets of varying complexity.

1) This is the cloud chamber with magnetic field used by Rochester and Butler and subsequently transferred to the Pic du Midi (2870 m) by members of the Manchester group (Armenteros et al., 1951). The set-up has already been described (Section 6). With this arrangement the Manchester group obtained decisive evidence for the $K^0_{\pi 2}$- and Λ^0 particles and the first Ξ-hyperon. The methods of analysis devised by various members of the group (Armenteros et al., 1951; Podolanski and Armenteros, 1954) set the pattern for much of the analytical work of later groups.

2) The double cloud chamber with magnetic field used at Pasadena (220 m) and at 1780 m by York et al. (1954). The chambers were rectangular and 30 cm in width. Much useful data on coplanarity and masses. Found the Σ^+-hyperon decay to a proton.

3) Chamber with magnetic field employed by Astbury, Newth and others from Manchester at the Jungfraujoch (3040 m) (Buchanan et al., 1954). The chamber was 50 cm in width. Notable for some of the most significant work on the lifetimes of V^0-particles.

4) Chamber with magnetic field employed by Thompson and his co-workers at Indiana (230 m). The chamber was 27 cm in width. Chamber system designed with

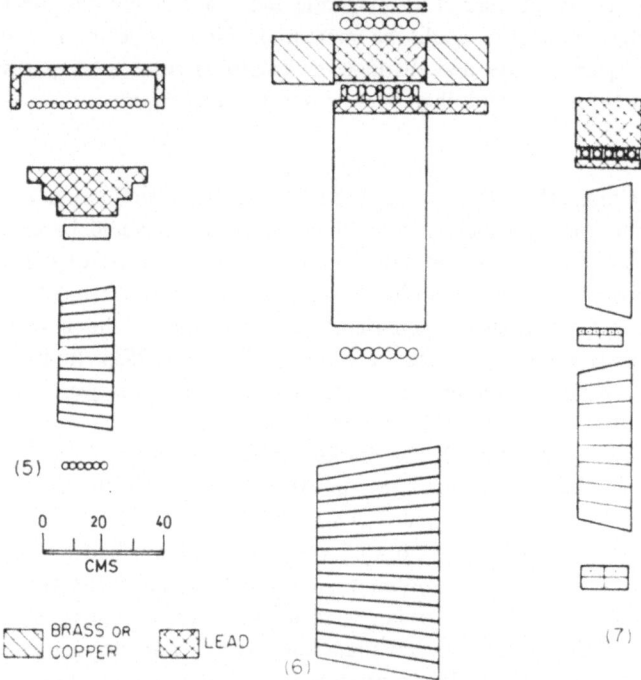

Fig. 4. (5) The multiplate chamber and counter set used by the MIT group at Echo Lake. (6) The double chamber arrangement used by the Ecole Polytechnique group at the Pic du Midi. (7) The complex chamber and counter arrangement used by the Princeton group at Echo Lake. Note the change in scale between Figs. 3 and 4.

great care to make it as free as possible from distortion resulting in some of the most accurate work ever done on the V^0-particles. This combined with an ingenious analytical method for separating the various classes of V^0 gave the best early value of the mass of the $K_{\pi 2}^0$, (now the K_s^0). Thompson et $al.$, also found unambiguous cases of $K^0 \to \pi^+ + \pi^- + \pi^0$, (now called K_L^0), a striking example of the so-called theta-tau (i.e. $\theta(K_s^0)$-τ) puzzle, which was an important element in the discovery of the non-conservation of parity.

5) Multiplate chamber used by the MIT group (Bridge et $al.$ 1954) at Echo Lake, Colorado (3230 m). The rectangular box above the chamber was a liquid scintillation counter. The width of the chamber was 48 cm. Used for a thorough study of S-events. Obtained clear evidence for the two main types of K-decay, the $K_{\pi 2}^{\pm} \to \pi^{\pm} + \pi^0$ and the $K_{\mu 2}^+ \to \mu^+ + \nu$.

6) Double chamber used on the Pic-du-Midi (2870 m) by The École Polytechnique group (EP), Gregory et $al.$ (1954), a set-up reminiscent of the Brode work on the mass of the muon. The top chamber had a magnetic field and the bottom chamber contained copper plates. The width of the chamber was 64 cm. With this set-up it was possible to measure the mass of the stopping K-meson and the range of its charged secondary. Gave convincing proof of the existence of the $K_{\mu 2}^+$-meson and the

proportion at which it occurred. Obtained valuable data on the $K_{\pi 2}^+$.

7) Experimental arrangement used by the Princeton group at Echo Lake (3230 m). Set up similar to (6). The rectangular boxes between and below the chambers were proportional counters (Hodson *et al.*, 1954). The upper and lower chambers were 40 cm and 50 cm in width. Noted for a remarkable photograph of a $K_{\pi 2}^+ \rightarrow \pi^+ + \pi^0 \rightarrow \pi^+ + e^+ + e^- + e^+ + e^-$ decay.

Of the cloud chambers not included in the above list mention should be made of the excellent multiplate chambers made by Fretter and his workers at Berkeley which produced striking photographs showing the creation of V^0-particles.

7.4 Charged K-mesons

The developments described in the previous paragraph led to considerable advances in the identification of the main modes of decay of charged K-mesons in the period 1951-54. The early work was due to O'Ceallaigh (Subsection 7.2) using the emulsion technique and to Bridge and Annis (1951) using a multiplate chamber. By 1954 emulsion physicists, Menon and O'Ceallaigh (1954), had found four decays of charged K-particles to pions of unique momentum and this was confirmed in many other examples by 'Bonetti *et al.* (1954), Bøggild *et al.* (1954), Baldo *et al.* (1954), Appa Rao and Mitra (1955), and the Padua Group. The decay was therefore assumed to be $K^{\pm} \rightarrow \pi^{\pm} + a$ neutral particle. The MIT, EP and Princeton cloud chamber groups found that the neutral particle was a π^0-meson. The decay scheme via this mode then became

$$K_{\pi 2}^{\pm} \rightarrow \pi^{\pm} + \pi^0.$$

For fuller details reference may be made to Powell *et al.* (1959) for emulsion work and to Bridge (1956) for the cloud chamber work.

The emulsion results on the decay of K-mesons to muons did not at first lead to a clear picture of this mode of decay but the multiplate work of the MIT group and the double chamber work of the EP group established that the commonest form of decay of the charged K-meson was to a muon of unique range. Thus both principal modes led to charged particles of unique range in one case to a pion and the other to a muon. Full accounts of the evidence on which these conclusions were based were given in Armenteros *et al.* (1955) and Bridge *et al.* (1955) where it was also shown that the full decay of the $K_{\mu 2}^+$ was

$$K_{\mu 2}^+ \rightarrow \mu^+ + \nu.$$

For the $K_{\mu 2}^+$ mode, especially, the double cloud chamber technique of the EP proved decisive since the top chamber was in a magnetic field and the lower chamber was a multiplate chamber. This meant that values of the masses of the primary particles and the ranges of the secondary particles could be obtained with reasonably high precision. The mass values proved to be close to the mass of the $K_{\pi 3}$-meson but not identical and it was therefore thought, at first, that K-mesons with different modes of decay had slightly different masses. It was soon shown that this was not the case, indeed, clear evidence emerged at the Bagnères Conference in 1953 that all reliable mass values were statistically distributed about the $K_{\pi 3}$-mass, at that time, 966 m.

Later emulsion results from the large stacks, e.g. the G-stack in cosmic rays, and stacks exposed at the machines confirmed the principal decay modes of the charged

K-meson as found in the multiplate chambers, namely, that roughly 63% of K-mesons decay via $K_{\mu 2}^{+}$ and 21% decay via $K_{\pi 2}^{\pm}$. The $K_{\pi 3}$ decay accounted for 5% and the $K_{\beta 3}$ and $K_{\mu 3}$ 5% and 3% respectively. The lifetime of the charged K-meson was found to be 10^{-8} s.

7.5 Neutral K-mesons

Whilst early cloud chamber studies of V^{0}-particles clearly indicated two main types, V_{1}^{0} and V_{2}^{0}, precise values of their masses had to await improvements in analytical procedures and cloud chamber techniques. Somewhat later the emulsion technique was able to improve on the accuracy of the cloud chamber determination of the mass of the V_{1}^{0}-particle. We shall return to this topic in Subsection 8.1 but here we shall be concerned with the V_{2}^{0}-particle. Throughout, a two particle decay mode will be assumed, but for lack of space the considerable body of the evidence for this cannot be given. We shall start with the Manchester group's treatment of the dynamics of two body decays and then proceed logically to the theoretical and experimental work of the Indiana group.

The main problem to be faced in cloud chamber studies of the V^{0}-particles stemmed from the fact that most of the primary particles and their decay products had values of $\beta \approx 1$ and that therefore relatively little information about their nature was available. Because of this the Manchester group made the first attempt to classify V^{0}-decays by purely dynamical methods. This group introduced a parameter, α, defined as $(p_{1L} - p_{2L})/(p_{1L} + p_{2L})$, where p_{1L} and p_{2L} were the longitudinal components of p_1 and p_2, by which it was possible to examine isotropy in the C.M. System. The other important parameter was the Q-value, that is, the mass decrement in the decay process. Thus for the V_{2}^{0}-particle we have,

$$V_{2}^{0} \rightarrow \pi^{+} + \pi^{-} + Q.$$

Armenteros et al. (1951) showed that α could be expressed in terms of the momenta and masses by the equation,

$$\alpha = \frac{p_{1}^{2} - p_{2}^{2}}{P^{2}} = \frac{m_{1}^{2} - m_{2}^{2}}{M^{2}} + 2p^{*}\cos^{*}\left(\frac{1}{M^{2}} + \frac{1}{P^{2}}\right)^{1/2}$$

where the symbols are defined in Figs. 5 and 6.

The average value of α, $\bar{\alpha}$, is

$$\bar{\alpha} = \frac{m_{1}^{2} - m_{2}^{2}}{M^{2}}$$

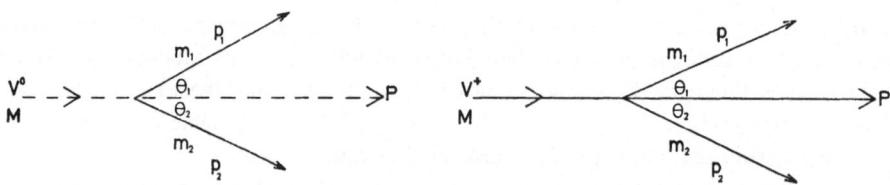

Fig. 5. V-particle decay schemes in the laboratory system.

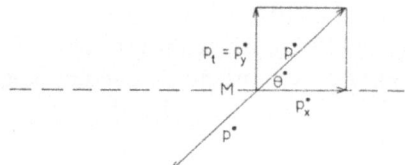

Fig. 6. Two-particle decay in the C.M. system.

and is a constant for a particular decay scheme. Resolving p^* into its two components p_x^* and p_y^* it can be shown that

$$p_x^* = \frac{(\alpha - \bar{\alpha})}{(2/\beta M)} .$$

Then, since $p_x^{*2} + p_y^{*2} = p^{*2}$

$$\frac{(\alpha - \bar{\alpha})^2}{(2p^* / \beta M)^2} + \frac{p_t^2}{p^{*2}} = 1$$

or

$$\frac{(\alpha - \bar{\alpha})^2}{(2p^*)^2 (M^{-2} + P^{-2})} + \frac{p_t^2}{p^{*2}} = 1$$

an equation originally derived by Thompson *et al.* (1953a).

For a given velocity, β, this equation describes a family of ellipses in the variables α and p_t. Fully to represent the equation requires a three dimensional surface, (the so-called Q-surface) but for our purpose since most of the V^0-decays have $\beta \approx 1$ the results can be represented on the (α, p_t) plane for $\beta = 1$. If a third particle is produced the observed points on the (α, p_t) plane scatter but lie within the ellipse. Typical Q-curves as given by Thompson *et al.* (*ibid*) for V_1^0 and V_2^0 are shown in Fig. 7. The important 'anomalous' cases shown within the V_2^0 curve were discussed in Subsection 7.3(c).

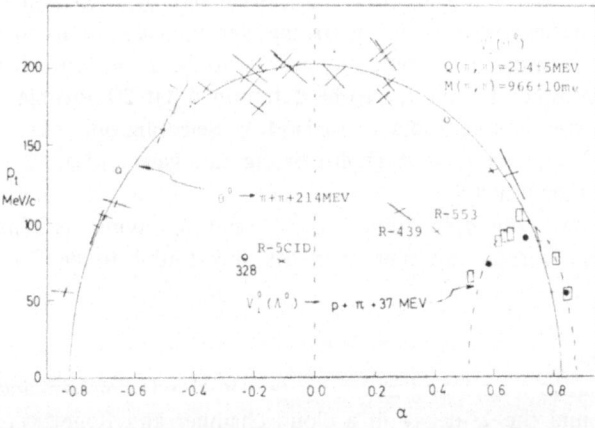

Fig. 7. Q-curve plots in the $p_t - \alpha$ plane for V_1^0- and V_2^0-decays, a representation introduced by Thompson. With accurate experimental data these curves delineated clearly V_1^0- and V_2^0-particles and enabled accurate values of the masses to be obtained.

A fuller treatment of the properties of the parameters α and Q was given in an excellent article by Thompson (1956), where the generality of the Q-surface method was pointed out. Indeed, the surface of constant Q-value unambiguously defines the decay scheme irrespective of the decay parameters and without any *a priori* assumptions as to the decay scheme.

The importance of the Thompson extension of the Manchester analysis was only realized when used with cloud chamber measurements of high accuracy such as presented by Thompson at the Bagnères Conference in 1953. The reason for the great improvement in precision was the great attention paid by Thompson to the reduction of track distortion through strict temperature control which led to an M.D.M. for long tracks of better than 50 BeV/c. From the Q-plots of Fig. 7 it is seen that the two main types of V^0-particle are clearly delineated. Later results, Thompson *et al.* (1953b). gave the Q value of the V_2^0-particle as (214 ± 5) MeV leading to a $K_{\pi 2}^0$ mass of $(966 \pm 10)m$.

Cloud chamber determination of the lifetime gave a value of 1.7×10^{-10} s, Gayther (1954).

8. Hyperons

8.1 The Λ^0-Hyperon
At the Bagnères Conference clear, additional, evidence was presented by several groups on the nature of the secondary particles of the V_1^0-decay. Several groups, particularly the MIT group, Bridge and Rossi (1953), and the Indiana group, Thompson *et al.* (1953a) found the Q-value to be close to 37 MeV. The decay could thus be written,

$$\Lambda^0 \rightarrow p^+ + \pi^- + 37 \text{ MeV}$$

leading to a mass of 2189 m.

The determination of the Q-value of the Λ^0-hyperon from the cloud chamber work led to a directed search in large emulsion blocks for two-pronged stars which simulated Λ^0-decays. The most extensive work was carried out by Friedlander *et al.* (1954), at Bristol, and the decays were found by tracing identified π^--mesons to two-pronged stars for which the other particle could be proved to be a proton. Measurement of the proton energy allowed the Q-value to be determined. Of 20 possible cases, 10 Q-values fell within a narrow interval $(35.5 - 38.5)$ MeV. Selecting only high quality Λ^0-decays gave $Q = (36.92 \pm 0.22)$ MeV. A Q-plot of the data gave a diagram very similar to the Thompson plot shown in Fig. 7.

No evidence has ever been found for Λ^+ and Λ^- hyperons. The lifetime of the Λ^0-hyperon from cosmic ray measurements was found to be 3.7×15^{-10} s (Page, 1954).

8.2 The Σ-triplet
The first Σ-hyperons were found in cosmic rays in cloud chambers and emulsions. York *et al.* (1953) found the Σ^+-decay in a cloud chamber and Bonetti *et al.* (1953) in an emulsion. The decay mode was

$$\Sigma^+ \rightarrow p^+ + \pi^0 + 110 \text{ MeV}.$$

The alternative mode was also found by Bonetti *et al.*

$$\Sigma^+ \to n + \pi^+ + 110 \text{ MeV.}$$

The Σ^-- and Σ^0-hyperons were found at the cosmotron. The lifetimes of the charged hyperons were found to be typically of the order of 10^{-10} s.

8.3 The $\equiv$-hyperon

This remarkable hyperon was first discovered by the Pic du Midi group in 1952 by Armenteros et al, its characteristic feature being a cascade, i.e. a V^- decay together with a V^0-decay below the point at which the charged V-particle decayed. The correct decay mode was identified by Cowan (1954) as $\equiv^- \to \Lambda^0 + \pi^- + 65$ Mev.

9. Postscript

Kinematical tables for all the principal two-body decays over a wide range of energies have been published by CERN, Salmeron (1963), and it is therefore possible to look again at the original V^0- and V^+-decays. Using the data given in Subsection 4.4 and the Salmeron Tables it is found that V^0 fits a $K_{\pi 2}^0$ closely with Λ^0 definitely excluded; V^+ fits $K_{\mu 2}^+$-decay closely but $K_{\pi 2}^+$ is not completely excluded. Σ^+ decays are excluded.

Thanks are expresseed to the following for permission to reproduce photographs or figures: Professor C.D. Anderson, Plate 3a; Dr. C.C. Butler, Plate 3b; Professor H.S.Bridge, The Editor and Publishers, The North Holland Publishing Co. Ltd., Amsterdam, Figs. 3 and 4; and Professor R.W. Thompson, The Editors and Publishers of the Physical Review, The American Institute of Physics, Fig. 7.

The writer is grateful to the Physics Department of the University of Durham for considerable technical help in the preparation of this article.

Drs. C.C. Butler and J.V. Major kindly read this article in MS and made useful comments.

Professor Thompson sent me a copy of his paper to the International Symposium on the History of Particle Physics held at the Fermi N. A. L. in May 1980, and for this paper and his elucidation of various points concerning the history of the V^0-particles, I am most grateful.

REFERENCES

Amaldi, E., Anderson, C.D., Blackett, P.M.S., Fretter, W.B., Leprince-Ringuet, L., Peters, B., Powell, C.F., Rochester, G.D., Rossi, B. and Thompson, R.W.: 1954, *Nuovo Cim.* **11**, 213.

Appa Rao, M.V.K. and Mitra, S.: 1955, *Proc. Ind. Acad. Sci.*, **41**, 30.

Armenteros, R., Barker, K.H., Butler, C.C. and Cachon, A.: 1951, *Phil. Mag.*, **42**, 1113.

Armenteros, R., Gregory, B., Hendel, A., Lagarrique, A., Leprince-Ringuet, L., Muller, F. and Peyrou, C.: 1955, *Nuovo Cim.*, **1**, 915.

Baldo, M., Belliboni, G., Sechi, B. and Zorn, G.T.: 1954, *Nuovo. Cim. Supp.* **12**, 220.

Barker, K.H. and Butler, C.C.: 1951, *Proc. Phys. Soc.*, A, **64**, 4.

Blackett, P.M.S.: 1936, *Proc. Roy. Soc.*, A, **154**, 564.

Blackett, P.M.S. and Lees, D.S.: 1932, *Proc. Roy. Soc.*, A, **132**, 658.

Bøggild, J.K , Hooper, J.E. and Scharff, M.: 1954, *Nuovo Cim. Supp.*, **12**, 223.

Bonetti, A., Levi Setti, R., Panetti, M.B. and Tomasini, G.: 1953, *Nuovo Cim.* **10**, 345, 1736.

Bonetti, A., Levi Setti, R., Panetti, M. and Tomasini, G.: 1954, *Proc. Roy. Soc. A*, **221**, 318.

Bostick, W.H.: 1942, *Phys. Rev.*, **61**, 557.

Bridge, H.S.: 1956, *Prog. Cosmic Ray Phys.*, **3**, 182.

Bridge, H.S. and Annis, A.: 1951, *Phys. Rev.*, **82**, 445.

Bridge, H.S., Courant, H., Dayton, B., De Staebler, H.C., Rossi, B., Stafford, R. and Willard, D.: 1954, *Nuovo Cim.*, **12**, 81.

Bridge, H.S., De Staebler, H., Rossi, B. and Sreekantan, B.V.: 1955, *Nuovo Cim.,* **1**, 874.

Bridge, H.S. and Rossi, B.: 1953, *Proc. Conf. Int. Ray. Cosmique, Bagnères,* A. 9.

Brown, R., Camerini, U., Fowler, P.H., Muirhead, H., Powell, C.F. and Ritson, D.M.: 1949, *Nature,* 163, 82.

Buchannan, J.S., Cooper, W.A., Millar, D.D. and Newth, J.A.: 1954, *Phil. Mag.,* **45**, 1025.

Butler, C.C., Rosser, W.G.V. and Barker, K.H.: 1950, *Proc. Phys. Soc., A,* **63**, 145.

Camerini, U., Fowler, P.H., Lock, W.O. and Muirhead, H.: 1950, *Phil. Mag.,* **41**, 413.

Cocconi, G.: 1941, *Phys. Rev.,* **60**, 532.

Conversi, M., Pancini, E. and Piccioni, O.: 1947, *Phys. Rev.,* **71**, 209.

Cowan, E.W.: 1954, *Phys. Rev.,* **94**, 161.

Daudin, J.: 1944, *Ann. de Phys.,* **19** (April–June).

Fowler, P.H.: 1950, *Phil. Mag.,* **41**, 169.

Friedlander, M.W., Keefe, D., Menon, M.G.K. and Merlin, M.: 1954, *Phil. Mag.,* **45**, 433.

Gayther, D.B.: 1954, *Phil. Mag.,* **45**, 570

Gregory, B.A., Legarrique, A., Leprince-Ringuet, L., Muller, F. and Peyrou, C.H.: 1954, *Nuovo Cim.,* **11**, 292.

Hamilton, J., Heitler, W. and Peng, H.E.: 1943, *Phys. Rev.,* **64**, 78.

Harding, J.B.: 1950, *Phil. Mag.,* **41**, 405.

Hazen, W.E.: 1944, *Phys. Rev.,* **65**, 67.

Herzog, G.: 1941, *Phys. Rev.,* **59**, 117.

Hopper, V.D. and Biswas, S.: 1950, *Phys. Rev.,* **80**, 1099.

Hodson, A.L., Ballam, J., Arnold, W.H., Harris, D.R., Rau, R.R., Reynolds, G.T. and Trieman, S.B.: 1954, *Phys. Rev.,* **96**, 1089.

Jánossy, L., McCusker, C.B.A. and Rochester, G.D.: 1941, *Nature,* **148**, 660.

Jánossy, L.: 1943, *Phys. Rev.,* **64**, 345.

Jánossy, L. and Rochester, G.D.: 1943, *Proc. Roy. Soc., A,* **182**, 180.

Jánossy, L., Rochester, G.D. Broadbent, D.: 1945, *Nature,* **155**, 142.

Lattes, C.M.G, Occhialini, G.P.S. and Powell, C.F.: 1947, *Nature,* **160**, 453, 486.

Leighton, R.B., Anderson, C.D. and Seriff, A.J.: 1949, *Phys. Rev.,* **75**, 1432.

Marshak, R.E.: 1980, *Int. Symposium on the History of Particle Physics,* May, 1980.

Marshak, R.E. and Bethe, H.: 1947, *Phys. Rev.,* **72**, 506.

Menon, M.G.K. and O'Ceallaigh, C.: 1954, *Proc. Roy. Soc., A,* **221**, 292.

O'Ceallaigh, C.: 1951, *Phil. Mag.,* **42**, 1032.

Page, D.I.: 1954, *Phil. Mag.,* **45**, 863.

Piccioni, O.: 1950, *Phys. Rev.,* **77**, 1.

Podolanski, J. and Armenteros, R.: 1954, *Phil. Mag.,* **45**, 13.

Powell, C.F.: 1953, *Phil. Mag.,* **44**, 219.

Powell, C.F., Fowler, P.H. and Perkins, D.H.: 1959, *The Study of Elementary Particles by the Photographic Method,* Pergamon Press, London.

Powell, W.M.: 1940, *Phys. Rev.,* **58**, 474.

Powell, W.M.: 1941, *Phys. Rev.,* **60**, 413.

Powell, W.M.: 1946, *Phys. Rev.,* **69**, 385.

Rochester, G.D.: 1946, *Proc. Roy. Soc., A.,* **187**, 464.

Rochester, G.D. and Butler, C.C.: 1947, *Nature,* **160**, 855.

Rochester, G.D. and Butler, C.C.: 1948a, *Proc. Phys. Soc., A,* **61**, 307.

Rochester, G.D. and Butler, C.C.: 1948b, *Proc. Phys. Soc., A,* **61**, 535.

Rochester, G.D. and Wilson, J.G.: 1952, *Cloud Chamber Photographs of the Cosmic Radiation,* Pergamon Press, London.

Rossi, B., Jánossy, L., Rochester, G.D. and Bound, M.: 1940, *Phys. Rev.,* **58**, 761.

Sakata, S. and Inoue, T.; 1946, *Prog. Theor. Phys.,* **1**, 143.

Salmeron, R.A.: 1963, *Kinematical Tables for Two-Body Decays,* CERN, Geneva.

Schein, M., Iona, Jr., M. and Tabin, H.: 1943, *Phys. Rev.,* **64**, 253.

Seriff, A.J., Leighton, R.B., Hsiao, C., Cowan, E.W. and Anderson, C.D.: 1950, *Phys. Rev.,* **78**, 290.

Thompson, R.W.: 1956, *Prog. Cosmic Ray Phys.,* **3**, 255.

Thompson, R.W., Cohn, H.O. and Flum, R.S.: 1951, *Phys. Rev.* 83, 175.

Thompson, R.W., Buskirk, A.V., Etter, L.R., Karzmark, C.J. and Rediker, R.H.: 1953a, *Phys. Rev.*, 90, 329.
Thompson, R.W., Buskirk, A.V., Cohn, H.O., Karzmark, C.J. and Rediker, R.H.: 1953b, *Proc. Conf. Int. Ray Cosmique*, Bagnères, B. 2.
Wollan, E.O.: 1941, *Phys. Rev.*, 60, 532.
York, C.M., Leighton, R.B. and Bjornerud, E.K.: 1953, *Phys. Rev.*, 90, 167.
York, C.M., Leighton, R.B. and Bjornerud, E.K.: 1954, *Phys. Rev.*, 95, 159.

DISCOVERY OF HYPERNUCLEI : THE BEGINNINGS

Jerzy PNIEWSKI[*]

Cosmic ray research has been viewed as a key to the initiation and development of hypernuclear physics. A retrospect of the early stages of hypernuclear studies is presented. Main steps in the development of hypernuclear physics are recollected and confronted with the progressing understanding of the elementary particle problems at that time. Some information about the continuation of the hypernuclear studies carried out with the aid of artificially constructed beams is also presented.

1. Introduction

The observation of the first nuclear fragment containing a bound Λ hyperon revealed the existence of a third nuclear component besides the two previously known constituents: the proton and the neutron. Nuclear structures containing the Λ hyperon have been named hypernuclei. The existence of hypernuclei was not anticipated before the first hypernucleus event had been observed. By this discovery cosmic rays gave rise to a new field of research, that of hypernuclear physics, a border region between nuclear and elementary particle physics. It is interesting to trace some of the correlations between these two fields of research, both initiated and, at the early stages, developed with the aid of cosmic rays.

2. The Contribution of the Hypernucleus Discovery to the Development of Elementary Particle Physics

At the turn of the forties the cosmic ray investigations carried out with the use of the cloud chamber and the nuclear emulsion techniques dominated in elementary particle physics. Subsequently, towards the end of the forties the decays of neutral and charged heavy mesons were observed (Rochester and Butler, 1947; Brown *et al.*, 1949; see also Leprince-Ringuet and l'Heritier, 1944).

In 1951 the existence of two neutral particles of different masses was established. They were called V_1^0, and V_2^0 particles. One of them was heavier than the proton and the other evidently lighter (Armenteros *et al.*, 1951). It was found that the V_1^0 particle decayed into a proton and a light particle, whereas V_2^0 decayed into two light particles. It was assumed from the beginning that the light particles were the π mesons although the ionization measurements were not precise enough to distinguish them from the muons. The problem was solved when some events with characteristic decay or interaction of the light secondaries had been found (e.g., Leighton *et al.*, 1951, 1953; Bridge *et al.*, 1952).

In this way the decay modes of the two kinds of the V^0 particles could be established: $V_1^0 \rightarrow p + \pi^-$, $V_2^0 \rightarrow \pi^+ + \pi^-$, and the V^0 particles were named later Λ

---

[*] University of Warsaw, Warsaw, Poland.

Y. Sekido and H. Elliot (eds.), Early History of Cosmic Ray Studies, 323–337.
© *1985 by D. Reidel Publishing Company.*

hyperons and K mesons, respectively. However, there seemed to be two contradictory facts. The V^0 particles were relatively frequently produced in fast processes of high energy interactions while their life times seemed to be rather long ($\tau \geq 10^{-10}$s) from the point of view of their production rate. Since it was accepted that the secondaries were strongly interacting particles, one might expect that the observed decays should be classified as fast processes occuring in 10^{-22}s. In fact, one was faced with the same problem in the case of the charged heavy meson decay ($\tau \to \pi^+ + \pi^- + \pi^+$) and, two years later, in the case of the charged hyperon decay (Bonetti et al., 1953, 28.1.53*).

In order to get out of the impasse, Pais proposed the mechanism of associated production of the newly discovered particles (Pais, 1952, 22.1.52). However, within the framework of his approach a pair production of two hyperons was also possible. Peaslee was unsuccessful when he was looking for a solution in the isospin and parity conservation (Peaslee, 1952, 12.2.52).

Meanwhile, the first hypernucleus event observed and interpreted in the second half of 1952 made the discussed problem even more exciting (Danysz and Pniewski, 1952, 20.10.52). As the latter observation revealed, the V_1^0 particle (a Λ hyperon), produced in fast processes of strongly interacting cosmic ray particles, could subsequently form bound states with the nucleons not suffering any fast decay.

Two additional hypernuclear events observed shortly afterwards confirmed the original interpretation (Tidman et al., 1953, 15.12.52; Crussard and Morellet, 1953, 5.1.53).

As things were, one had to look for a better understanding of the peculiar features of the new particles.

In the middle of 1953 the existence of hypernuclear structures was for the first time taken into account as a new ingredient in the theoretical attempts to understand the elementary particle problems (Peaslee, 1953, 10.6.53). Peaslee who was studying the problem of the peculiar binding of the V_1^0 particle in an atomic nucleus came to the conclusion that only a special restriction may inhibit a fast decay of this particle. Accordingly, he introduced a new quantum number thus being, in fact, quite near to the correct solution of the problem.

Shortly afterwards first Gell-Mann (21.8.53), and next Nakano and Nishijima (16.9.53) made a plucky attempt to solve the problem by ascribing an integer isospin value to the V_1^0 fermion and a half integer value of the isospin to the V_2^0 boson (Gell-Mann, 1953; Nakano and Nishijima, 1953). They assumed that fermions might be isobosons and bosons - the isofemions. In fact they were able to achieve substantial progress in understanding the mechanism of the observed decays in which the strong and electromagnetic interactions were inactive.

Meanwhile, at the end of 1953 clear examples of the associated production of K mesons together with neutral or charged hyperons were observed (Lal et al., 1953, 3.10.53; Fowler et al., 1953, 10.11.53, see also 1954).

In the next year, at the Glasgow Conference (13-17.7.54) Gell-Mann and Pais (1954) proposed various models to explain the unsolved problem. One of the models, coinciding with an explanation proposed at the same time by Nishijima (1954, 13.7.54), was based on a concept of a new charge and the corresponding quantum

* The date of admission for publication.

number. As a matter of fact it was equivalent to the concept of strangeness.

At the same time the first associated production of the K^+ ($\tau \rightarrow \pi^+ + \pi^- + \pi^+$) and the Λ hyperon bound in a hypernucleus was observed (Debenedetti et al., 1954, 8.8.54). Moreover, examples of the hypernuclei produced first in the K^- captures and next in the Σ^- captures were also found (Naugle et al., 1954, 29.6.54; Ceccarelli et al., 1955, 9.7.55; Schein et al., 1955, 25.8.55).

Again these events were observed in an emulsion irradiated by cosmic rays and the K^- mesons and the Σ^- hyperons were found to be only the secondaries produced by the primary cosmic radiation.

All these observations helped to find the straight road to the concept of strangeness.

At the Pisa Conference (12-18.6.55) Gell-Mann (1955) stated precisely the idea of strangeness conservation which has remained valid ever since in strong interaction processes. An independent and equivalent solution was also presented by Nishijima (1955, 11.2.55).

The rules governing the hyperon and heavy meson production or decay and the phenomenon of the hypernucleus existence were made clear once the strangeness concept was accepted.

A relation between the strangeness, S, the electric charge, Q, the third isospin component, I_3, and the baryonic number, B, turned out to be $Q = I_3 + \dfrac{B+S}{2}$. It shows that the conservation of strangeness in strong and electromagnetic interactions is thus a result of the conservation of all the other quantities. In fact, strangeness different from zero informs us only that the strange particle isospin multiplets are shifted by a half integer. This fact itself makes it possible to ascribe proper isospin values to the strange particles and in this way to extend the conservation rules of the isospin to these new particles. By means of strangeness, one can easily show which are the allowed and forbidden reactions in strong and electromagnetic interactions.

3. Production of Hypernuclei

Hypernuclei with the same strangeness as hyperons and K^- mesons ($S = -1$) are copiously produced by K^- mesons. If non strange particles produce hypernuclei, the latter are accompanied by particles with the opposite strangeness ($S = +1$) e.g., K^+ or K^0 mesons.

The first one and many subsequently observed hypernuclear events were produced in the interactions of cosmic rays with the nuclei of the photographic emulsion irradiated in the stratosphere. It was C.F. Powell and his collaborators from Bristol who initiated the baloon flights and developed the nuclear emulsion technique. Emulsion with its unique spatial resolution was well suited to become the basic tool for hypernuclear studies and this remained the case for many years to follow.

On the other hand already in 1953 high momentum π^- meson beams formed in Brookhaven Cosmotron Laboratory were used in the experiments devised for studying elementary particle phenomena. In 1954 the first hypernuclear event produced by a meson from an artificially constructed beam was observed (Hill et al., 1954, 30.1.54). Nevertheless, for two years that followed cosmic radiation was, as before, the main source of the particles used for the production of hypernuclei. Only at the end of

1955 some hypernuclear events resulting from interactions of the π^- mesons from an artificial beam were found. Finally, by the end of 1956 hypernuclei were copiously produced with a low momentum K^- beam first constructed at Berkeley Bevatron Laboratory (Barkas *et al.*, 1957, 26.12.56, Fry *et al.*, 1957, 29.3.57). Since then hypernuclear physics has entered the accelerator laboratories for good.

4. Terminology and Hypernuclear Symbols

Hypernuclei formed as fragments of desintegrated target nuclei were first referred to as unstable or excited nuclear fragments. For the first two years following their discovery these structures had no name and it was only in 1954 when M. Goldhaber suggested the term hyperfragment. At the end of 1953 new names for the elementary particles were accepted by the collective decision of a committeee of ten physicists (Amaldi *et al.*, 1954, 12.12.53). The proposed name — hyperfragment — was a consequence of the name hyperon chosen for the V_1^0 particle. In 1955 the name suggested by Goldhaber was generally accepted (Fry *et al.*, 1955a, 16.2.55). As the years went by, the observed hyperfragments were formed also from the whole target nuclei and in this case it was more proper to call them hypernuclei rather than hyperfragments. Since any nuclear fragment is, in fact, also a nucleus, the name hyperfragment is no longer much in use.

At first, hypernuclear symbols did not differ from the ordinary nuclear symbols and the equation $H^3 = \pi^- + He^3 + Q$ was an example of the notation used to describe a hypernucleus decay. However, shortly after a star was added to the hypernuclear symbol $H^{3*} = \pi^- + He^3 + Q$. Only in 1955 the symbol Λ was first used as a special suffix to distinguish a hypernucleus from an ordinary nucleus: e.g., $_\Lambda H^3 \rightarrow \pi^- + He^3$, $^3H_\Lambda \rightarrow \pi^- + {}^3He$ or $_\Lambda^3H \rightarrow \pi^- + {}^3He$ (Schneps *et al.*, 1955, 22.10.55). It is recommended to use the notation of the last example with the Λ mark and the mass number A preceding the hypernucleus name in agreement with the notation of the nuclear symbols proposed by the IUPAP Commission (SUN Commision, 1978). In the $_\Lambda^A Z$ hypernucleus the symbol A denotes the baryonic number, i.e., the total number of baryons.

The hypernuclear names have been adopted from the corresponding chemical symbols of nuclei e.g., the $_\Lambda^3 H$ hypernucleus is called the hyperhydrogen 3 or the hypertriton, $_\Lambda^5 He$ — · the hyperhelium 5. In 1958 Dalitz and Downs anticipating the possible binding of two Λ hyperons in one hypernuclear structure simply used the symbol $_{\Lambda\Lambda}^6 He$ for the double hyperhelium 6 (Dalitz and Downs, 1958). It is customary to call all the nucleons bound in a hypernucleus a hypernuclear core, or a parent nucleus, e.g., the hypernuclear core in $_\Lambda^5 He$ is an α particle, in the case of $_\Lambda^9 Be$ it is a ^8Be nucleus and in $_{\Lambda\Lambda}^6 He$ — an α particle etc.

5. The First Hypernucleus Event

The first hypernucleus event found in Warsaw (Fig. 1) originated from the interaction of a cosmic ray particle with a bromine or a silver nucleus in the photographic emulsion (Danysz and Pniewski, 1952, 20.10.52 and 1953, 1.12.52).*

* The emulsion plates were irradiated and processed in Bristol.

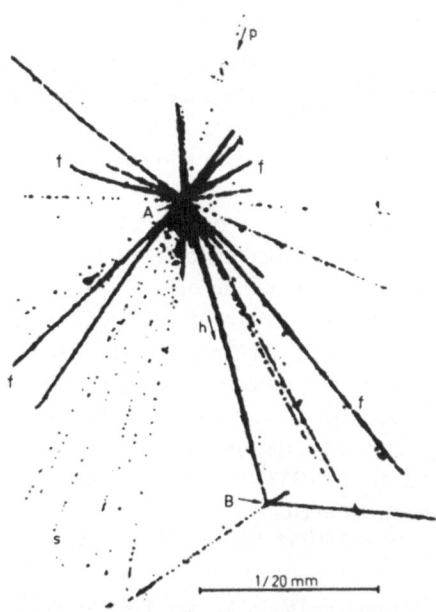

Fig. 1. First hypernucleus event: p-primary cosmic ray particle, A-target nucleus desintegraration, h-hyperfragment, B-hypernucleus decay. Danysz and Pniewski (1952).

The track of the fragment finally recognized as a hypernucleus was evidently tapering and the fragment appeared to stop and decay at rest or just before being stopped. A rough estimate of its charge gave the Z value of about 5. Four tracks coming out of the end of the fragment track could only belong to secondary particles produced in a spontaneous decay of the fragment. The life time deduced from the time of flight of the fragment could not be less than 3 picoseconds.

A chance coincidence of two independent events was considered to be very improbable in the case of the emulsion stack employed. The kinematical analysis limited to a single 600μ thick pellicle could not be carried out completely. One of the four secondary tracks might belong to a proton or a pion, which would prevent the identification. On the assumption that the event represented an unstable nuclear fragment the unidentified particle had to be a proton. However, the estimated energy released in the decay, i.e., the sum of the kinetic energies of all the charged particles and the neutrons balancing the momenta together with their binding energies in the fragment, exceeded 140 MeV. Long life time ($\tau > 3$ps) and high energy release classified the event as being decidedly unusual. One could easily reject an interpretation in terms of a highly excited long lived nuclear fragment ($Q \geqq 140$ MeV, $\tau > 3$ps).

However, it could be noted that the lowest limit of energy release roughly coincided with the energy corresponding to the rest mass of the pion. Was the pion really involved in the process?, – and where could it come from? Only two explanations were worth considering (i) the hypothesis that it was a π^- mesic atom ($Q \approx 140$ MeV) or (ii) that it was a peculiar nuclear fragment containing a bound V_1^0 particle i.e., a Λ

hyperon ($Q \approx 170–180$ MeV). The first interpretation, attractive as it was, did not seem to be likely from the point of view of the production mechanism of the mesic atom in the high energy interaction. One may estimate at present that the π^- meson would have to be bound in the 3d or the 4f orbit, for a more strongly bound pion would not have survived the time of flight, whereas a pion from a higher state would be lost in flight.

Being left with the interpretation of a bound Λ hyperon one had to recall again the previous statement that the unidentified particle might be a proton or pion. A rough and not very reliable analysis performed according to the present identification criteria shows that it might be a mesonic decay of $^{12}_{\Lambda}C$, $^{13}_{\Lambda}C$ or $^{11}_{\Lambda}B$ or else a nomensonic decay of a hypernucleus with atomic and mass numbers between $^{9}_{\Lambda}B$ and $^{14}_{\Lambda}N$.

Two similar hypernuclear events observed subsequently finally excluded a chance coincidence, and a π meson track observed first by Crussard among the secondary tracks proved explicitly that the interpretation in terms of a bound V^0_1 particle was certainly correct (Tidman $et\ al.$, 1953, 15.12.52; Crussard and Morellet, 1953, 5.1.53). The first theoretical attempt to exclude the pionic atom interpretation was already done in 1953 (Cheston and Primakoff, 1953, 29.6.53).

6. Hypernuclear Decay Modes, First Identifications

Kinematical analysis of hypernuclear decays makes it possible to identify individual hypernuclear events. In such an analysis the tracks of decay particles have to be followed up to the end of their ranges. Starting in 1953 stacks of stripped emulsion have been exposed to cosmic radiation and subsequently to the beams constructed in the laboratories. With the stripped emulsion one could follow the tracks through many pellicles and in this way in 1954 one of the observed hypernuclear events was definitely identified. It was the hyperhydrogen 3 observed via the two body mesonic decay: $^{3}_{\Lambda}H \rightarrow \pi^- + ^3He$ (Fig.2), (Bonetti $et\ al.$, 1954, 12.1.54). For the first time the Λ binding energy(B_{Λ}) was calculated from the deficiency of the energy released in the hypernucleus decay. In fact it was the first information pertaining to the strength of the Λ-core or the ΛN interaction responsible for the binding of a Λ hyperon.

There are two mesonic decay channels for a free Λ hyperon: 1) $\Lambda \rightarrow p + \pi^-$ (64.2%) and 2) $\Lambda \rightarrow n + \pi^0$ (35.8%). The π^--decays of a bound lambda hyperon were observed at a very early stage, whereas the first π^0-decay of a hypernucleus was not reported till 1958. It was a π^0 mode of a two body decay of hyperhelium 4: $^{4}_{\Lambda}He \rightarrow \pi^0 + ^4He$ and the π^0-meson could be detected by observing a Dalitz pair from the decay of the π^0 : $\pi^0 \rightarrow \gamma + e^+ + e^-$ (Levi Setti and Slater, 1958, 1.5.58). Three months later the first π^+ decay of a hypernucleus was found and the π^+ meson was identified owing to its characteristic decay mode $\pi^+ \rightarrow \mu^+ \rightarrow e^+$. However, the identification of the hypernucleus was not unique: $^{7}_{\Lambda}Li$ ($^{7}_{\Lambda}Be$) $\rightarrow \pi^+ + ^6He$ (^6Li) $+ n$ (Schneps, 1958, 31.7.58). The observation of other hypernuclear decays and specifically of neutron deficient hypernuclei ($^{4}_{\Lambda}He$, $^{7}_{\Lambda}Be$) was very important since the π^+ mode is forbidden for a free Λ hyperon. Various theoretical models which were devised to explain these decays were confronted with experiment.

Very common nonmesonic decays have been observed from the very beginning of

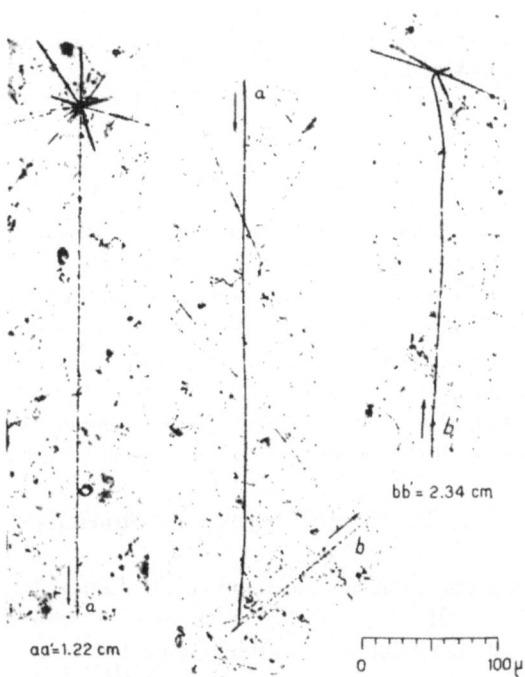

Fig. 2. The first uniquely identified hypernucleus event $^3_\Lambda H \to \pi^- + {}^3 He$. One can distinguish: the hypernucleus production star being a result of a cosmic ray particle interaction, the hypernucleus decay star and the pion capture star, Bonetti *et al.* (1954).

hypernuclear studies. It was noticed that their decay rates increased strongly with the baryonic mass number and it was realized that it could not be a result of the π^- meson absorption only. One can consider this decay as a stimulated Λ decay being a result of the ΛN weak interaction with the neighbouring nucleons: $\Lambda + N \to n + N$. The emission of neutrons makes in general the identification of non-mesonic events unfeasible. A nonmesonic decay of the hyperboron 11: $^{11}_\Lambda B \to {}^8 Li + {}^2 H + {}^1 H$ represents an exceptional example. The absence of neutrons and the emission of a $^8 Li$ fragment recorded as a "hammertrack" made it possible to identify the hyperfragment (Evans *et al.*, 1959). Another decay $^9_\Lambda Be \to {}^8 Li + {}^1 H$ represents a similar example (Ammar, 1963).

All examples of π^0, π^+ decay modes and the nonmesonic decay presented here were observed in emulsion stacks irradiated by K^- beams.

The attempts to detect leptonic hypernuclear decays or mesonic decays with accompanying γ radiation have not been successful. The presented examples were unreliable and not confirmed by any events found in systematic studies of large samples of hypernuclear decays. Most likely they occur only in a small percentage of all hypernuclear decays as should be expected since the known decay rates of a free Λ hyperon are: $pe^-\bar{\nu}_e$: $(8.07 \pm 0.28) \times 10^{-4}$; $p\mu^-\bar{\nu}_\mu$: $(1.57 \pm 0.35) \times 10^{-4}$; $p\pi^-\gamma$: $(8.5 \pm 1.4) \times 10^{-4}$.

The calculations of the rate of the leptonic decays of hyperhydrogen 4: $^4_\Lambda H \rightarrow {}^4He + e^-$ (μ^-) + $\bar{\nu}_e$ ($\bar{\nu}_\mu$) resulted in values not much different from the free Λ decay rates (Lyulka, 1963; McNamee and Oakes, 1966).

Detection and recognition of these rare decay modes are very difficult since in the presence of neutrinos or photons the kinematical analysis cannot be done directly.

Varfolomeev *et al.* (1956a) have recognized one of the observed cosmic ray events as a mesonic decay of a $^4_\Lambda H$ accompanied by a γ emission: $^4_\Lambda H \rightarrow \pi^- + {}^4He + \gamma$, $E_\gamma = (20 \pm 9)$. However, the authors themselves admitted that in case of unbalanced momentum an alternative interpretation with the emission of a neutron might be accepted: $^7_\Lambda He \rightarrow \pi^- + {}^6Li + n$. The alternative interpretation is certainly much more convincing.

The first summary review of the results obtained with cosmic rays in hypernuclear physics investigations was presented by Danysz at the Pisa Conference (12-18.6.1955). The list of all observed hypernuclear events was presented at the Budapest Conference (28-31.8.1956) by Filipkowski *et al.*, (1956). Similar summary reviews and new lists of the identified hypernuclei were subsequently presented by other physicists.

7. The Simplest Hypernuclear Structures

The Λ particle does not form bound states with a single nucleon and dibaryonic hypernuclear structures, $^2_\Lambda H$ and $^2_\Lambda n$, do not exist.

The hypertriton $^3_\Lambda H$ containing three different constituents Λ, p and n is the simplest hypernucleus and the only existing representative of tribaryonic hypernuclear systems.

Early experimental attempts undertaken to find a hyperdeuteron revealed some events only tentatively interpreted as examples of the looked for decays $^2_\Lambda H \rightarrow \pi^- (\pi^0)$ + $p(n)$ + p. However, they probably represented decays of the Σ^+ hyperons in flight or simply the hypertriton decays (George *et al.*, 1956). Similar observations carried out with cloud chamber technique could probably be explained by overlapping of the lambda V track on an incidental proton track (Alexander *et al.*, 1955). The other decay $^2_\Lambda n \rightarrow \pi^- + p + n$, could not have been distinguished from a common free Λ decay. Further systematic investigations gave no sign of the existence of these structures (Tripp, 1960) and the independent theoretical considerations showed also that a search for these structures was futile. In fact the binding energies of $^3_\Lambda H$ and other light hypernuclei indicated that the Λ particle interaction with nucleons was weaker than the NN interaction. Using the Λ binding energy as an input, the deduced ΛN potentials predicted no ΛN bound state (Blatt and Butler, 1956; Derrick, 1956).

At an early stage $^3_\Lambda H$ was considered to be a member of the isospin triplet comprising $^3_\Lambda n$, $^3_\Lambda H$ and $^3_\Lambda He$, making it reasonable to search for the two remaining hypernuclei, $^3_\Lambda n$ and $^3_\Lambda He$, stable under this assumption (Dalitz, 1955). However, all experimental investigations since have given no indication of their existence. In one case the suggested decay $^3_\Lambda n \rightarrow \pi^- + d + n$ turned out to be based on a not too certain estimate of the deuteron mass (Friedlander and Bercha, 1956). By the end of the fifties the problem was solved, since $^3_\Lambda H$ was recognized as an isospin singlet and the tribaryonic systems with isospin 1 were predicted not to form particle stable bound states (Downs and Dalitz, 1959).

8. Uniquely Identified Hypernuclei

In the first summary review of hypernuclear events presented in 1956 at the Budapest Conference on Cosmic Radiation the total number of events was equal to 72 and among them about 60 were produced by cosmic rays (Filipkowski *et al.*, 1956). However, only 27 events were considered to be uniquely identified. They represented 5 different hypernuclides: $^{3}_{\Lambda}\mathrm{H}$, $^{4}_{\Lambda}\mathrm{H}$, $^{4}_{\Lambda}\mathrm{He}$, $^{5}_{\Lambda}\mathrm{He}$ and $^{7}_{\Lambda}\mathrm{Li}$. The contemporary identification criteria were not very rigorous, nevertheless at least some of these hypernuclides were well recognized and the binding energy (B_{Λ}) of the Λ hyperon measured.

In the late fifties experimental physicists from all parts of the world started to form large collaborating teams, for instance the European K^- Collaboration with successively increasing number of member laboratories from Bristol, London, Brussels, Dublin, Warsaw, Berlin and Belgrade or in the United States the Enrico Fermi Institute and North Western University from Chicago. In this way the collected experimental data could be based on a statistically significant material. Enormous effort was spent by these teams and especially by the laboratories of the European K^- Collaboration. Hundreds of thousands individually analyzed hypernuclear events recorded in emulsion made it possible to identify more than ten thousands hypernuclei, all of them according to very stringent criteria. After 15 years of the work twenty two different hypernuclides were discovered. All of them have been observed in their ground, particle stable states. The list of these hypernuclides is presented in Fig. 3.

Z \ A	3	4	5	6	7	8	9	10	11	12	13	14	15	16
1	$^{3}_{\Lambda}\mathrm{H}$	$^{4}_{\Lambda}\mathrm{H}$												
2		$^{4}_{\Lambda}\mathrm{He}$	$^{5}_{\Lambda}\mathrm{He}$	$^{6}_{\Lambda}\mathrm{He}$	$^{7}_{\Lambda}\mathrm{He}$	$^{8}_{\Lambda}\mathrm{He}$	·							
3					$^{7}_{\Lambda}\mathrm{Li}$	$^{8}_{\Lambda}\mathrm{Li}$	$^{9}_{\Lambda}\mathrm{Li}$	·		·				
4					$^{7}_{\Lambda}\mathrm{Be}$	$^{8}_{\Lambda}\mathrm{Be}$	$^{9}_{\Lambda}\mathrm{Be}$	$^{10}_{\Lambda}\mathrm{Be}$	·	·	·			
5							$^{9}_{\Lambda}\mathrm{B}$	$^{10}_{\Lambda}\mathrm{B}$	$^{11}_{\Lambda}\mathrm{B}$	$^{12}_{\Lambda}\mathrm{B}$	·	·	·	·
6								·	·	$^{12}_{\Lambda}\mathrm{C}$	$^{13}_{\Lambda}\mathrm{C}$	$^{14}_{\Lambda}\mathrm{C}$	·	·
7											·	·	$^{15}_{\Lambda}\mathrm{N}$	·
8												·	·	·

Fig. 3. Hypernuclides uniquely identified with the emulsion technique.

The dots in this figure indicate species so far not identified, although certainly existing as particle stable hypernuclides. Recently hypernuclides were also identified with the use of the counter technique, at first only in the excited states and next in the ground states as well ($^{16}_{\Lambda}\mathrm{O}$, $^{32}_{\Lambda}\mathrm{S}$, $^{40}_{\Lambda}\mathrm{Ca}$ and others; see e.g., Povh, 1978)

9. Cryptofragments, Discovery of Heavy Nuclei

The Λ hyperon produced in the strange particle capture by a heavy emulsion nucleus may be bound with the residual part of the target nucleus. A heavy hypernucleus obtained in this way cannot be detected easily, because the production star and the hypernucleus decay star practically coincide in space: However, Friedlander *et al.* (1955) and also Goldsack and Lock (1966) noticed that heavy hypernuclei produced by the hyperons should manifest their existence in the energy balance considerations. The sum of two energies released in the production and nonmesonic decay processes is always much higher than the energy observed in the production of a free lambda hyperon only. The K^- Stack Collaboration confirmed these expectations and the existence of heavy hypernuclei was thus proved (Bonetti, 1957).

Directly undetectable heavy hypernuclei have been called cryptofragments. They were produced in Σ^- and K^- captures and the production rates for the K^- capture was estimated in 1961 (Davis *et al.*, 1961; Filipkowski *et al.*, 1962).

An attempt to detect apparent heavy hypernuclear events among the fission fragments induced by K^- mesons has failed (Levi-Setti and Slatter, 1959). They were clearly observed only when 800 MeV/c K^- mesons were used (Jones *et al.*, 1962). K^- mesons interacting in flight transferred sufficiently high momenta to the residual recoils with bound Λ particles to separate hypernuclear decay stars from the production stars. Systematic investigation of heavy hypernuclei produced by high momentum K^- beams were continued by the European K^- Collaboration between 1961 and 1968. The binding energies of heavy hypernuclei obtained in those investigation have led to estimates of the Λ binding energy in an infinite nuclear matter. However, none of the observed heavy hypernuclei could be identified uniquely in the emulsion and their Z and A numbers were known only approximately.

At present some heavy hypernuclei obtained in two body strangeness exchange reactions might be identified with the use of the counter technique.

10. Double Hypernuclei

So far only two double hypernuclei, i.e., with two lambdas bound in a single hypernuclear structure, have been identified, both found in emulsion irradiated with high momentum K^- beams producing doubly strange Ξ^- hypersons. The first event was a double hyperberylium 10 or 11 found in 1963 in Warsaw in an emulsion stack of the European K^- Collaboration (Danysz *et al.*, 1963) and the second one, the double hyperhelium 6, was discovered three years later in Los Angeles (Prowse, 1966). Some other examples have not been uniquely identified, or else interpreted as short lived double hypernuclear resonances decaying into two separate hypernuclei (Wilkinson *et al.*, 1969).

Some physicists discussed a possible existence of triple hypernuclei, which would manifest the triple strangeness. In principle they could be produced by the Ω^- hyperons captured e.g., by carbon or oxygen nuclei (Dalitz, 1963). Using a shell model description, two of the Λ particles can be put in the s-shell but the third lambda owing to the Pauli principle should be bound in the p state. Accordingly, one cannot look for very light triple hypernuclei if their ground states have to be particle stable. So far such structures have not yet been observed.

11. Are There Any Other Hypernuclear Structures?

Initially at every stage of hypernuclear studies the possible existence of hypernuclear structures containing hyperons other than the lambda was thoroughly considered and looked for. In these investigations people showed interest in analysing all peculiar hypernuclear events whenever the estimated energy was incompatible with the decay energy of the Λ hyperon. However, the suggested interpretations were never unique and sometimes even inconsistent with the subsequently established basic features of the elementary particles. Most probably the identification of the decay particles and the estimate of their kinetic energies could be questioned, in particular nonmesonic decays might be mistaken for mesonic ones. The analysed events were found in emulsion stacks irradiated at first by cosmic rays and later on by the artificially constructed beams.

Today, one can easily show by simple energy conservation arguments that in general no other strange baryon can form a new variety of hypernuclear structures stable with respect to a fast decay.

In Table 1A we list fast elementary processes resulting from strong hyperon-nucleon interaction. In all these processes the hyperons heavier than the Λ hyperon are transformed into a lambda. Thus the Ω^- hyperon cannot form any bound hypernuclear structure with nucleons, however, one cannot exclude the binding of the Σ and the Ξ hyperons in some very simple systems like $\Sigma^- n$, $\Xi^- n$, $\Sigma^- nn$, $\Xi^- nn$ or in their charge symmetric counterparts. A discussion of this problem was initiated by Holladay in 1955. Since the energy conservation considerations were not conclusive in the case of simple ΣN structures, many theoretical papers assuming various ΣN potentials were published in the late fifties. The conclusion reached at that time was that these dibaryonic systems probaly did not exist as bound structures or they were only loosely bound. Some theoreticians have come back to the problem recently and experiments have been started with the heavy hydrogen bubble chamber irradiated by low momentum K^- beams. If there is no ultimate answer with this method one might still try to solve the problem by carrying through the hypernuclear spectroscopy experiments planned for the investigations of the Σ hypernuclear resonances.

Another interesting example is the ΞN dibaryonic system with the total electric charge equal to 0. In principle two states $^1S_0$ and $^3S_1$ might be considered and each of these states should be a mixture of two isospin states with $I = 0$ and $I = 1$ (Dalitz, 1969). The $^1S_0(\Xi N)$ state should decay in a fast process into $\Lambda + \Lambda$, however, owing to the Pauli principle, the $^3S_1(\Xi N)$ state may only decay via the electromagnetic interaction, $^3S_1(\Xi N) \rightarrow \Lambda + \Lambda + \gamma$. It would be interesting to prove the existence of this structure with the life time of 10^{-19}s.

Table 1

A. List of fast elementary processes ($Y + N$).

$\Sigma^+ + n = \Lambda + p + 75.1$ MeV, $\Xi^0 + n = \Lambda + \Lambda + 23.3$ MeV,

$\Sigma^0 + N = \Lambda + N + 76.9$ MeV, $\Xi^- + p = \Lambda + \Lambda + 28.4$ MeV,

$\Sigma^- + p = \Lambda + n + 80.5$ MeV.

$\Omega^- + n = \Xi^- + \Lambda + 174.9$ MeV, $\Omega^- + p = \Xi^0 + \Lambda + 180.0$ MeV.

B. Hyperons interacting with bound mucleons.

$\Xi^- + \alpha = {}^3H + 2\Lambda + 8.6$ MeV, $\Xi^0 + \alpha = {}^3He + 2\Lambda + 2.7$ MeV,

$\Xi^- + \alpha = {}^4_\Lambda H + \Lambda + 10.6$ MeV, $\Xi^0 + \alpha = {}^4_\Lambda He + \Lambda + 5.1$ MeV.

More complex structures were considered by Podgoretski in 1957. He suggested a possible binding of a Ξ hyperon to some light nuclear cores and in particular to the α particle, to be expected from the low values of the energy released in the fast processes of interaction with bound nucleons as indicated in Table 1B. The answer to the problem depends strongly on the Ξ binding energy and the existence or nonexistence of the $_{\Lambda\Lambda}^{5}\mathrm{He}$ and $_{\Lambda\Lambda}^{5}\mathrm{H}$ double hypernuclei.

The possible existence of hypernuclear structues composed of hyperons only may also be considered, however, for the present one cannot find any experimental approach as to how one should produce or investigate such structures.

In astrophysics it is argued that in a contraction process of some very dense stars ($d \sim 10^{15}$ g cm^{-3}) protons and electrons are spontaneously transformed first into neutrons and then into hyperons. The highest hyperon concentration is expected to be in the star centre and for the leptons at its periphery. The whole star might be called a "giant hypernucleus". The hyperon stability or their decay processes are governed by the energy equilibrium of the star as a whole. The hyperon decay might be inhibited in those stars more or less as the decay of a bound neutron is inhibited in a nucleus.

12. Supernuclei = Charmed Nuclei ?

The discovery of the charmed particles encouraged physicists to consider the possible binding of charmed baryons with nucleons and to look for a new variety of nuclear structures. this problem is being discussed currently by many theoreticians (Tyapkin, 1975; Iwao, 1977; Dover and Kahana, 1977). Some people suggested calling the charmed baryons – superons, and the new nuclear structures – supernuclei.

The experimental efforts made to observe the supernuclei have not so far led to an explicitly positive result. Some events tentatively ascribed to supernuclei were found in nuclear emulsion irradiated by 70 and 250 GeV protons. (Batusov *et al.*, 1981).

Certainly one should expect that supernuclei do really exist, although there is no easy way to produce and observe these structures. In the case of interacting hadrons the production rates are rather low. The short lived charmed mesons could not be used in a smiliar way as K^- mesons have been used for the hypernucleus production. It would be interesting, if high energy leptons, e.g., neutrinos, were more effective to produce these structures.

It is not out of place to mention here once more the early investigations of the odd hypernuclear events observed in emulsion irradiated by cosmic rays. Among those odd events some hypernuclear decays with the emission of K mesons were reported (Fry *et al.*, 1955b, c; Varfolomeev *et al.*, 1956 b; Grigoreev *et al.*, 1957; Bannik *et al.*, 1958 see also the events produced by pions, Azimov *et al.*, 1960).

Two interesting facts have been noticed. The energy balance was nearly the same, at least for some of those events, and in four cases the primaries were neutral. Moreover, no such events have subsequently been reported with K^- mesons interacting at rest or in flight.

May we say now that they were the first examples of charmed nuclei produced, let us say, by neutrinos, or just that the identification of secondly K mesons was incorrect?

The only argument used for the identification of K mesons was the rough estimate of their masses based on the ionization measurements. This could not be confirmed by the chracteristic decay or any interaction of the assumed K mesons. The number of the observed events would not find a justification in the intensity and cross sections of high energy cosmic ray neutrinos. Most probably the standardization of the emulsion was not too good for reliable ionization measurements and then it would be certainly extremely risky to consider those odd events as the first supernuclear events.

Recapitulating this retrospective review it should be stressed that the cosmic rays made it possible to discover the first hypernuclear events and to create thereby a new field of research in physics. Cosmic rays also enabled us to start and develop systematic investigations in hypernuclear physics and to continue these studies until the accelerators could replace them.

In the 30 years of hypernuclear investigations many basic features of these structures have been established and used for the study of lambda nucleon interaction.

In the sixties hypernuclear spectroscopy investigations were started with the use of the emulsion technique, the possible existence of hyperhelium 7 isomer was discussed and next the existence of the hypercarbon 12 resonant state was established, the latter being the first step in pion spectroscopy experiments. In the seventies enormous progress in hypernuclear spectroscopy was made with the use of the counter technique first in the gamma and next in the pion spectroscopy.

REFERENCES

Alexander, G., Ballario, C., Bizarri, R., Brunelli, B., De Marco, A., Michelini, A., Moneti, G.C., Zavattini, E., Zichichi, A. and Astbury, J.P.: 1955, *Nuovo Cimento* 2, 365.

Amaldi, E., Anderson, C.D., Blackett, P.M.S., Fretter, W.B., Leprince-Ringuet, L., Peters, B., Powell, C.F., Rochester, C.D., Rossi, B. and Thomson, B.W.: 1954, *Nuovo Cimento* 11, 213.

Ammar, R.G.: 1964, *Proc. Int. Conf. on Hyperfragments, St. Cergue, 1963, CERN* 64-1, p.7.

Armenteros, R., Barker, K.H., Butler, C.C., Cachon, A. and Chapmann, A.H.: 1951, *Nature* 167, 501.

Azimov, S.A., Gulyamov, U.G., Karimova, R. and Rakhimbayev, B.G., 1960, *Zh. Eksp. Teor. Fiz.* 38, 697.

Bannik B.P., Gulyamov, U.G., Kopylova, D.K. and Nomofilov, A.A., 1958, *Zh. Eksp. Teor. Fiz.* 34, 286.

Barkas, W.H., Dudziak, W.F., Giles, P.C., Heckman, H.H., Inman, F.W., Mason, C.J., Nickols, N.A. and Smith, F.N.: 1957, *Phys. Rev.* 105, 1417.

Batusov, Ya.A., Bunyatov, S.A., Lyukov, V.V., Sidorov, V.M., Tyapkin, A.A. and Yarba, V.A.: 1981 JETP Lett 33, 56.

Blatt, J.M. and Butler, S.T.: 1956, *Nuovo Cimento* 3, 409.

Bonetti, A.: 1957, *K-Stack Collaboration, Int. Conf. on Mesons and Recently Discovered Particles,* Padova-Venezia, Com. II 79.

Bonetti, A., Levi-Setti, R., Panetti, M. and Tomasini, G.; 1953, *Nuovo Cimento* 10, 345.

Bonetti, A., Levi-Setti, R., Panetti, M. and Tomasini, G.: 1954, *Nuovo Cimento* 11, 210, 330.

Bridge, H., Courant, H. and Rossi, B.: 1952, *Phys. Rev.* 85, 159.

Brown, R., Camerini, U., Fowler, P.H., Muirhead, H., Powell, C.F. and Ritson, M.D.: 1949, *Nature* 163, 82.

Ceccarelli, M., Dallaporta, N., Grilli, M., Merlin, M., Saladin, G., Sechi, B. and Ladu, M.: 1955, *Nuovo Cimento*, 2, 542.

Cheston, W. and Primakoff, H.: 1953, *Phys. Rev.* 92, 1537.

Crussard, J. and Morellet, D.: 1953, *C.R. Acad. Sci. Paris* 236, 64.

Dalitz, R.H.: 1955, *Phys. Rev.* **99**, 1475.

Dalitz, R.H.: 1964, *Proc. Int. Conf. on Hyperfrahments, St. Cergue, 1963, CERN* 64-1, 201.

Dalitz, R.H. and Downs, B.W.: 1958, *Phys. Rev.* **111**, 967.

Danysz, M.: 1956, *Proc. Pisa Conf., 1955, Nuovo Cimento Suppl.* **4**, 609.

Danysz, M. and Pniewski, J.: 1952, *Bull. Acad. Polon. Sci.* **3**, 42.

Danysz, M. and Pniewski, J.: 1953, *Phil. Mag.* **44**, 348.

Danysz, M., Garbowska, K., Pniewski, J., Pniewski, T., Zakrzewski, J., Fletcher, E.R., Lemmone, J., Renard, P., Sacton, J., Toner, W.T., O'Sullivan, D., Shah, T.P., Thompson, A., Allen, P., Heeran, Sr.M., Montwill, A., Allen, J.E., Beniston, M.J., Davis D.H., Garbut, D.A., Bull, V.A., Kumar, R.C. and March, P.V.: 1963, *Nucl. Phys.* **49**, 121.

Davis, D.H., Csjethey-Barth, M., Sacton, J., Jones, B.D., Sanjeevaiach, B. and Zakrzewski, J.: 1961, *Nuovo Cimento* **22**, 275.

Debenedetti, A., Garelli, C.M., Tallone, L. and Vigone, M.: 1954, *Nuovo Cimento* **12**, 466.

Derrick, G.H.: 1956, *Nuovo Cimento* **4**, 565.

Dover, C.B. and Kahana, S.H.: 1977, *Phys. Rev. Lett.* **39**, 1506.

Downs, B.W. and Dalitz, R.H.: 1959, *Phys. Rev.* **114**, 593.

Evans, D., Jones, B.D. and Zakrzewski, J.: 1959, *Phil. Mag.* **4**, 1255.

Filipkowski, A., Gierula, J. and Zieliński, P.: 1957, *Proc. Int. Conf. on Cosmic Radiation, 1956*, Hungarian Acad., Sci, Budapest, p. 145; 1956. *Acta Physica Polonica* **16**, 141.

Filipkowski, A., Marquit, E., Skrzypczak, E. and Wróblewski, A.: 1962, *Nuovo Cimento* **25**, 1.

Fowler, W.B., Shutt, R.P., Thorndike, A.M. and Whittemore, L.: 1953, *Phys. Rev.* **91**, 1287; 1954, **93**, 861.

Friedlander, M.W., Fujimoto, Y., Keefe, D. and Menon, M.G.K.: 1955, *Nuovo Cimento* **2**, 90.

Friedlander, E. and Bercha, S.: 1956, *Dokl. Akad. Nauk SSSR* **107**, 51.

Fry, W.F., Schneps, J. and Swami, M.S.: 1955a, *Phys. Rev.* **99**, 1561.

Fry, W.F., Schneps, J. and Swami, M.S.: 1955b, *Nuovo Cimento* **2**, 346; 1955c, *Phys. Rev.* **97**, 1189.

Fry, W.F., Schneps, J., Snow, G.A., Swami, M.S. and Wold, D.C.: 1957, *Phys. Rev.* **107**, 257.

Gell-Mann, M.,: 1953, *Phys. Rev.* **92**, 833.

Gell-Mann, M.: 1956, *Proc. Pisa Conf., 1955, Nuovo Cimento Suppl.* **4**, 848.

Gell-Mann, M. and Pais, A.: 1955, *Proc. Glasgow Conf. on Nuclear and Meson Physics, 1954*, Pergamon Press, p.342.

George, E.P., Hertz, A.J., Noon, J.H. and Solntseff, N.: 1956, *Nuovo Cimento* **3**, 94.

Goldsack, S.J. and Lock, W.O.: 1956, *Nuovo Cimento* **3**, 600.

Grigoreev, A.P., Toporkova, Ye. P. and Fiesenko, A.J.: 1957, *Zh. Eksp. Teor. Fiz.* **32**, 1589.

Hill, R.D., Salant, E.O., Widgoft, M.W., Osborn, L.S., Pevsner, A., Ritson, D.M., Crussard, J. and Walker, W.D.: 1954, *Phys. Rev.* **94**, 797.

Holladay, W.G.: 1955, Private suggestion, e.g., cited by Sachs, R.G., *Phys. Rev.* **99**, 1573.

Iwao, S.: 1977, *Lett. Nuovo Cimento* **19**, 647.

Jones, B.D., Sanjeevaiah, B., Zakrzewski, J., Csjethey-Barth, M., Langnaux, J.P., Sacton, J., Beniston, M.J., Burhop, E.H.S. and Davis, D.H.: 1962, *Phys. Rev.* **127**, 236.

Lal, D., Yash, Pal and Peters, B.: 1953, *Proc. Ind. Acad. Sci.* **A33**, 398.

Leighton, R.B., Wanlass, S.D. and Alford, W.L.: 1951, *Phys. Rev.* **83**, 843.

Leighton, R.B., Wanlass, S.D. and Anderson, C.D.: 1953, *Phys. Rev.* **89**, 148.

Leprince-Ringuet, L. and l'Heritier, M.: 1944, *C.R. Acad. Sci. Paris* **219**, 618.

Levi-Setti, R. and Slater, W.: 1958, *Phys. Rev.* **111**, 1395.

Levi-Setti, R. and Slater, W.E.: 1959, *Nuovo Cimento* **14**, 895.

Lyulka, V.A.; 1963, *Zh. Eksper. Teor. Fiz.* **45**, 164.

McNamee, P. and Oakes, R.J.: 1966, *Phys. Rev.* **149**, 1157.

Nakano, T. and Nishijima, K.: 1953, *Prog. Theor. Phys.* **10**, 581.

Naugle, J.E., Ney, E.P., Freier, P.S. and Cheston, W.B.: 1954, *Phys. Rev.* **96**, 829, 1383.

Nishijima, K.: 1954, *Prog. Theor. Phys.* **12**, 107.

Nishijima, K.: 1955, *Prog. Theor. Phys.* **13**, 285.

Pais, A.: 1952, *Phys. Rev.,* **86**, 663.

Peaslee, D.C.: 1952, *Phys. Rev.* **86**, 127.

Peaslee, D.C.: 1953, *Prog. Theor. Phys.* **10**, 227.

Podgoretski, M.I.: 1957, *Hypernuclei, Conf. on Use of Thick Emulsion*, Dubna.

Povh, B.: 1978, *Ann. Rev. Nucl. Part. Sci.* **28**, 1.

Prowse, D.J., 1966, *Phys. Rev. Lett.* **17**, 782.

Rochester, G.D. and Butler, C.C.: 1947, *Nature* **160**, 855.

Schein, M., Haskin, D.M. and Leenow, D.: 1955, *Phys. Rev.* **100**, 1455.

Schneps, J.: 1958, *Phys. Rev.* **112**, 1335.

Schneps, J., Swami, M.S. and Fry, W.F.: 1955, *Phys. Rev.* **100**, 1263.

SUN Commission: 1978, *Symbols, Units and Nomenclature in Physics, Doc. UIP* 20.

Tidman, D.A., Davis, G., Hertz, A.J. and Tennet, R.M.: 1953, *Phil. Mag.* **44**, 350.

Tripp, R.D.: 1958, *Proc. Annual Conf. on High Energy Physics at CERN* p.184.

Tyapkin, A.A.: 1975, *Yad. Fiz.* **22**, 181.

Varfolomeev, A.A., Gerasimova, R.J. and Karpova, L.A.: 1956a, *Doki. Akad. Nauk SSSR* **110**, 758.

Varfolomeev, A.A., Gerasimova, R.J. and Karpova, L.A.: 1956b, *Dokl. Akad. Nauk SSSR* **110**, 959.

Wilkinson, D.H., StLorant, S.J., Robinson, D.K. and Lokanathan, S.: 1959, *Phys. Rev. Lett.* **3**, 397.

PARTICLE THEORY – COSMIC RAYS – ACCELERATORS CONFLICTS AND RECONCILIATIONS

Eugene L. FEINBERG*

1. Introduction

Half a century ago D.V.Skobeltzyn placed a cloud chamber in a magnetic field and discovered the cosmic ray particles with energies greatly exceeding those of natural radioactivity (Skobeltzyn 1927, 1929). Almost immediately W.Bothe and W.Kolhörster confirmed this conclusion using their famous counter coincidence method (Bothe and Kolhörster, 1928, 1929). From the standpoint of theory this meant the birth of high energy particle physics. From the experimental standpoint, for many years to come, two basic tools of research have been established. During the two decades preceeding the appearance of accellerators of superrelativistic particles, cosmic ray physics was united with particle theory. However when this third partner – powerful and rapidly growing – stepped in the inter-relations within the triangle became complicated. Competition and alternating sympathies, neglection, and disbelief of many theoretician in the former companion, a disbelief which many years later turned out to be unfounded, replaced one another. I shall illustrate this history by two examples: for the period of unclouded union of particle theory and cosmic rays – by considering the history of the meson problem; for the subsequent decades – by the changing views concerning the mechanism of multiple production.

2. The Plight of the Meson Concept

Up to 1932 ideas about the raw material which God had at hand for constructing the world was simple: electrons, protons and photons. The discovery of positron was a triumph for the Dirac theory and produced no confusion. It was a pure joy and a source of inifinite fantasies about matter and antimatter. The discovery of neutron as a neutral version of proton, vaguely anticipated by Rutherford, in a very natural way solved some puzzles concerning nuclear properties. And together with it as a central problem of nuclear physics the problem of nuclear forces arose.

In 1934 Fermi proposed his brilliant theory of beta decay and at once it was realized that it meant the possibility of new forces between nucleons mediated by the exchange of pairs of particles – electron or positron plus neutrino or antineutrino. I.Tamm was one of the two or three men who immediately understood it. Moreover he produced a quantitative theory, only to find out that these "β-forces" were far too weak to explain the stability of nuclei (Tamm, 1934). He found four possible modifications of the Fermi scheme and tried all of them, but in vain. I remember his disappointment being so great that he could not force himself to write the full paper after the extremely short letter mentioned above. Only in the summer of 1936 did he

* P.N. Lebedev Physical Institute, Acad., Sci., Moscow, U.S.S.R.

Y. Sekido and H. Elliot (eds.), Early History of Cosmic Ray Studies, 339–353.
© *1985 by D. Reidel Publishing Company.*

do so (Tamm, 1936).

In the meantime two events took place. One was a misleading zigzag, the other a great step in science. I mean by the first the conclusion of experimentalists that the original Fermi "Ansatz" for the $e\nu N$ (electron-neutrino-nucleon) interaction gave the wrong shape for the β-spectra. In order to improve the situation E.Konopinski and G.Uhlenbeck proposed a new formula for the interaction Hamiltonian constructed not of the ψ_e and ψ_ν wave operators but of their derivatives over four-coordinates x_α, namely

$$H_{int} = ig \frac{\partial^s \psi_e^+}{\partial x_\alpha{}^s} \gamma_\mu \frac{\partial}{\partial x_\mu} \frac{\partial^s \psi_\nu}{\partial x_\alpha{}^s}, \tag{1}$$

g being the coupling constant, s - some integer (Konopinsky and Uhlenbeck, 1935). For several years some experimentalists were enthusiastic: the theory agreed with measured spectra much better. This Hamiltonian enchanted even many theorists, although there were very serious warnings against it. In fact, when introducing higher order time derivatives we violate the basic principle of quantum mechanics, since in that case the time evolution of a system governed by the Schrödinger equation should be determined not only by a wave function fixed at some moment $t = t_0$ but also by some of its time derivatives at the same moment. Nevertheless even Heisenberg could not resist the temptation (I shall come to this point later again).

It very soon turned out, however, that disagreement of β-spectra with the initial Fermi Hamiltonian was fictitious and to a considerable extent resulted from the superposition of several spectra of successive decays. Accordingly the Konopinski-Uhlenbeck version was discarded. Now it is forgotten. Were it correct, then for high energy exchanged electrons and neutrinos each differentiation would give a large extra energy-momentum factor and we could somehow hope to obtain sufficiently strong β-forces within a nucleus.

The second event was a real revolution. I mean the appearance of the remarkable and fantastic meson idea by H.Yukawa (Yukawa, 1935). Since β-forces according to Tamm were too weak, Yukawa argued, then there should exist quanta of some new field with a strong coupling to nucleons, their mass being determined by the known range of nuclear forces. This idea was almost unbearably bold. Even the simple connection of the range of forces with the mass of the field particle noticed by Yukawa, which in our times seems obvious, was not easily digestable. Only three years later when G.Wick (Wick, 1938) explained it plainly by indeterminacy relation it became conceivable to everybody. But Yukawa, later together with his collaborators (S.Sakata and M. Taketani), had not been marking time. They were developing the idea by adding new bold hypotheses. The meson was strongly coupled to nucleons, but wherefrom does originate β-decay? Already in 1935 Yukawa did not hesitate to introduce the rare ("weak") decay of the meson into an electron and a neutrino. Therefore the meson spin became an integer. By choosing a reasonable coupling constant and, therefore, life time of the meson, it turned out possible to fit the β-decay data. The spin dependance of nuclear forces seemed to indicate a nonzero spin. Accordingly H. Fröhlich, W. Heitler and N. Kemmer (Fröhlich *et al.*, 1938) and N. Kemmer (Kemmer, 1938) constructed the spin one theory of meson and of meson nuclear forces and this theory was worked out in all details by various theoriests. By and by physicists all over the world were becoming influenced by the meson theory.

The Yukawa meson with the mass of some 200-300 electron masses was to be searched for in cosmic rays where energies were encountered sufficiently large for creating such mesons in matter (from the lattitude effect of cosmic ray flux it had already been known that the spectrum of charged particles coming from space had extended at least to some 30 GeV). And they *were* found. Of course it could be anticipated even earlier due to the existence of two components of cosmic rays in the atmosphere – the soft and the hard ones – with drastically different absorptions. They could be seen in various experiments even from a penetrating ability curve presented as far back as 1929 (Bothe and Kohlhörster, 1929). But "there are none so deaf as those who won't hear". Many people, for instance, preferred to believe that somehow at the energy above $137\ m_e c^2 \sim 10^8$ MeV (m_e is the electron mass) the electromagnetic interactions were not governed by the then known quantum electrodynamics. Why? This is very interesting psychologically. On the one hand theorists firmly believed that an elementary particle necessarily must be pointlike. Otherwise, if it is for example as hard as a classical electron, then within it the relativistic limitations on the signal velocity would be violated. You could find such a statement even in some superior textbooks, like the one by L. Landau and E. Lifshitz. On the other hand, the classical idea of the electron radius $r_e = e^2/m_e c^2$ was still influential, as a valid bound for the Lorentz electrodynamics (at smaller wave length the radiation reaction term in the Lorentz theory was impermissibly large). One did not realize that such a limitation could pertain at best to the *transferred* energy (even here it was unfounded, as became clear much later), while in a process like Bremsstrahlung and pair creation, as had been shown already in 1934 by C.F. von Weizsäcker (Weizsäcker, 1934) and by E.J.Williams (Williams, 1934), the transferred energy usually is much smaller than $137\ m_e c^2$ at arbitrarily high energy of colliding particles. Thus there might be no such forbidding threshold. Nevertheless some vague mixing of point-like particle and classical electron radius ideas (erroneous, as we know at present) was a serious obstacle and physicists became *en mass* "wanting to hear" about mesons only about 1938.

Nobody who recollects those times can forget the impression produced by the photograph by S.H.Neddermeyer and S.D.Anderson of a cloud chamber with a counter inside from which a thick track of a slow meson emerged (Neddermeyer and Anderson, 1938). Of course, no less important were their earlier studies of the energy loss of relativistic charged cosmic ray particles in a metal plate: the ones accompanied by other particles, i.e. belonging to "showers", were loosing their energy as prescribed by electrodynamics for electrons. But the unaccompanied particles were not (Neddermeyer and Anderson, 1937). Now people could put two and two together: the heavy particles could not radiate γ-quanta efficiently and thus the unaccompanied particles were interpreted as mesons. The "normal" group, on the countrary, consisted of electrons.

Of course there were many other important experiments which contributed to the same conclusion, and many celebrated names originated herefrom. Their story may be found, for instance, in Chapter I of the well known S.Hayakawa's book (Hayakawa, 1969) or in the collected papers edited by W.Heisenberg (Heisenberg, ed., 1943).

This first triumph of Yukawa theory came at a time when it became comprehensible to everybody particularly due to Wick (see above). However, it was only the first part of the complex scheme, namely, the very existence of the "semiheavy" particle, the meson. To complete the picture first, its decay into e + ν, second, its integer spin, almost definitely equal to one, and third, the strong coupling to nucleons, securing nuclear stability had to be cleared. Here new problems appeared.

The decay was actually found and the meson life time was measured only much later, in 1941–1943 (Rasetti, 1941; Rossi and Neresson, 1942, 1943). But long before that, the idea of decay had been indirectly used for a very elegant explanation of cosmic ray flux temperature variations by P.M.S.Blackett (Blackett, 1938) and in a particularly extensive and convincing way by H.Euler and W.Heisenberg (Euler and Heisenberg, 1938) who in a review paper succeeded in showing how the meson decay helped to understand some cosmic ray effects. The decay was accepted as an established fact but the meson life time of about 10^{-6} sec needed in these analyses was two orders of magnitude larger than the one according to the refined Yukawa scheme of β-decay (Yukawa and Sakata, 1938, 1939). The discrepancy was persistent and embarassing.

The second nail sticking out within the boot was the spin problem. For spin 1 the calculations led to electromagnetic effects (Bremsstrahlung etc.) increasing with energy which is not permissible (Smorodinsky, 1940; Booth and Wilson, 1940, etc.). Moreover Tamm proved that a spin one charged particle had no stable state in a coulombian field. For instance it should fall into the nucleus (Tamm, 1940). This impressed L.Landau to such an extent that he concluded that the meson must have a finite size (Landau 1940). Another idea was due to Ginzburg who noticed that a particle with a magnetic moment different from that of the Dirac particle had a magnetic moment radiation reaction which dominated over the electric charge radiation reaction. Accordingly he looked for the solution of the puzzle by constructing a theory of particle capable of being in states with different spins (Ginzburg, 1941, 1943, 1944; sea also a general theory in Ginzburg and Tamm, 1947).

It was only later that burst experiments indicated the half integer spin of the penetrating cosmic ray mesons (Christy and Kusaka, 1941) and made these theoretical complications unnecessary. However these experiments contradicted the integral spin necessary for the nuclear forces.

And, finally, the third and most striking puzzle of all: If mesons interact strongly enough to secure nuclear stability, then why do cosmic ray mesons penetrate so easily through the atmosphere and various thick screens? This contradiction was still amplified by speculations on the mechamism of meson production in the upper layers of the atmosphere. For example the latitude effect observed in the lower part of the atmosphere was too small for the penetrating particles of rather low energy which were found here and should be very sensitive to the Earth's magnetic field. The answer was that a high energy primary particle produces several lower energy mesons in the top layers. This and similar other arguments led some theorists, like L.W.Nordheim and M.H.Hebb (Nordheim and Hebb, 1939) to the idea that one high energy interaction gives some three or four mesons. But in the usual quantum electrodynamics (QED) this is practically impossible. This again led to the idea that QED is incorrect at high energies.

The problem of the validity of QED has already been mentioned above. The full theory of Bremsstrahlung and pair creation had been built as far back as 1933-34 (Heitler, 1933; Bethe and Heitler, 1934). All suspicions concerning the use of the strange "hole theory" by Dirac died away when W.Pauli and V.F.Weisskopf demonstrated that very similar formulas followed for production of pairs of electrically charged Bose particles which needed no filled vacuum and no holes (Pauli and Weisskopf, 1934). Nevertheless people could not even understand the possibility of multiplication of electrons which seemingly followed from the shower like phenomena which had already been noticed by Skobeltzyn (Skobeltzyn, 1929) and were clearly demonstrated by many experiments, particularly by the B.Rossi multicoincidence reasearches (Rossi, 1933) and by the discovery of Auger showers (Auger *et al.*, 1938), now called extensive air showers. Even in 1936 W.Heitler believed that QED fails at $E > 137 \, m_e c^2$ and here disagrees with experiment (Heitler, 1936, see section 24.2). It was only in 1938 that Heisenberg gave another criterium for its validity based on the magnitude of *transferred* 4-momentum. It moved the bound farther away (Heisenberg, 1938; at present it is also known to be fictitious). But already in 1937 Heitler himself together with Bhabha (Bhabha and Heitler, 1937), as well as J.F.Carlson and J.R.Oppenheimer (Carlson and Oppenheimer, 1937) constructed the cascade theory of electromagnetic showers, and the problem became clarified.

At present it may seem unbelievable that before that time the cascade multiplication was considered as inefficient, but this can be easily proved. In 1936 in an attempt to solve the puzzle of multiplication Heisenberg proposed the theory of *simultaneous* multiple production of electrons (and neutrinos) on the basis of the Konopinski-Uhlenbeck theory of β-decay mentioned above. He used the fact that in Eq. (1), as has already been mentioned, for a sufficiently large number of differentiations s at high energies the weak $e\nu N$ interaction can become strong. In such a situation perturbation theory breaks down, all successive approximations contribute equally and true multiple production becomes possible. However, during the next year this theory was discarded together with the Konopinski-Uhlenbeck Hamiltonian. Moreover, it turned out to be unnecessary as the cascade theory was put forward. It explained rather satisfactorily the "Pfotzer curve" — the increase of cosmic ray flux with altitude up to a maximum close to the top of the atmosphere and the subsequent decrease down to the primary cosmic ray flux — as a pure electron–photon cascade process. The necessary primary electron energy (7÷10 GeV) nicely fitted the estimate according to the latitude effect. Moreover, it left enough place for penetrating mesons at low and medium altitudes (when translating the book by Heitler for the Russian edition in 1939 I had already the opportunity to write and include into it two additional sections describing the Bhabha-Heitler cascade theory and the successful explanation of the Pfotzer curve). However the Heisenberg paper contained some very interesting ideas which we shall discuss in the next section.

Going back to the meson problem I should like to close at this stage. As we have seen in 1938-39 almost everything was ready for the final solution of the meson puzzle and cosmic ray physicists were inevitably approaching the true explanation by means of the two meson scheme — the "nuclear active" pion (discovered some ten years later by C.M.G.Lattes, G.Occhialini and C.F.Powell (Lattes *et al.*, 1947)) plus the penetrating leptonic muon. It was exactly in those years that decisive steps were taken: validity of electrodynamics beyond the false limit 137 $m_e c^2$ was established; meson decay was indirectly but safely demonstrated, its life time some 100 times exceeding that of the Yukawa "nuclear" meson; necessity of multiple production of mesons (with average multiplicity of some 3-5 per event at the energy of 10-30 GeV) was becoming consistently recognized. The stream of discoveries was increasing. But then World War II broke out and the final decision was postponed for almost a decade. It was only in USA, still enjoying peace, that for a few years the investigations on meson problem were still in progress.

My aim was not to describe this final decision. I intended merely to show the entangled history, the unconceivable misunderstandings and the triumphant victories of the interpenetrating particle theory and cosmic ray experiment.

3. Mechanism of Multiple Production

Now we shall switch over to another subject which is some 45 years old and nevertheless cannot be considered as being entirely clarified. However its history itself is very instructive.

As has been told above, at the end of thirties it was realized that there were two possible mechanisms of multiplication of high energy particles, — the cascading for electromagnetic forces weak enough for perturbation theory to hold, and multiple production in the case of strong interaction where no strict calculations were (and are even presently) possible, only models can be tried. The necessity of the latter mechanism for the particles which we now call hadrons was experimentally proved and accepted unanimously only at the end of the forties. But already in the papers of the prewar period (e.g. Nordheim and Hebb, 1939) it was anticipated, and Heisenberg in his "unhappy" paper (Heisenberg, 1936) partially discussed above made the first attempt to treat the process of such a kind theoretically. The idea of thermodynamical description was put forward for the first time. To be more exact, he pointed out that when *many* strongly interacting particles were produced in a single encounter, then we might use a quasiclassical approach and these particles when flying away from the point of their common origin should proceed to interact with each other and even produce in these interactions new particles until their energy dropped below the threshold at which interaction ceased to be strong. In this situation we may use the temperature concept and the final energy spectrum must be of the Planck type. This paper seems to have been forgotten, at least it was never referred to. But actually the idea of thermodynamical treatment was later restored or rediscovered by several authors many years later in various versions, and in the first place we should mention G.Wataghin (Wataghin, 1943), E.Fermi (Fermi, 1950) and I.Ya. Pomeranchuck (Pomeranchuck, 1951). Heisenberg himself preferred to follow a different course. He assumed as a basis the quasiclassical treatment of a wave field (particles being its quanta) which arises on the collision in two primary high energy particles as a wave packet of small size. Heisenberg assumed further some new nonlinear Hamiltonian of the field seflinteraction. Such a selfinteraction is strong for high "matter densities" (large local field amplitudes) and vanishes when the field expands in space (and its local amplitudes become small). The wave equation enables one to trace the process of expansion of the initial wave packet until the interaction can be completely neglected. At this stage the Fourier expansion of the wave packet gives the final momentum spectrum with some definite multiplicity (Heisenberg, 1949, 1952). The two lines —

thermodynamical (or statistical) treatment and nonlinear equation treated quasi-classically — were developing in parallel.

Everyone knows the famous Fermi model: two relativistically contracted nucleons collide, overlap and stop still releasing all their energy in the form of heat in this overlap volume, in the CMS system it is $V_F = \frac{4\pi}{3} r_0^3 m_N / E$, $r_0 \sim m_\pi^{-1}$ being the hadron radius, m_π and m_N the pion and nucleon masses, 2E — their combined CMS energy. The temperature is high, $T_F \approx 1.7 (E/m_N)^{1/2} m_\pi \approx 1.2 (\frac{E_L}{m_N})^{1/4} m_N$, where E_L is the laboratory energy. With these V_F and T_F the spectra of produced Bose (pions) and Fermi (kaons, nucleons) particles can be easily found. When temperature exceeds all masses (i.e. at $E_L \gg 10^3 \, \text{GeV}$), the composition of produced particles is governed by the numbers of their internal quantum states, e.g. $n_\pi : n_K : n_N = 3:4:8$.

We may speculate that Fermi was led to this idea by the success of his β-decay theory in which the momentum spectrum was determined mainly by the phase space factor in the formula for differential probability, $dW \sim |M|^2 p_e^2 p_\nu^2 \, dp_e dp_\nu \delta (E_0 - E_e - E_\nu)$ (M is matrix element). If very many particles are produced then the phase space factor should be even more essential and prescribes the microcanonical distribution thus leading to thermodynamics.

Now we must stop for a moment and recollect the world situation at this time — the end of the forties and the first half of the fifties. In the first place the emulsion technique had already demonstrated the existence of high energy jets with tens of charged relativistic particles produced at an estimated energy $E_L \sim 10^3$ GeV and above. Study of extensive air showers already gave definite indications of the approximate constancy of interaction cross section, of its peripheral character (inelasticity coefficient smaller than 0.5) and of the energy dependence of multiplicity within the tremendous range of energies, up to $E_L \sim 10^5$ GeV (Zatsepin et al., 1947; Zatsepin, 1949 and subsequent investigations of various cosmic ray physicists).

In the second place, the first accellerators based on new ideas proposed by V.Veksler in USSR and independently one year later by E.McMillan in USA already existed but their energy was low. They started with 2÷3 GeV protons, and only in 1956 appeared the Dubna machine (10 GeV). Of course here multiplicities were quite small. It took more than 20 years to reach 10^3 GeV energy already familiar in cosmic ray experiments both in emulsion studies and in extensive air showers investigations. This difference produced the lasting difference in psychology between cosmic ray and accellerator physicists.

In the third place normal scientific exchanges between the various countries were not still restored after the war and this led to many misunderstandings. One example has been mentioned above: the Veksler papers became known in USA with a few years' delay.

The Fermi model could not then be extensively checked in cosmic rays: one could still not distinguish experimentally various kinds of particles produced in high energy jets or measure their transverse momenta, etc. But the average multiplicity of charged particles, $\langle n_s \rangle \sim E_L^{1/4}$, was reasonable, according to the analysis of the extensive air showers mentioned above. Therefore the theorists, Fermi himself among them, paid more attention to advantages of accellerator studies. A rather ludicrous fact: his model initially intended for very high energy collisions, did not work too badly in accelerator particle production of very *low* (some $2 \div 3$ new particles) multiplicities (later we learned that for higher energies the model drastically contradicted the experiment: composition predicted disagreed with the experimental one which is $n_\pi : n_\kappa : n_N$ $\approx 9:1:0.1$; average transverse momenta did not follow the predicted increase with E_L being instead remarkably constant up to $\sim 10^5$ GeV, etc.). This strengthened the enthusiasm of those who believed that the thorough study of low multiplicity collisions in the accellerator was more efficient for discovering the mechanism of multiple production than the cosmic ray researches. A rather strange idea indeed. The result was that during the few decades when the accellerator energy slowly increased in parallel with multiplicity (early sixties: $E_L \lesssim 30$ GeV, $\langle n_s \rangle \lesssim 4.5$; late sixties: $E_L \leqslant 76$ GeV, $\langle n_s \rangle \leqslant 6$; seventies: $E_L \sim 400 \div 2000$ GeV, $\langle n_s \rangle \leqslant 12$), the models proposed on the basis of accellerator were permanently changing. Models popular in cosmic rays and since fifties based on truly high multiplicity events ($\langle n_s \rangle \gtrsim 10$) were not held in esteem by the accelerator minded theorists. However in our days we observe that these models, particularly fireballs, became more and more popular although under a different name — "clusters". We shall follow this process later in more detail.

For the time being, let us return to the 1950 Fermi model. When the Progress of Theoretical Physics issue containing his paper arrived in Moscow it caused great excitement and appreciation. But two weak points were noticed immediately. Firstly, as has been told, we had already known from an extensive study of air showers that dominating collisions up to $E_L \sim 10^5 \div 10^6$ GeV were not catastrophic. The low inelasticity coefficient $\langle K \rangle \leqslant 0.5$ meant that no common Fermi pot could be formed during a collision of two nucleons. The idea of "peripherality" was already, although vaguely, anticipated. Secondly, Pomeranchuck at once noticed an internal contradiction in the Fermi scheme: if the particles interact so strongly that they can come to a standstill when two primaries collide, then the produced multitude of particles cannot escape from the small overlap volume V_F independently. They must interact and suffer transformations until their mutual distance exceeds the range of force, $\sim m_\pi^{-1}$. Thus the final volume is $V_P = \dfrac{4\pi}{3} r_0^3 \langle n \rangle = \dfrac{E}{m_N} \langle n \rangle V_F$. This leads to a low final temperature, $T_P \sim m_\pi$, and high multiplicity, $\langle n \rangle \sim E_L^{1/2}$ (Pomeranchuck, 1951; one can easily recognize here the close similarity to the Heisenberg (1936) idea).

The necessity of nonzero impact parameter b was independently recognized by several theorists, and all of them, with slight differences, tried to modify accordingly the head on collision scheme. In a peripheral collision, $b \neq o$, the Lorentz contracted nucleons overlap though partially. Fermi took this into consideration very simply (Fermi, 1951): he assumed that even this partial overlap is sufficient to form a single pot having however an angular momentum proportional to b and the Planck distribution of produced particles had to be correspondingly modified in a standard way. But as was immediately noticed (Feinberg and Chernavsky, 1951) this meant that the thermodynamical equilibrium was supposed to propagate in the *transverse* direction over a distance $b \gtrsim m_\pi^{-1}$, beyond the small overlapped part of the colliding nucleons' substance, during the very short collision time $\Delta t \sim r_0 m_N / E$, i.e. to propagate with the superlight velocity $v \sim E/m_N$, thus violating relativity.

Heisenberg within his quasiclassical wave packet expansion picture speculated differently (Heisenberg, 1952): the larger is b, the smaller is the energy ΔE carried within the overlapping parts of nucleons. It is only this ΔE which can form the wave packet and produce new particles. Their number n is accordingly smaller the larger b is. Integrating over all b up to the ones for which $n = 2$, Heisenberg obtained the total cross section $\sigma_{tot} = (\pi/4)r_0{}^2(\ln(2E_L/m_N))^2$. Two decades later, when it was found from accelerator studies that beyond $E_L \sim 100\,\text{GeV}$ the cross section increases, $\sigma_{tot} \approx 40 + 0.5\,(\ln(E_L/m_N))^2$ millibarns, some physicists pointed to this Heisenberg result as to a prediction of the newly found energy dependance. But this opinion can hardly be considered correct. The actual ln squared term is relatively quite small while in the Heisenberg 1952 result it led to an absurd energy increase of σ_{tot}: for E_L increasing from few GeV up to some 10^3 GeV, σ_{tot} was predicted to grow from 40 mb up to 1000 mb while the real increase attains some 10%. As has been mentioned above, already in those times it was known from extensive air shower studies that even at $E_L \sim 10^6$ GeV the cross section differs from its value at, say, 10 GeV not more than by a factor $1/2 \div 2$ (instead of being some 10^3 mb according to the Heisenberg formula).

This inconsistency in the Heisenberg result with cosmic ray data was stressed immediately and the source of the error was also indicated (Feinberg and Chernavsky, 1953): when b becomes large and ΔE correspondingly small we come to a violation of indeterminacy relation for the actual collision time Δt and ΔE. For the quasiclassical treatment of Heisenberg to be valid we must have $\Delta E \cdot \Delta t \gg 1$. Thus beyond some impact parameter, since ΔE is very small (actually for a rather small one, but slightly exceeding $b \sim m_\pi^{-1}$), the quasiclassical approach is erroneous. The authors could not withstand the temptation to end their paper by the remark that "the neglect of indeterminacy relation as in paper 1 (this is the Heisenberg 1952 paper; E.F.) is not permissible".

It is worth mentioning that the quasiclassical treatment along the same lines (thus also violating indeterminacy relation) was tried also by H.J. Bhabha (Bhabha, 1953) and by some other theorists.

It follows, herefrom, that the peripheral collision is to be treated as a quantum field process of interaction leading possibly to a formation of excited substates which may decay in many particles according to thermodynamical (statistical) scheme of Fermi or Pomeranchuck as was observed still in 1951 (Feinberg and Chernavsky, 1951). The model proposed was as follows: peripherally colliding nucleons exchange a field quantum (a pion), become excited and decay. Such a "thermodynamically-peripheral scheme" in various versions (later as a multiperipheral one, including fireballs) was the basis of further theoretical models which were being developed by the Lebedev Institute group for decades.

The reader will no doubt observe that this scheme is almost exactly the one independently proposed by S. Takagi (Takagi, 1952) and by W.L. Kraushaar and L.J. Mark (Kraushaar and Mark, 1954), more than once quoted in the literature (see e.g. Hayakawa, 1969), while the Soviet papers just mentioned did not attract such attention. It is but natural and presents another example of the insufficiency of scientific contracts in those times. These papers as well as some others (e.g. Smorodinsky, 1940; Laudau, 1940; Zatsepin et al., 1947; Zatsepin, 1949; Pomeranchuck, 1951) were published only in Russian and for many years remained unknown outside of the USSR.

The same is true for the next step — for the hydrodynamical model of multiple hadron production by L.D. Landau (Landau, 1953). This was a natural generalization of the Pomeranchuck work necessary for such a high energy that the pressure and accelleration within the expanding cloud could not be neglected. According to this theory the process within a strongly contracted nucleon is treated classically, and this is (within the assumed model) well founded from the point of view of the conditions of applicability of a non-quantum approach. This macroscopic relativistic hydro-dynamics "within a nucleon" attracted anyone who learned of it by its elegance. Besides it gave the natural explanation to many properties of multiple hadron production gradually accumulated in cosmic ray researches in the fifties:, to the drastic predominance of pions, to the small and constant transverse momentum, to $<n> \sim E_{\mathrm{L}}^{1/4}$ multiplicity law, etc.

Unfortunately Landau considered only head on ("central") collisions. He was motivated to this by the weak dependence of multiplicity on the energy of the overlapping parts of colliding nucleons which he believed, would make it unnecessary to differentiate $b \neq 0$ collisions from the head on ones. He did not pay due attention to the fact that it influenced, at any rate, the angular distribution. Only much later the Landau model was consistently included into the "thermodynamically-peripheral scheme" (Zhirov and Shuryak, 1975). However, actually similar ideas can be recognized in the Hagedorn model (Hagedorn, 1965, 1967, 1968) if its formulas are properly reinterpreted and freed from the temperature bound introduced by Hagedorn (Feinberg, 1972).

Although the Landau theory was quite soon published in English (Belenky and Landau, 1956) and was developed by many Soviet and Japanese theorists it did not attract due attention in USA and Western Europe until the middle of the seventies. Theorists were mainly enthusiastic about accelerators with their low average multiplicities insufficient for extensive application of thermodynamical models. But those who were concerned with cosmic rays could not leave it aside. In this respect one torturing puzzle persisted: besides the Landau model with its $E_{\mathrm{L}}^{1/4}$ multiplicity law there existed the quasi-classical nonlinear wave theory by Heisenberg. It was seemingly quite well founded but prescribed a different multiplicity, $\sim E_{\mathrm{L}}^{1/2}$. The origin of difference remained obscure until a young theorist G.A. Milekhin clarified the situation (Milekhin, 1959). He demonstrated that if the 4-gradient of the wave function in the Heisenberg theory is in a definite way connected with the velocity of the substance of the expanding wave packet, then the theories by Heisenberg and Landau may be given an identical shape, in which a definite choice of the state equation of matter in the Landau theory corresponds to a definite choice of the nonlinear selfinteraction Lagrangeian in the Heisenberg theory, and vice versa. The difference in predicted multiplicities was actually due to the difference in assumed state equations (or of selfinteraction Lagrangeians). I remember how glad was Heisenberg to learn this when during the International Conference on High Energy Physics in Kiev (1959) he met Milekhin and the situation could be discussed thoroughly.

But all this did not interest accelerator physicists. In the early sixties they got finally fairly large multiplcities at 30 GeV and in 1968 enough data were accumulated for drawing the basic characteristics of multiple production. When summarized (Turkot, 1968) they turned out, within this rather small energy range (E_L = 10÷30 GeV), to agree remarkably well with the same regularities which had been deduced from cosmic ray studies long before and summarized by C.F. Powell (Powell, 1959) as valid within the incomparably larger energy range, $E_L \leqslant 10^5$ GeV.

When the 30 GeV accelerator data were obtained people tried to explain them within the Fermi model. Of course the result was unsatisfactory (the model predicted too many kaons etc.). Were the CERN and Brookhaven physicists familiar with the Pomeranchuck statistical model they could have seen how well it worked for not too low multiplicities, the more so if fitted into the "thermodynamical-peripheral scheme". But this model was almost unknown, and still for many years to come by statistical model most of people meant the Fermi one.

At this stage theorists proposed the famous Regge pole method. For few years, when low multiplicities were paid predominant attention, its success was complete. The colliding hadrons were supposed to exchange some Regge pole, to be excited and to decay into some 1÷3 hadrons each (within limitations imposed by the pole quantum numbers). Of course some constants (vertices in the Feynman-Regge graphs etc.) were to be fixed to fit the experiment, their number increasing along with further progress of experiment, but everything was consistent and elegant. Many theorists believed that such "Reggeistics" can in general be a substitute for the quantum field theory. No doubt, for such binary or quasibinary processes this model is of high value and even now keeps its position invariably, at least as a parameterization of experimental data. However when the accelerator energy increased further it became clear that quasibinary processes in no way exhaust the situation. Cosmic ray physicists had known it long ago. Since their experiments covered a tremendously larger range of energies and multiplicities they could distinguish, among the products of collision, both the products of excitation and subsequent decay (generally into few particles) of primary colliding hadrons, called "isobar" or fragmentation particles ("leading" ones among them), and the more numerous and less energetic (in CMS) "pionization" particles. In 1958 the heavy fireball model was put forward by the Cracow M. Miesowicz group, by K. Niu and by G. Cocconi (Ciok et al., 1958, Niu, 1958, Cocconi, 1958), as a mechanism for producing pionization particles. Fireballs were found by emulsion technique at $E_L \gtrsim 10^3$ GeV when sufficiently high multiplicity events were selected ($n_s \geqslant 6$). Two fireballs having masses m_f = 2÷5 GeV were declared to arise in each collission. One year later, with a different technique (Wilson chamber in magnetic field together with ionization calorimeter) similar fireballs were observed by the N.A. Dobrotin group (Grigorov et al., 1959; Dobrotin et al., 1961) in Pamir mountain laboratory at lower energy, $E_L \sim 200$ GeV, with only one fireball per event. Fireball decay agreed with the Pomeranchuck statistical model.

However the accelerator people, both experimentalists and theorists, were not willing to listen to the necessity of a special procedure for singling out fireballs. In fact, even now the selection of clusters from the accelerator data ($E_L \leqslant 2.10^3$ GeV) is a complicated task. Accordingly statements of the cosmic ray people could not be considered safe enough. They were not confirmed by accelerator data summarized over all n_s and the fireball idea was far from being convincing.

Insufficiency of the quasibinary Regge pole model was the reason for turning attention to the AFS multiperipheral model (Amati *et al.*, 1962) based on the one pion exchange with production in each triple vertex of a ρ-meson (or, in modern version, of some other light meson resonances as well), decaying into pions each. This very important model led, in principle, to a field theoretical understanding of the vacuum Regge pole (of a "Pomeron") and also had some other very attractive features. The resulting logarithmic energy dependence of multiplicity agreed with the experiment rather well in the accelerator energy range (with some corrections this holds even now, for $E_L \lesssim 10^3$ GeV until the difference from $E_L^{1/4}$ law becomes noticeable). The model became even more popular and attractive when at the end of the sixties Feynman proposed his parton model with its remarkable scaling law (Feynman, 1969) and the multiperipheral model could be understood as its elegant quantum field countterpart. It dominated the psychology of physicists beyond the middle of the seventies. However very soon analysis of various authors demonstrated (Salzman, 1963; Koba and Krzywicki, 1963; Feinberg and Chernavsky, 1964; Kobayashi *et al.*, 1964) that such a model with production of only light mesons (ρ, etc.) could not agree with high multiplicity data within the entire energy range studied in cosmic rays. For example, it gives drastically contradicting total cross section (which is prescribed to decreased with energy as $E_L^{-0.7}$), and absolute value of multiplicity, disagrees in rapidity distribution, etc. It was also pointed out that agreement is restored if vertices in the multiperipheral chain are supposed to be heavy fireballs similar to those declared by cosmic ray experimentalists, while the advantages of the original multiperipheral model, like the field theoretical explanation of the vacuum Regge pole, persist (the calculated slope of the Pomeron trajectory was few times smaller than adopted in those times by the accellerator physicists; however it fitted better with later accelerator experiments as became clear at the end of the seveties).

This idea formed the basis of the model developed in the following years in the Lebedev Institute by D.S. Chernavsky and his collaborators (Dremin *et al.*, 1965; 1970). For brevity it is called the "multifireball model" since it assumes the presence (within the multiperipheral chain) also of heavy nonresonant clusters of fireballs besides the usual rather light meson and baryon resonances. The average size and number of fireballs per event (i.e. per chain) increases with energy. After being fitted to agree with the experiment and thus with fixed arbitrary parameters the model gave a consistent description both of elastic (Pomeron slope value etc.) and inelastic processes. A detailed comparison of this model (its parameters fixed in 1972) with accelerator data on pp and πp collisions at E_L = 25÷400 GeV demonstrated the agreement with all measured distributions (inclusive, semiinclusive, correlations etc.), both known then and obtained later (Volkov *et al.*, 1973, 1975, 1976; a compendium is given by Dremin *et al.*, 1978). Specifically, according to this model the average number of heavy fireballs is close to one per event at $E_L \sim 200$ GeV and increases further with energy tending roughly to two at $E_L \sim 2000$ GeV, thus not contradicting the older cosmic ray data quoted above. Obviously this model is a compromise between the original AFS multiperipheral model (Amati *et al.*, 1962) and the simple cosmic ray fireball model, which contained only fireballs and no light resonances.

The reader can recognize in this theoretical model a rather close similarity to the less sophisticated model deduced at the very beginning of sixties by S. Hasegawa (Hasegawa, 1961, 1963) from the analysis of pseudorapidity distributions in individual cosmic ray emulsion chamber events. Here multiple production proceeds via formation of chain of identical hadronic objects called H-quanta and having the mass of some 2 GeV (i.e. markedly smaller than in the Lebedev Institute model where it is some 2-5 GeV but moreover light meson and baryon resonances are directly produced).

Unfortunately, this line of progress did not interest accelerator physicists until they had accumulated enough detailed data for $E_L = 400 \div 2000$ GeV. And then there appeared an essential disagreement with the simple scaling predicted both by the multiperipheral AFS model and by the Feynman parton model. An active CERN theoretician said: "If these results had been known a few years earlier, maybe "scaling in the central region" or even "multiperpheral model" would not be an extremely popular concept at present. On the contrary we might all be very familiar for example with Landau's hydrodynamical model!" (Le Bellac, 1976).

The existance of clusters, although as yet rather small ones, was admitted. In a review talk, C.De Tar (De Tar, 1978) claimed that multiple production goes via clusters with average mass of ~ 1.5 GeV decaying into $3 \div 4$ pions. Therefore they could not be reduced only to the well known ρ, ω, ... etc. resonances. Actually, as was shown by a detailed analysis of correlations, when the "multifireball model" is taken into account (Volkov et al., 1976; Dremin et al., 1978), in a hadron collision at $E_L = 30 \div 1000$ GeV some $3 \div 4$ intermediate unstable objects are produced. Within this energy range their average number does not change essentially. However at the lower end of the range they are almost exclusively the usual light resonances while when E_L increases the fireballs appear more frequently and their mass increases until at $E_L \sim 10^3$ GeV it becomes of the same order as indicated two decades ago by cosmic ray observations. The De Tar estimate mentioned above may be considered as the result of averaging over all intermediate objects, i.e. over light resonances and heavy fireballs.

Thus we observe an interesting situation. When accelerators approached the energy range studied long time ago in cosmic rays the theoretical schemes neglecting fireball ideas (qussibinary reactions with Regge pole exchanges, as a dominating process; multiperpheral AFS chain containing only light resonances; parton scaling model) became inadequate and had to be modified somehow in the direction of a picture akin to the one worked out long ago in cosmic rays. The important element of these modifications is the presence of heavy clusters or fireballs.

The problem has not yet come to an end in many respects. First of all, the very existence of heavy clusters is not unanimousely accepted, their identification among the accelerator data asks for a complicated and not altogether unquestionable procedure, and even if accepted then it is still not proved that they are not simply some new heavy resonances. However, it was already shown that in the quantum field theory there exists quite a natural place for a heavy nonresonant cluster. It might be simply a hadron far from its mass shell (displacement in time like direction) (Dremin and Feinberg, 1978, 1979). When various correlations among produced particles are taken into account, the presence of fireballs becomes more and more probable. Secondly, their thermodynamical way of decay is in no way established, they could instead decay by successively emmitting hadrons ("tree like graphs") as is supposed to be the case for hypothetical higher resonances in the dual resonance model (Makarov et al., 1973). Finally, if the heavy cluster is a reality we do not know what happens for $E_L > 10^3 \div 10^4$ GeV: either the number of fireballs, with some bounded mass, per event increases or their mass is unbounded and may increase further. In particular, the multifireball model by the Lebedev Institute group (the multiperipheral chain with both, light resonances and fireballs) is in no way recognized widely. Although it has not met any contradiction with experimental data, it is described here mainly as an example of a possible and plausible synthesis of cosmic ray fireball ideas with multiperipherality of AFS model. In the near future colliding beam accelerators with the equivalent energy of $E_L \gtrsim 10^5$ GeV will appear and we may hope that the problem will be solved.

4. Conclusion

Here we presented the history of two problems most important for the particle theory. This very instructive history of entangled mistakes and triumphs, of blindness and striking insights in which comic ray previsions played an outstanding part should not be forgotten. One can recollect the lines of the celebrated Russian poetesse Anna Akhmatova: " One cannot imagine the trash from which poetry grows knowing no shame." Our knowledge in the sphere of particle physics, with all its tremendous growth during half a century when cosmic rays contributed to it, also proceeded, as I tried to show, to considerable extent from the trash of misunderstandings, unfair neglections and unconceivable errors. Nevertheless it *did* grow up.

REFERENCES

Amati, D., Stanghellini, A. and Fubini, S.: 1962, *Nuovo Cim.* 26, 896; 1962, *Phys. Lett.* 1, 29.
Auger, P., Mase, R. and Grivet-Meyer, T.: 1938, *Comp. Rend.* 206, 1721.
Belenky, S.Z. and Landau, L.D.: 1956, *Suppl. Nuovo Cim.* 3, 15.
Bethe, H. and Heitler, W.: 1934, *Proc. Roy. Soc.* 146, 83.
Bhabha, H.J. and Heitler, W.: 1937, *Proc. Roy. Soc.* 159, 432.
Bhabha, H.J.: 1953, *Proc. Roy. Soc.* 219, 293.
Blackett, P.A.S.: 1938, *Phys. Rev.* 54, 973.
Booth, F. and Wilson, A.H.: 1940, *Proc. Roy. Soc.* 175, 483.
Bothe, W. and Kolhörster, W.: 1928, *Naturwiss.* 16, 1045; 1929, *Z. Physik* 56, 751.
Carlson, J.F. and Oppenheimer, R.J.: 1937, *Phys. Rev.* 51, 220.
Christy, R.F. and Kusaka, S.: 1941, *Phys. Rev.* 59, 405, 414.
Ciok, P., Coghen, T., Gierula *et al.*: 1958, *Nuovo Cim.* 8, 166.
Cocconi, G.: 1958, *Phys. Rev.* 111, 1699.
De Tar, C.: 1976, *Proc. 18th Inter. Conf. High Energy Phys. Tbilisi*, Vol. 1, A3–4.
Debrotin, N.A. *et al.* : 1961, *Nucl. Phys.* 35, 152.
Dremin, I.M. and Feinberg, E.L.: 1978, *Proc. IXth Inter. Symp. High Energy Multipart. Dynamics, Tabor.*
Dremin, I.M. and Feinberg, E.L.: 1979, *Phys. Elem. Chast. Atomn. Yadra*, 10, 996 (1979, *Sov. Phys. Elem. Part. Nucl. Phys.*, 10, N 5).
Dremin, I.M., Orlov, A.M. and Volkov, E.I.: 1978, *Lebedev Inst. Preprint*, N 247.
Dremin, I.M., Royzen, I.I., White, R.B. and Chernavsky, D.S.: 1965, *J. Exp. Theor. Phys.* 48, 952.
Dremin, I.M., Royzen, I.I. and Chernavsky, D.S.: 1970, *Usp. Phyz. Nauk* 101, 385 (1971, *Sov. Phys. Uspekhi* 13, 438, the early review paper).
Euler, H. and Heisenberg, W.: 1938, *Erg. Exakte Naturwiss.* 17, 1.
Feinberg, E.L. and Chernavsky, D.S.: 1951, *Doklady Akad. Nauk SSSR* 81, 795; 1953, 91, 511 (in Russian).
Feinberg, E.L. and Chernavsky, D.S.: 1964, *Usp. Fiz. Nauk* 82, 3 (1964 *Soviet Phys. Usp.*, July–August.)
Feinberg, E.L.: 1972, *Phys. Rep.* 5, 237.
Fermi, E.: 1950, *Progr. Theor. Phys.* 5, 570.

Fermi, E.: 1951, *Phys. Rev.* **81**, 683.

Feynman, R.: 1969, *Phys. Rev. Lett.* **23**, 1415.

Fröhlich, H., Heitler, W. and Kemmer, N.: 1938, *Proc. Roy. Soc.* **166**, 154.

Ginzburg, V.L.: 1941, *J. Phys. (Moscow)* **5**, 47; 1944, **8**, 33; 1943, *Phys. Rev.* 1.

Ginzburg, V.L. and Tamm, I.E.: 1947, *J. Exp. Theor. Phys. USSR* **17**, 227 (in Russian).

Grigorov, N.L. *et al.*: 1959, *Proc. Inter. Conf. Cosmic Rays, Moscow,* Vol. 1.

Hagedorn, R.: 1965, *Suppl. Nuovo Cim.* **3**, 147; 1968, **6**, 311; 1967, *Nuovo Cim.* **52A**, 1336.

Hasegawa, S.: 1961, *Progr. Theor. Phys.* **26**, 150.

Hasegawa, S.: 1963, *Progr. Theor. Phys.* **29**, 128.

Hayakawa, S.: 1969, *Cosmic Ray Physics,* Wiley-Interscience.

Heisenberg, W.: 1936, *Z. Phys.* **101**, 533; 1938, *Z. Phys.* **110**, 251.

Heisenberg, W.: 1949, *Z. Phys.* **126**, 569; 1952, *Z Phys.* **133**, 65.

Heisenberg, W.: 1943, *(Herausgeb.) Kosmische Strahlung,* Springer- Verlag.

Heitler, W.: 1933, *Z. Phys.* **84**, 145.

Heitler, W.: 1936, *The Quantum Theory of Radiation,* Oxford, (see § 24.2).

Kemmer, N.: 1938, *Proc. Roy. Soc.* **166**, 127.

Koba, Z. and Krzywicki, A.: 1963, *Nucl. Phys.,* **46**, 471, 485.

Kobayashi, T., Namiki, M. and Ohba, I.: 1964, Waseda University preprint.

Konopinski, E. and Uhlenbeck, G.: 1935, *Phys. Rev.* **48**, 107.

Kraushaar, W.L. and Mark, L.J.: 1954, *Phys. Rev.* **93**, 326.

Landau, L.: 1940, *J. Exp. Theor. Phys. USSR* **10**, 718 (in Russian).

Landau, L.D.: 1953, *Izv. Akad. Nauk SSSR, Ser. Fiz.* **17**, 51 (in Russian).

Le Bellac, M.: 1976, *Proc. VIIth Inter. Colloquium Multipart, Reactions, Tutzing.*

Milekhin, G.A.: 1959, *Inter. Conf. Cosmic Rays, Moscow,* Vol. 1, p.220.

Makarov, V.I., Miransky, V.A., Shelest, V.P., Struminsky, B.V. and Zinovjev, G.M.: 1973, *Phys. Lett.* **43B**, 72; 1973, *Lett. Nuovo Cim.* **8**, 151.

Neddermeyer, S.H. and Anderson, S.D.: 1937, *Phys. Rev.* **51**, 884; 1938, **54**, 88.

Niu, K.: 1958, *Nuovo Cim.* **10**, 994.

Nordheim, L.W. and Hebb, M.H.: 1939, *Phys. Rev.* **56**, 494.

Pauli, W. and Weisskopf, V.F.: 1934, *Helv. Phys. Acta* **7**, 709.

Pomeranchuk, I.Ya.: 1951, *Doklady Akad. Nauk SSSR* **78**, 889 (in Russian).

Rasetti, F.: 1941, *Phys. Rev.* **60**, 198.

Rossi, B.: 1933, *Z. Phys.* **82**, 151.

Rossi, B. and Nereson, N.: 1942, *Phys. Rev.* **62**, 417; 1943, **64**, 199.

Salzman, F.: 1963, **131**, 1786.

Smorodinsky, Ya.: 1940, *J. Exp. Theor. Phys. USSR* **10**, 840 (in Russian).

Skobeltzyn, D.V.: 1927, *Z. Physik* **43**, 354; 1929, **54**, 687.

Takagi, S.: 1952, *Progr. Theor. Phys.* **7**, 123.

Tamm, I.: 1934, *Nature* **133**, 981.

Tamm, I.: 1936, *Phys. Zs. Sowjetunion* **10**, 567.

Tamm, I. 1940, *Dokl. Akad. Nauk USSR (Compte Rendus)* **29**, 551.

Volkov, E.I., Dremin, I.M., Dunaevskii, A.M., Royzen, I.I. and Chernavsky, D.S.: 1973, *Sov. J. Nucl. Phys.* **17**, 407; 1973, **18**, 437; 1975, **20**, 149.

Volkov, E.I., Kanarek, R.I., Royzen, I.I. and Chernavsky, D.S.: 1976, *Sov. J. Nucl. Phys.* **24**, 1212.

Wataghin, G.: 1943, *Phys. Rev.* **63**, 137; 1944, **66**, 149.

Weizsäcker, C.F., von.: 1934, *Z. Phys.* **88**, 612.

Wick, G.: 1938, *Nature* **142**, 993.

Williams, E.J.: 1935, *Kgl. Danske Vid. Selskeb* **13**, N 4.

Yukawa, H.: 1935, *Proc. Phys. Math. Soc. Japan* **17**, 48.

Yukawa, H. and Sakata, S.: 1938, *Proc. Phys. Math. Soc. Japan* **20**, 720.

Yukawa, H. and Sakata, S.: 1939, *Proc. Phys. Math. Soc. Japan* **21**, 138.

Zatsepin, G.T.: 1949, *J. Exp. Theor. Phys. USSR* **19**, 1104 (in Russian).

Zatsepin, G.T., Miller, V.V., Rosental, I.L. and Eidus, L.Kh.: 1947, *J. Exp. Theor. Phys. USSR* **17**, 1125 (in Russian).

Zhirov, O.V. and Shuryak, E.V.: 1975, *Sov J. Nucl. Phys.* **21**, 861.

PART 8 FURTHER IN SPACE PHYSICS

FROM BALLOONS TO SPACE STATIONS

Sergei N. VERNOV*, Naum L. GRIGOROV*,
George B. KHRISTIANSEN*, Agasi N. CHARAKHCHAN**
and Alexander E. CHUDAKOV

In this paper the authors present the results obtained by academician D.V. Sko-beltzyn's school in high-altitude cosmic-ray observations.

The event which has brought a genuine revolutionary improvement to the study of the atmosphere took place on January 30, 1930 when Prof. P.A. Moltchanov was the first to launch a balloon which could radio-transmit the results of meteorological measurements to the Earth. Nobody could suggest at that time that the event would have such an outstanding impact on the development of cosmic-ray physics as well.

It was known in the early thirties that a weak latitudinal cosmic-ray effect (10% while travelling from high latitude to the equator) was observable at sea level. So, as to the minor part of primary cosmic ray particles, it was possible to state that they had an electric charge and were deviated by the Earth's magnetic field, i.e. that their energy was below 15 GeV. As to the major part of the cosmic rays there was no information at all.

In order to obtain such information, it was necessary to measure the cosmic ray intensity at high altitudes, in the stratosphere, and at various geomagnetic latitudes. The apparatuses applied at that time for the cosmic ray intensity measurements in the stratosphere (Regener, Pfotzer) were taken up by balloons and the measurement results were fixed by means of an automatic recorder or camera. Thus, an apparatus which would not miss any information had to be found. It is evident that such a kind of apparatus could be applied only in densely populated areas and even in this case one had to rely on good luck. To study the latitudinal effect in sparsely populated areas of the Earth, new means of investigation had to be sought.

It was a balloon with a radio-transmitting device that became the new means of cosmic ray measurements (Vernov, 1934, 1935a, b). It was based on Moltchanov's meteorological balloon, and the first apparatus for cosmic ray studies was designed in close cooperation with Prof. Moltchanov.

The first apparatus for cosmic ray study was launched on April 1, 1935 at the Aerological Observatory (Fig. 1). The equipment was a counter telescope and weighed 18.6 kg. The second model with a single counter was as heavy as 9.8 kg and made its ascent in the same year of 1935.

Prof. Moltchanov also took part in the ascent of the first apparatus and the reception of its signals. Later on, he wrote (Moltchanov, 1938): "The signal received from the apparatus for cosmic ray study is of special interest. While ascending, the number of signals increases gradually. At low altitude it takes 1~2 minutes to hear the

* Moscow State University, Moscow, U. S. S. R.
** Lebedev Physical Institute, Acad. Sci., Moscow, U. S. S. R.
*** Institute for Nuclear Research, Acad. Sci., Moscow, U. S. S. R.

Y. Sekido and H. Elliot (eds.), Early History of Cosmic Ray Studies, 357–374.
© 1985 by D. Reidel Publishing Company.

Fig. 1. Launching the first radiosonde for cosmic ray measurements on April 1, 1935. On foreground
– Prof. A P. Moltshanov.

sound indicating the hit of a cosmic ray[†]. The higher the apparatus, the more frequent
the hits. At an altitude of 8~9 kilometers they are falling like drops of heavy rain.
Higher than 9 km, the hits become still more frequent and we have to apply special
automatic instruments to get correct hit counting".

The balloon method of cosmic ray detection opened up numerous possiblities for
designing various devices to study the phenomena caused by cosmic rays. The radio-
transmission made it possible to carry out the research in the areas most suitable for
revealing physical processes wherever they are, i.e. in fact it turned the method into a
universal means of investigation.

The first cosmic ray balloons were used to study the cosmic ray latitudinal effect in
the stratosphere. Every apparatus comprised two miniature G-M counters for registering
particles. One of them (1 cm long) was permanently connected to a pulse amplifier
while the other, a larger one (3 cm long), was connected to the amplifier by a
barograph pen. Abrupt changes in the particle counting rate defined the atmospheric

† In the first apparatus, the information was taken by ear. Every event of particle registration
switched on a relay which, in its turn, switched on the transmitter. It was fixed as a high tone click
in a telephone.

pressure, i.e. the measurement altitude. The battery of Clark normal elements were used as a high voltage source (about 1400 V). In order to maintain a constant temperature inside the apparatus (fluctuations were allowed within ± 5°, while the outdoor temperature fell down to 40°~50°C below zero) chamical temperature controller was applied. As the temperature fell down, a valve opened and concentrated sulphuric acid started dropping from a glass bulb into a vessel with alkali. The heat emitted by the reaction raised the temperature in the apparatus up to +8~10°C. To reduce heat loss, the apparatus was covered with eiderdown (Vernov, 1945).

The first measurements at the Leningrad and Yerevan latitudes were taken in 1935–36. The equipment proved to operate properly and 12 identical instruments were made to carry out the measurements at the equator. A picture of one of them is shown in Fig. 2. All the instruments were thoroughly checked and the participants of the planned equatorial expedition thought it to be quite sufficient to provide, every apparatus with a battery, an amplifier and a transmitter and to pour acid and alkali into a thermoregulator that the apparatus be ready for ascent (S.N. Vernov and N.L. Grigorov were among the participants of the expedition).

It was in 1937 when, under the general scientific supervision of D.V. Skobeltzyn and with the great assistance of S.I. Vavilov, an expedition onboard tanker "Sergo" sailing from the Black Sea to Far East was organized. The expedition consisted of four persons. Its aim was to carry out cosmic ray measurements at high altitudes in equatorial region.

The preparation of the first apparatus to flight was started when sailing in the Red Sea. With great precautions the instrument was taken from the box and the eiderdown cover was removed. A careful inspection showed no signs of damage. The laboratory supply sources were turned on but the apparatus kept silent. The second one – the same result. The third, the fourth – none of the 12 apparatus worked. There were no evident reasons which could cause any damage to the instruments which worked safely in Moscow.

Fig. 2. Photograph of the instrument for measuring the cosmic ray latitude effect in the stratosphere in 1937–38.

It was a real bad luck for the expedition just because the only meters were two magnetoelectric devices and an electrometer. The devices were used to charge the battery, and the electrometer to control the operations of the counters, so that there did not remain any means to measure the counter voltage and the resistance in the circuits of the counters (thousands of meg) and the amplifier (tens of meg).

In order to find the "culprit" responsible for the damage of all apparatuses one had above all to measure the battery voltage and the parameters of the resistors. A tea glass was turned into a static voltmeter. In the lid covering the glass an amber insulator was adjusted. A wire with two leaves of thin foil was put through the insulator. It was an electroscope known to all schoolchildren. Graph paper glued to one side of the glass served as a scale. The voltmeter thus made was graded with the help of dry batteries and proved to operate reliably during the entire period of the expedition. As for the resistance, it was measured according to the time of capacitor discharge for a given value of charge.

When a chance appeared to measure the battery voltage feeding the counters, it turned out that they did not work. The reason was the almost 100% air humidity. A thin water film covered the glass into which the counters, getinax frameworks, with Clark elements, were soldered. The water film created strong leakage currents which were equivalent in the Clark elements to the currents of short-circuit. The surface leakage was so great that it changed all the high-ohmic resistors whereupon the amplifiers stopped operating. In order to counter this trouble, an effective rosin-talcum glue was invented to cover all the units of the equipment. All the instruments were dismounted and boiled in the glue. A thin film of cold glue proved to be a perfect insulating material even in the highly humid air.

Only after the restoration of insulating properties of all materials by the described method did the instruments start working safely, whereupon they were taken up near the equator.

During the expedition of 1937–38, simultaneously with but independently of Millikan an energy spectrum of the main flux of primary cosmic rays was obtained. The primary cosmic rays were found to be electrically-charged particles which, when interacting with the atmospheric matter, lose their energy much more intensively than it should be expected from energy loss for ionization (Vernov, 1945).

Having answered some questions, the results of the equatorial expedition raised a number of others, namely

(1) what is the origin of the primary particles?

(2) what is the mechanism of the intensive energy loss?

The World War II interrupted the experiments. They were resumed in 1945 on a considerably higher technical level.

First of all the replacement of the medium wavelengths by ultrashort waves has permitted the noise immunity of radio reception to be raised immensely. The heavy and complex systems of chemical temperature control and heat insulation with expensive eiderdown were replaced by solar heating. With this purpose, the equipment was painted with black and white strips. To avoid cooling by atmospheric air during its ascents and descends, the equipment was put into a transparent cellophane envelope. This improvement decreased the mass of the apparatus by several kilograms and made their operation much easier.

The high-voltage batteries of Clark normal elements were replaced by batteries of miniature dry elements, thereby increasing the reliability and decreasing the size and weight of the high-voltage source.

Finally, the radio signal recording system was improved greatly. The reception by ear as was done with the first balloons was replaced by photoregistration. The signals from the receiver entered the oscilloscope; its display being projected by a lens to a continuously moving film.

The system made it possible to transmit a great number of coded signals, including the amplitudes of electric pulses if necessary.

The cosmic ray balloon research based on the new instrumentation was rapidly developed in many directions to seek the solution for two interrelated problems, namely

a) the study of the nature of primary cosmic ray and the origin of secondary components in the stratosphere;

b) the study of the mechanisms of primary cosmic ray interactions with matter and the processes of production of secondary components.

Various instruments were developed to study the origin of secondary components.

The electron-photon component in the stratosphere was studied using the effect of production of multiparticle showers in lead by a high-energy primary electron or photon. In this case the particle number was measured by a counter of as small area as $1{\sim}1.5$ cm^2 in the atmosphere and inside a lead sphere alternatively. The lead sphere wall thickness in different instruments varied from 1 to 4 cm. To eliminate the effect of unpredictable variations in the sensitivity of the instruments on the measurement results, the counter was alternatively placed for several minutes inside and outside (at a 30 cm distance) the lead sphere using a spring mechanism in the early instruments and a small electric motor in their later versions. The position of the counter was transmitted by a special radio signal. The results of the measurements carried out in 1947–48 (Brikker *et al.*, 1947) and presented in Fig. 3 showed that the mean energy of electrons and photons was $(2{\sim}3){\cdot}10^8$ eV and did not in practice vary with altitude up to the highest altitudes of 26 km reached in the experiments.

The measurements showed also that the major flux of primary cosmic rays did not consist of electrons since the primary electrons should have been of energy of about 3 GeV and should have given a nearly 10 times as high multiplication factor in the atmosphere).

In the experiments, the role of μ-mesons in the production of the electron-photon component in the stratosphere was appraised and it was demonstrated that not all of the observed intensity could be explained by their decays. It was also asserted that some new process of the electron-photon component generation by primary particles, presumably by protons was effective (Brikker *et al.*, 1948).

The measurements with the counter were not fully reliable since they involve a particle counting loss and hence a decrease in the multiplication factor. Despite the necessary measures taken to minimize the effect (the counters were small and the sphere size was large compared with the size of the counters), such an adverse effect was still possible. The effect could have been removed entirely by measuring the ionization produced by particles under and outside a lead screen. However, was very difficult to realize.

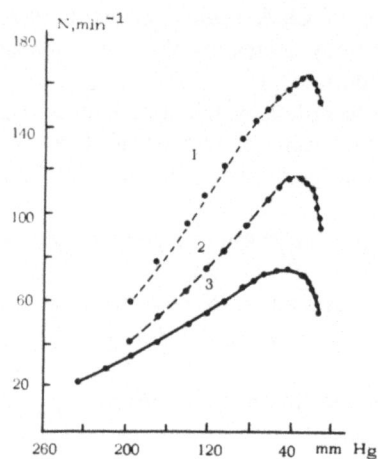

Fig. 3. The cosmic ray transition effect (air-lead) versus the residual atmospheric pressure over the instrument (Brikker *et al.*, 1948): 1 — lead screen of 1- and 2-cm thickness, 2 — same with the screen of 4-cm thickness, 3 — without a screen.

First of all, the ionization chamber should be small, and the weight of the lead filter of several cm thickness surrounding the chamber from all sides be within several kilograms.

Secondly, the chamber walls should be thin not to absorb the low-energy electrons emitted from lead and yet to withstand a gas pressure of several atmospheres.

Thirdly, it was necessary to work out a method of radio transmission of the ionization current measurement results. There was no experience in making such measurements since up till then the weak currents from the ionization chambers were measured with electrometers whose filaments were photographed.

To solve all these problems, a miniature electrometer was developed. In the instrument, a 1 mm² mirror and a cross rocker were fastened to a 20 micron diameter glass thread in the middle part of a thin quartz thread of some 1.5 cm length. The assembly as a whole was balanced by small counter weights in such a way that the movable system did not react on tilts.

After that, the threads were coated with a thin film of gold and got conductive. An electrometer of the size of a thimble was used as one of the electrodes of a small spherical ionization chamber with a glass window in its wall. A thin light beam was directed through the window to the mirror and, having been reflected, came back through the same window to an opaque screen with a system of slits. A photocell was placed behind the screen. The photocurrent was amplified by a valve circuit and, after that, was used to modulate the sonic frequency of the transmitter.

The ionization measurements in the stratosphere were carried out in 1947. Similarly to the experiments with counters, the ionization was alternatively measured under and outside lead filters (Vernov *et al.*, 1947). The measurements confirmed the low multiplication factor for the electron-photon component particles.

The measurements of the particle number and the ionization in the stratosphere made it possible to determine the mean ionizing power of cosmic ray particles which proved to be nearly twice as high as that of the particles at sea level. This fact meant that about a half, of the ionization in the stratosphere was caused by the strongly ionizing particles produced in the nuclear disintegrations of the atmospheric atoms (Grigorov *et al.*, 1951). The result was unexpected. To exclude any assumption about instrumental errors, check experiments were made. The same technique as in the stratosphere was used to measure the particle number and the ionization in the Moscow metro tubes. Ionization currents as weak as 10^{-16} A had to be measured, which proved to be quite possible with the electrometers developed for the stratospheric measurements.

The presence of the particles with increased ionizing power at high altitudes in the atmosphere and their absence at sea level indicated that they were closely associated with the primary cosmic radiation.

Thus, it became clear towards the end of 1947 that the intensive electron-photon component in the stratosphere and the particles of increased ionizing power were the products of the interaction of the primary cosmic rays with the atmosphere. The fact that the penetrating component muons is also composed of secondary particles was known long before; namely, they cannot reach the earth because of their instability. To clarify the processes of generation of secondary aprticles it was, first of all, necessary to find out how the primary particles were absorbed in the atmosphere.

Having assumed that the primary particles can generate electron-nuclear showers (Veksler *et al.*, 1947), the authors of (Vernov *et al.*, 1950) decided to study the altitude dependence of such showers.

Two types of devices were used to detect the showers. In one of them, a lead absorber of 8 cm thickness inserted between three counters arranged in a telescope. Several counters connected to the circuit of coincidences with the telescope pulses were placed around the absorber. In such a way, the traversal of a penetrating particle through the telescope resulting in the responce of even a single side counter (i.e. shower generation in the thick lead block) was recorded and radio-transmitted.

With that device, the dependence of shower intensity on residual atmospheric pressure was measured (Vernov *et al.*, 1950). It was found that at 51° latitude the component producing the electronnuclear showers was absorbed according to a power law with the absorption path $\lambda_\alpha = 120$ g.cm^{-2}.

The second device for studying the showers initiated by primary particles was much more complicated (Vernov *et al.*, 1950). It also comprised a three-counter telescope monitoring the vertical particle flux. The telescope counters were interlayed with 10 cm Pb in one series of experiments and with 10 cm Al in another. The shower particles were detected in two trays of counters placed under the telescope as shown in Fig. 4. The counters were interconnected to form a hodoscope, and the information about the response of the counters to each passage of a particle through the telescope was radio-transmitted. Thus the device, though roughly, allowed one "to see" the picture of showers (the black circles in Fig. 4 indicate the counters responding to one of the showers).

It was found using this device that the interactions of primary particles with lead and aluminium nuclei initiated showers. But in contrast with the showers generated in lead, size of the showers produced in aluminium was smaller. The showers generated in aluminium were narrower and did not comprise the strongly scattered particles charac-

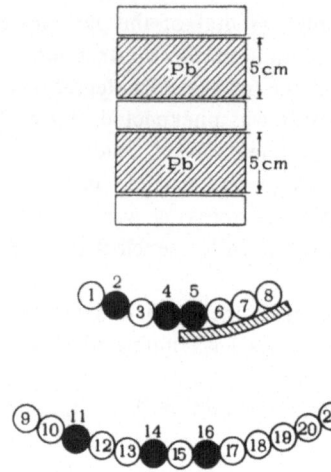

Fig. 4. Arrangement of the counters in the instrument for measuring the secondary showers (Vernov *et al.*, 1950). The black circles are the counters responding to one of the showers; lead absorbers are dashed.

teristic of the electron-photon showers developing in lead.

The comparison of showers in lead and alminium compelled the authors to conclude (Vernov, 1949) that the electron-photon component was generated in the interaction of primary particles with nuclei through the intermediate decay of a particle whose life time was shorter than 10^{-10} sec. The conclusion pre-empted the discovery of π^0-mesons.

In order to clarify the production mechanism of the strongly ionizing particles in the stratosphere, an instrument for recording nuclear disintegrations, the so called "ionization bursts", was developed.

The device consisted of an ionization chamber and an electric pulse amplifier. The number of pulses exceeding a given value (about 1 mV at the amplifier input) was radio-transmitted to the earth.

This method was used in 1948 to measure altitude dependence of the number of nuclear disintegrations. The measurements showed that the secondary particles produced by cosmic rays in the atmosphere played the major role in the production of nuclear disintegrations (Vernov *et al.*, 1948).

In the same period (1947–51), S.N. Vernov and A.E. Chudakov studied cosmic rays on rockets beyond the atmosphere (Vernov *et al.*, 1958). In their experiments, the primary cosmic ray flux was measured, and the generation of electrons and photons by primary cosmic rays in their interactions with matter inverstigated. The primary low-energy photon flux and the ionizing power of the total primary cosmic ray flux were measured.

To facilitate the study A.E. Chudakov developed the method for measuring the ionization without electrometers and the technique of radio transmission of electric pulse amplitudes. Both methods were extensively used later in the stratospheric measurements (Grigorov and Murzin, 1953; Grigorov, 1956).

Thus, the first stage of the stratospheric studies of cosmic rays onboard balloons and rockets had given the following qualitative results (Vernov *et al.*, 1950) by the end of 1948.

(1) In their interactions with atomic nuclei, the primary particles (protons) produce showers of particles. As a result, the electron-photon and penetrating components of cosmic rays are generated.

(2) The secondary particles in these showers are nuclear-active and can produce showers similar to the primary particle-generated showers and cause nuclear disintegrations. The same experiments brought forward new problems demanding their examinations, namely

(i) what are the quantitative characteristics of proton interactions?

(ii) how does the interaction character depend on primary energy?

(iii) what is the reason for the absence of azimuthal asymmetry of cosmic rays in the stratosphere if the primary particles are protons?

Concerning the last problem, it was assumed in the literature of that period that the absence of the asymmetry could be associated with strong quasi-isotropic scattering of secondary particles in the stratosphere. However, the measurements of the angular distribution of particles up to altitudes of 27 km (Vernov and Kulikov, 1950) showed that the assemption was incorrect because the scattering of secondary particles is insignificant and cannot account for the Johnson-Barry result (too small effect of the East-West asymmetry in the stratosphere at the equator).

In 1949, Vityaz, the oceanographic research vessel of the Academy of Sciences, of the USSR took off from the Black Sea to Far East and the second equatorial expedition was on her board. One of the main aims of the expedition was to study the east-west asymmetry in cosmic rays. With that purpose two types of instruments were designed. One of the instruments comprised a set of telescopes arranged at various zenith angles to measure the counting rates from all particles incoming at various angles to a vertical and at various azimuths. When ascending, the instrument spinned around its vertical axis, while a system of several photocells fixed an exact orientation of the instrument with respect to the Sun (Vernov and Kulikov). (Fig. 5).

Fig. 5. The last check-up of the apparatus for the measurements of cosmic ray azimuthal asymmetry before the launching in equator area. Motor ship "Vityaz", 1949.

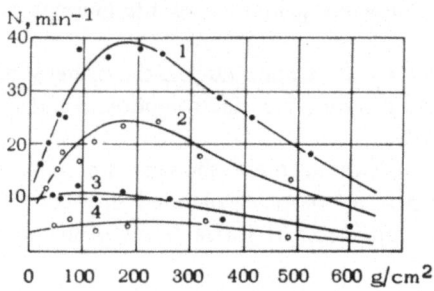

Fig. 6. Cosmic ray azymuthal asymmetry in the stratosphere (Vernov *et al.*, 1950). *X*-axis is the residual atmospheric pressure; *Y*-axis is the particle intensity at 60° to a vertical; the points and crosses are the measurements in the western and easter directions respectively.
1 and 2 – all particles. 3 and 4 – penetrating particles.

The second instrument was a telescope inclined at 60° to the vertical and mounted on a platform the orientation of which was fixed in the east-west direction by photocells independently of the orientation of the instrument proper. The direction of the telescope was automatically reversed from east to west at an interval of several minutes (Vernov *et al.*, 1949).

In the first run, the azimuthal asymmetry of all particles was measured. In the next run a lead absorber was inserted between the telescope counters to measure the azimuthal asymmetry of the penetrating component. The results of the measurements presented in Fig. 6 showed a great east-west asymmetry in the stratosphere.

By these experiments the assumption that the protons are the main constituent of primary cosmic ray flux was finally confirmed.

Besides the above research during the expedition, the altitude dependence of the electron-nuclear showers (Vernov *et al.*, 1953) and nuclear disintegrations (Grigorov and Murzin, 1953) produced by primary particles with mean energy about 40 GeV was measured.

When preparing the second expedition, the experience with humidity trouble during the first equatorial expedition in 1937 was taken into account. The instruments were designed in such a way that even a 100% humidity would not affect operations. Due to this precaution expedition had no dramatic incidents. During the 30 days at the equator there were launched 16 instruments and all of them, except for one, were acting satisfactorily safely.

The quantitative characteristics of the elementary event of cosmic ray proton interactions were studied in 1949–52 (Grigorov and Murzin, 1953) also in the stratosphere. The experiments were based on the so called energy approach proposed by D.V. Skobeltzyn's school. The main idea of the approach to measure the distribution of the primary particle energy between the secondary components. With that purpose thorough measurements of the energy fluxes of the electron-photon component at various atmospheric altitudes (by measuring the air-lead transition effect), of the altitude dependence of the nuclear disintegrations rate, and of the spectrum of energy release in the disintegrations were carried out in various geomagnetic latitudes at altitudes from 5 to 25 km. Use was made of new and improved devices which required neither electrometers nor spring shock absorbers when measuring the ionization bursts. In the same period, I.D. Rapoport measured the ionizing power of individual cosmic ray particles at altitudes from sea-level to 20 km (Grigorov et al., 1953). A pulsed ionization chamber was used to measure the energy transferred to π^0-mesons in a single event of ~20 GeV primary proton interaction with lead nuclei.

All the experiments have given the quantitative characteristics of the elementary interaction event which may be outlined as follows (Grigorov and Murzin, 1953; Vernov and Grigorov, 1956; Grigorov, 1956):

(1) the fraction of the energy lost by primary particles for generation of neutral pions in their first collision with nucleus is some 10~15% and depend little on primary particle energy in the 3~20 GeV range. Therefore, the energy loss for pion production is about 30~45%;

(2) the primary particles with energies of about 20 GeV lose the major portion of their energy (80%) for pion production in the atmosphere. The production of π-mesons in the atmosphere by high-energy particles (~20 GeV) does not occur in a single event but in several events (more than three), i.e. the nuclear-cascade process takes place the idea of which G.T. Zatsepin set forth in his analysis of extensive air showers in 1949;

(3) Analysis of the experimental data on nuclear disintegrations has shown that the mean energy loss, for a single event of disintegration of a light nucleus is about 400 MeV and does not depend on the energy of the primary in the 3~20 GeV range. Disintegration of a light nucleus produces on the average three neutrons capable of initiating further nuclear disintegrations.

These energy characteristics have made it possible to quantitatively describe the cosmic rays propagation in the atmosphere and the production of secondary components at various latitudes (Grigorov, 1956) and were confirmed later in the accelerator experiments.

In the mid-fifties, a new trend associated in a certain way with the International Geophysical Year (IGY, 1957) appeared in the altitude experiments. The cosmic ray studies were an essential part of the IGY program which laid special emphasis on examination of the relationships of cosmic rays to various geophysical and solar events. It was decided, therefore, that continuous balloon-borne observations of cosmic rays in the stratosphere should be carried out simultaneously with the ground-based observations.

The valuable experience in the stratospheric cosmic ray observations has helped A.N. Charakhchyan to develop a light, handy, and reliable instrument for measuring cosmic rays. The Charakhchyan device became a conventional type and was put in quantity production. The daily launchings of the standard instruments carried out at various latitudes (since 1957 in Murmansk and Alma-Ata, and since 1962 in Mirny in the Antarctics) have yielded unique experimental material. The results of the stratospheric measurements demonstrating an 11-year circle of cosmic ray intensity are presented in Fig. 7. It has been established (Charakhchyan and Charakhchyan 1968; Stozhkov and Chrakhchyan, 1970) that the modulation of galactic cosmic rays in 1959–70 was in a good correlation with the number helio latitude distribution of sunspots.

The continuous many-year stratospheric observations of cosmic rays have also shown that the empirical correlation was violated in 1971–73 when cosmic ray intensity did

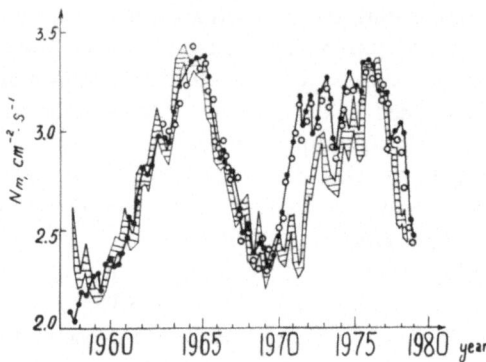

Fig. 7. Time variations of cosmic ray intensity in the stratosphere in the maximum of the absorption curve. The black circles are the Arctic measurements, the open circles are the Antarctic measurements. The dashed curve shows the calculations according to solar activity.

not follow solar activity. The anomaly coincided with the sign reversal in the Sun's general magnetic field. The explanation of the anomaly was given by J. Jokipii (Jokipii and Kopriva, 1979) and B.A. Tverskoy (1980).

Before the systematic observations of cosmic rays in the stratosphere, the general opinion was that the cosmic ray generation on the Sun was a rare occurrence at an interval of several years and took place during extremely powerful flares on the Sun. The cosmic rays with energies of up to tens of GeV observable at mountains and sea level are supposed to be generated in such flares. The continuous high-altitude measurements of cosmic rays organized by A.N. Charakhchyan have revealed the important fact that in practice each solar flare of class 3+ is accompanied by generation of solar cosmic rays (Charakhchyan and Charakhchyan, 1962).

Returning to the study of the interaction characteristics it is worth while to note that knowledge of the primary particle energy was of major importance to the research. In the stratospheric measurements the Earth's magnetic field (i.e. the cosmic ray latitude effect) was used to get such knowledge.

At the same time, the study of the particles with energies of and above hundreds of GeV which are actually unaffected by the Earth's magnetic field necessitated a new experimental method which would make it possible, as before, to deal with particles of known energies when studying the interactions.

The solution was found at the Moscow University where it was suggested to measure the energy of a primary particle by measuring the total ionization effect produced in a dense matter (Grigorov *et al.*, 1958). The instrument designed on the basis of this principle was called ionization calorimeter.

We think that two circumstances promoted the idea of the ionization calorimeter in D.V. Skobeltzyn's school.

The first was that the numerous stratospheric studies had been based on the "energy approach" in which cosmic ray energy influx was determined using the total ionization effect in the atmosphere.

The second was associated with the extensive use of ionization chambers in the

Moscow University studies which gave impetus to the development of various electronic devices for ionization measurements and of numerous technological methods of making the integral and pulsed ionization chambers for various applications. So a leap from that experience gained to the making of the ionization calorimeter required but one step. In 1957 the first instrument was made (Grigorov et al., 1958).

The ionization calorimeter has proved to permit the construction of large-area arrays of (up to tens of m^2) and, thereby, the study of the very high-energy particles. Besides that, the combination of the ionization calorimeter with the known instruments for visual observations of particles (cloud chamber, nuclear photoemulsions, emulsion chambers) proposed, by the author of the ionization calorimeter has made it possible to realize the idea of studying the interactions of cosmic ray hadrons of known energies in the $10^{11} \sim 10^{13}$ eV range (Grigorov et al., 1959a; Grigorov and Rapoport, 1961).

In the USSR, the hadron interactions at mountain-levels had been studied for a long time (see the review by N.A. Dobrotin et al.). We shall mention here but some of the experiments carried out at the Moscow University because they had promoted the development of cosmic ray experiments onboard heavy satellites. The program of the works proposed in 1956 by N.L. Grigorov (Grigorov et al., 1959b) consisted of the following three stages.

The first stage was to study the hadron interaction characteristics in the $10^{11} \sim 10^{12}$ eV energy range using a cloud chamber in combination with ionization calorimeter. This stage was realized by the Pamir collaborated in 1958–59 (See the review by N.A. Dobrotin et al.).

The second stage was to study the particle interaction characteristics in the $10^{12} \sim 10^{13}$ eV energy range using nuclear emulsion plates combined with ionization calorimeter (the method of controllable nuclear emulsions). This stage was realized in the Aragatz experiment in 1960's.

The third stage was to study the high-energy particle interactions in an emulsion pile combined with ionization calorimeter. This stage was realized in the experiment onboard the Intercosmos-6 satellite in 1972.

The second stage of the program was realized in the experiments with a large ionization calorimeter of 10 m^2 area. The energy spectrum of π^0-mesons generated in a carbon target by hadrons of energies above 10^{12} eV was studied. The experiments have revealed the existence of a new interaction process in which 60~80% of the energy of a primary particle is transferred to several π^0-mesons, while more than 25% of this energy goes to a single π^0-meson. The process is realized within a 10% probability (Babajan et al., 1962). In the center-of-mass system, the Lorentz-factor of a cluster whose decay may have been responsible for the π^0-meson production in such interactions was some $20 \sim 35$ (Babajan et al., 1965a) at the primary particle energies in the $10^{12} \sim 10^{13}$ eV range.

The characteristics of the newly found interactions proved to be drastically different from those of the ionization processes, so the authors defined the former to be a specific class of interaction (Grigorov and Shestoperov, 1964a). It was found out that the new class was responsible for generation of high-energy secondary gamma-rays and muons (Grigorov and Shestoperov, 1964b; Grigorov et al., 1973).

Some 10 years later, when the $10^{11} \sim 10^{12}$ eV accelerators were put into operation,

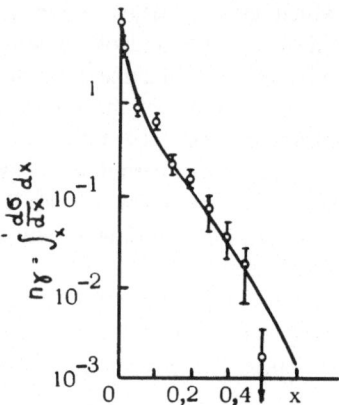

Fig. 8. The integral spectrum of gamma-rays produced in the high-energy hadron-air collisions. X-axis is the Feynman variable X, Y-axis is the mean number of gamma-quanta produced in a single collision. The points are the results of cosmic ray measurents (Grigorov and Shestoperov, 1980), the solid curve presents the accelerator data.

the new process (now it is called the hadron fragmentation process) was found anew. In Fig. 8, the old data obtained in the modern variables in cosmic rays (Grigorov and Shestoperov, 1980) are presented together with the new accelerator data on the fragmentation process.

In the mountain-level experiments a great number of hadrons with energies above 10^{12} eV were recorded. This fact made it possible to study the character of energy loss by hadrons in the ionization calorimeter, to measure the absolute intensity of hadrons of various energies at mountain-level, and to measure the intensity of the high-energy hadrons which are not accompanied by secondary particles, i.e. the intensity of so called single hadrons. Assuming that the single hadrons at mountain-level are primary protons which traversed the atmosphere without interactions, one can obtain the lower limit of the cross section of high-energy proton inelastic interaction with air nuclei (the method was developed as early as 1957 (Grigorov *et al.*, 1957). The intensity measurement of single hadrons of energies above 10^{12} eV made it possible to obtain the data on "the decrease of the interaction path of nuclear-active particles with increasing the particle energy" (Babajan *et al.*, 1965b). But the quantitative character of the dependence of the cross section on energy could not be established in the experiments.

Due to the development of the space technique in the mid 60's the possibility arose to launch heavy satellites with scientific equipment weighing several tons. It was decided to use this possibility when studying the high-energy primary cosmic rays and some of their interaction characteristics.

The series of the satellites carrying cosmic ray laboratories were called Proton. The main instrument for detecting the high-energy particles onboard Proton was the ionization calorimeter. Together with the targets of carbon and polyethylene and the detectors discriminating the protons and α-particles from all the primary particles, the calorimeter has made it possible to measure the effective cross section of the primary proton inelastic interactions with protons and carbon nuclei in the 20~600 GeV energy range (Grigorov *et al.*, 1967), the energy spectra of primary protons and α-particles (Grigorov *et al.*, 1969) in the $3 \times 10^{10} \sim 10^{13}$ eV energy range, and the spectrum of all cosmic-ray particles in the $3 \times 10^{10} \sim 3 \times 10^{15}$ eV energy range (Grigorov *et al.*, 1969; Akinov *et al.*, 1971).

As a result of the Proton measurements, not only was the increase of the effective inelastic cross section confirmed, but the quantitative dependence of the cross section on energy was also obtained.

The measurements of the primary cosmic ray energy spectrum onboard Proton satellites by the direct methods up to the energies of the particles measured until then only by indirect methods in extensive air showers (EAS) were of great importance to the study of the ultrahigh-energy particles by the EAS method. The comparison between the energy spectra obtained by indirect and direct methods allows us to verify our concepts of the EAS origin and development, i.e. the mode of the interactions of the ultrahigh-energy particles with matter.

The direct succession and relevance of the Proton experiments to all the previous studies can most clearly be demonstrated by the fact that, though the development of the equipment to be flown on the Proton series of space stations was initiated in August, 1963, Proton-1 was launched as early as July, 1965. In fact, but for the experience gained with the stratospheric balloons and rockets and in satellite measurements, the complex scientific equipment could not have been developed, manufactured, and checked.

The cosmic rays include not only the high-energy particles but also the superhigh-energy species (up to 10^{20} eV). It can hardly be expected to obtain such superhigh-energy particles (in the $10^{16} \sim 10^{20}$ eV range) in laboratories before the end of this century. It seems to us expedient therefore, to briefly discuss here some of the studies of EAS initiated by the supperhigh-energy cosmic rays.

It was noted above that the direct measurements of the energy spectrum of primary cosmic rays on the Proton satellites up to 10^{15} eV had confirmed the results obtained at the Moscow State University by recording EAS.

The interest in the detection and study of the EAS-producing superhigh-energy cosmic rays arose immediately after the discovery of the nuclear-cascade process in the superhigh-energy range and is ever growing especially because of the recent advances in the high-energy physics. The array at the Moscow State University and then the Yakutsk array were used to study the primary cosmic ray energy spectrum in the $10^{15} \sim 10^{20}$ eV range (Vernov *et al.*, 1968; Diminstein *et al.*, 1975) with special emphasis on the events caused by the nuclear-cascade process in the atmosphere (the fluxes of electrons, muons, and Cerenkov radiation). The hodoscope of 10,000 underground and ground-based Geiger counters of the Moscow State University array was used to study the EAS electron and muon size spectra. The essential differences in the sensitivities of the EAS electron and muon components to the parameters of the hadron-nucleus interactions have made it possible to show (Vernov and Khristiansen, 1967) on the basis of the set of the muon and electron data that the power-law exponent of the primary energy spectrum changes abruptly in the energy range $E_0 = (2 \sim 4) \times 10^{15}$ eV. This conclusion was confirmed later by measuring the EAS Cerenkov light densities with the Yakutsk array (Efimov *et al.*, 1979a) and can quantitatively be formulated as follows. If the primary energy spectrum of cosmic rays is presented by the power law $\sim E^{-\gamma}$, the power-law exponent γ will be 1.65 at $E_0 < 2 \times 10^{15}$ eV and

2.3 at $E_0 \sim 4 \times 10^{15} \sim 10^{17}$ eV.

The measurements of the Vavilov-Cerenkov radiation produced in EAS have permitted an absolute calibration of the detected showers with respect to their primary energy to within not worse than 25%. The idea of using the Cerenkov radiation flux as a measure of the energy of the EAS-generating primary particles goes back to D.V. Skobeltsyn's classical works in which the total energy loss for ionization in a cascade was proposed to use as a measure of the cascade energy. The fact is that the Cerenkov radiation flux in the $10^{14} \sim 10^{17}$ eV EAS near sea level is proportional to the total ionization loss in EAS cascades, with the proportionality factor being practically independent of EAS model.

The first systematic studies of the EAS Cerenkov light initiated in 1957 (Chudakov *et al.*, 1959) permitted an absolute primary energy calibration of EAS. The studies were continued with the Yakutsk array which comprises 43 points to detect EAS electrons and 33 points to detect EAS Cerenkov light over an area of about 18 km². The relationship between the EAS electron size and Cerenkov radiation flux may be used to get the mean absolute primary energy calibration of EAS.

The data on the primary energy spectrum obtained with the EAS arrays at the Moscow State University and Yakutsk (Vernov and Khristiansen, 1967; Efimov *et al.*, 1979a) and from the Proton satellites (see Fig. 9) have changed the earlier concept of a simple power-law form of the primary energy spectrum. The results of some studies (Dyakonov *et al.*, 1979; Khristiansen *et al.*, 1980) were indicative of new irregularities in the primary energy spectrum in the 3×10^{17} and 10^{19} eV ranges. Though the irregularity may be considered to be safely established in the 3×10^{15} eV range only, the detailed study of the form of the primary energy spectrum in the $10^{17} \sim 10^{20}$ eV range is one of he most urgent tasks of the superhigh-energy cosmic ray astrophysics.

The studies of the EAS electrons, muons, and Cerenkov light with the EAS arrays at the Moscow State University and Yakutsk have also yielded interesting information about the nature of the superhigh-energy hadron interactions ($10^{15} \sim 10^{18}$ eV). It has been shown (Kalmykov and Khristiansen, 1975, 1976), that the extrapolation of the experimental parameters of hadron interactions obtained at the accelerator energies ($\lesssim 2 \times 10^{12}$ eV) to the superhigh-energy range (above 10^{15} eV) is impossible because of the violation of the scaling.

The results shown in Fig. 7 and the data on the scaling violation have been obtained by the classical method of EAS study, namely by detecting the lateral distribution of particles at a given observation level in an individual shower (see, for example, in (Khristiansen *et al.*, 1980). Recently, the Yakutsk EAS array was used (Efimov *et al.*, 1973) to realize the method proposed in (Fomin and Khristiansen, 1971) which gives the pattern of the longitudinal development of an individual shower by measuring the time-differential density of the Cerenkov photon flux, i.e. the Cerenkov pulse shape, at a great distance from the shower axis. At present, this method has yielded the information about the position of the cascade maximum in individual EAS (Efimov *et al.*, 1979b) and about the shape of the cascade curve.

An attempt was made in this review to briefly describe the branch of the cosmic ray physics which began in the USSR with the launchings of balloons and developed into the scientific trend of the high-altitude cosmic ray studies. The immanent logic of the studies can be traced from the first balloonborne experiments to the sophisticate works

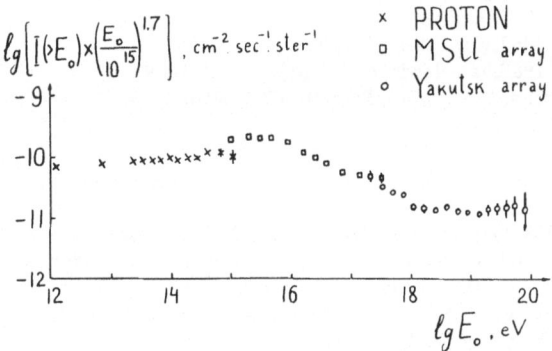

Fig. 9. The primary cosmic ray energy spectrum inferred from the EAS measurements and Proton experiments. The accuracy of the derivation of the absolute flux from the EAS data is within a factor of 1.7~2.0.

with automatic equipment and has led to the modern space research and to the giant EAS arrays.

REFERENCES

Akimov, V.A. *et al.*: 1971, *Izv. Acad. Sci. USSR, Phys.* **35**, 2434.

Babajen, K.P. *et al.*: 1962, *Izv. Acad. Sci. USSR, Phys.* **26**, 558.

Babajan, K.P. *et al.*: 1964, *Nucleonike* **IX**, 291.

Babajan, K.P. *et al.*: 1965a, *Izv. Acad. Sci. USSR, Phys.* **29**, 1648.

Babajan, K.P. *et al.*: 1965b, *Izv. Acad. Sci. USSR, Phys.* **29**, 1652.

Brikker, S.I., Vernov, S.N., Evreinova, I.M., Sokolov, S.P. and Charakhehjan, T.N.: 1947, *Doklady USSR Acad. Sic.* **57**, N 2.

Brikker, S.I., Vernov, S.N., Grigorov, N.L., Evreinove, I.M. and Charakhchjan, T.N.: 1948, *Doklady USSR Acad. Sci.* **61**, 629.

Charakhchyan, A.N. and Charakhchyan, T.N.: 1968, *Izv. Acad. Sci. USSR, Phys.* **32**, 1869.

Charakhchyan, A.N. and Charakhchyan, T.N.: 1962, *Geomagn. Aeron.* **2**, 829.

Chudakov, A.E. *et al.*: 1959, *6 ICRC, Moscow*, Vol. 2.

Diminstein, O.S. *et al.*: 1975, *14 ICRC, Munich*, **12**, 4318.

Dyakonov, M.N. *et al.*: 1979, *16, ICRC, Kyoto* **8**, 168.

Efimov, N. N. *et al.*: 1973, *13 ICRC, Denver*, Vol. 4, 2378.

Efimov, N. N. *et al.*: 1979b, *16 ICRC, Kyoto* **9**, 73.

Efimov, N. N. *et al.*: 1979a, *16ICRC, Kyoto* **8**, 152.

Fomin, Yu.A. and Khristiansen, G.B.: 1971, *Sov. Nucl. Phys.* **14**, 647.

Grigorov, N.L.: 1956, *Uspekhy Phys. Nauk* **58**, 599.

Grigorov, N.L., Everinova, I.M. and Sokolov, S.P.: 1951, *Doklady USSR Acad. Sci.* **81**, 379.

Grigorov, N. L. and Murzin, V.S.: 1953, *Izv. Acad. Sci. USSR Phys.* **17**, 21.

Grigorov, N.L., Murzin, V.S. and Rapoport, I.D.: 1958, *JETP* **34**, 506.

Grigorov, N.L. and Rapoport, I.D.: 1961, *Certificat* N 724308/26.

Grigorov, N.L. and Shestoperov, V.Ya.: 1964a, *Nucleonika* **IX**, 307.

Grigorov, N.L. and Shestoperov, V.Ya.: 1964b, *Izv. Acad. Sci. USSR, Phys.* **28**, 1778.

Grigorov, N.L. and Shestoperov, V.Ya.: 1980, *Izv. Acad. Sci. USSR, Phys.* **44**, 512.

Grigorov, N.L., Rapoport, I.D. and Shipulo, G.P.: 1953, *Doklady USSR Acad. Sci.* **91**, 491.

Grigorov, N.L. *et al.*: 1959a, **6 ICRC, Moscow.**

Grigorov, N.L. *et al.*: 1959b, **6 ICRC, Moscow.**

Grigorov, N.L., Rapport, I.D. and Shestoperov, V.Ya.: 1973, *High Energy Particles in Cosmic Rays*, Moscow.

Grigorov, N.L. *et al.*: 1957, *JETP* **33**, 1099.

Grigorov, N.L. *et al.*: 1967, *Space Res.* **V**, 420.

Grigorov, N.L. *et al.*: 1969, *Izv. Acad. Sci. USSR, Phys.* **33**, 1469.

Jokippii, J.R. and Kopriva, D.A.: 1979, *Proc. 16 ICRC, Kyoto* **3**, 7.

Kalmykov N. N. and Khriotiansen G. B. : 1975, *Lettens to JETP*, **21**, 666.

Kalmykov, N.N.and Khristianson, G.B.: 1976, *Lett. JETP* **23**, 595.

Khristiansen, G.B., Kulikov, G.V. and Fomin, Yu.A.: 1980, *Cosmic Ray of Superhigh Energy, Munich, Tiemig Verlag.*

Moltchanov, P.A.: 1938, *Radio Atmos. Res. "Nauka"* **3**, (in Russian).

Stozhkov, Yu.I. and Charakhchyan, T.N.: 1970, *Izv. Acad. Sci. USSR, Phys.* **34**, 2439.

Tverskoy, B.A.: 1980, *23 COSPAR Meeting, Budapest.*

Veksler, V. I. Kurnosova and Lubimov A.: 1947, *JETP,* **17**, 1026.

Vernov, S.N.: 1934, *Phys. Rev.* **46**, 822.

Vernov, S.N.: 1935a, *Nature.* **135**, 1072.

Vernov, S.H.: 1935b, *Proceeding USSR Conf. Stratosphere Research,* 423 (in Russian).

Vernov, S.N.: 1945, *Lebedev Institute Proceedings.*

Vernov, S.N.; 1949, *JETP* **19**, N 7.

Vernov, S.N. and Grigorov, N.L.: 1956, *Suppl. N. Cim.* **4**, 879.

Vernov, S.N. and Kulikov, A.M.: 1950, *Doklady USSR Acad. Sci.* **73**, 483.

Vernov, S.N. and Khristiansen, G.B.: 1967, *10 ICRC, Calgary,* **1**, 345.

Vernov, S.N. and Khristiansen, G.B. *et al.*: 1968, *Can. J. Phys.* **46**, 137.

Vernov, S.N., Grigorov, N.L. and Charakhchyan, A.N.: 1950, *Izv. Acad. Sci. USSR, Phys.* **14**, 51.

Vernov, S.N., Kulikov, A.M. and Charakhchyan, A.N.: 1953, *Izv. Acad. Sci. USSR, Phys.* **17**, 13.

Vernov, S.N., Grigorov, N.L. and Savin, F.D.: 1947, *Doklady USSR Acad. Sci.* **57**, 137.

Vernov, S.N., Grigorov, N.L. and Savin, F.D.: 1948, *Doklady USSR Acad. Sci.* **61**, 815.

Vernov, S.N., Dobrotin, N.A., Grigorov, N.L., Savin, F.D. and Sokolov, S.P.: 1949, *Doklady USSR Acad. Sci.* **68**, 253.

Vernov, S.N., Vakulov, P.V., Gorchakov, E.V., Logachev, Yu.I. and Chudakov, A.E.: 1958, *Artif. Earth Satell.,* N **2**, 61.

COSMIC RAY INTENSITY VARIATIONS AND
THE MANCHESTER SCHOOL OF COSMIC RAY PHYSICS

Harry ELLIOT[*]

1. Introduction

My first fleeting contact with cosmic ray research in Manchester was made early in
1939 when, together with a few other undergraduate physics students, I was taken on
a tour of the research laboratories by J.M. Nuttall, a senior member of the teaching
staff who in an earlier epoch had worked on radioactivity as a colleague of Rutherford
and together with Geiger had been joint discoverer of the well known relationship
between the range of an α-particle and the half life of its parent nucleus.

In the course of this tour we visited the large cosmic ray laboratory on the ground
floor of the Schuster Building in Coupland Street. This room with its white and brown
tiled interior typified the entire building which bore a striking resemblance in its
decorative style to the public lavatories of the time — so much so that it was popularly
believed by the student population that the building had come about as the result of a
misprint in the architect's brief, a misprint which was now spelled out for the benefit
of posterity in bricks and mortar and brown glazed tiles.

It was in this particular laboratory at this time that Jánossy, Lovell, Rochester,
Rossi, Wilson and others were unravelling the complexities of the secondary cosmic
rays using cloud chambers and a variety of ingenious arrangements of Geiger counters,
valves and lead shielding. I well remember that on this occasion Blackett too was to be
seen at work in the laboratory examining cloud chamber photographs with the aid of a
stereoscope from which activity he spared but a brief glance in our direction. P.M.S.
Blackett had arrived in Manchester from Birkbeck College, London, in 1937 as the new
occupant of the Langworthy Chair following the departure of W.L. Bragg to the
National Physical Laboratory. In the short space of two or three years Blackett had
established a powerful cosmic ray research group in the Manchester Physics Department which was the equal of any in the world at that time.

However, the darkening political situation in Europe already indicated the near
certainty of war in the not very distant future and in 1939 Blackett who, as a member
of the Air Defence Committee, had already been deeply involved in such matters left
Manchester to work full time on the scientific and technical aspects of warfare. From
that time until 1945 he was able to visit the Manchester Laboratories no more than
two or three times a year. Meanwhile work on cosmic rays continued under the
direction of Jánossy and Rochester who had at the same time to contend with a
greatly increased burden of undergraduate teaching through the absence of their
colleagues on war work.

[*]Blackett Laboratory, Imperial College, London, UK.

375

Y. Sekido and H. Elliot (eds.), Early History of Cosmic Ray Studies, 375–384.
© 1985 by D. Reidel Publishing Company.

2. Duperier's Work at Imperial College

In the immediate pre-war years the political and social upheavals in continental Europe had led to the exodus of many distinguished scientists from their native lands and as a result of strenuous efforts by Blackett and others a number of these exiles had been found positions in the UK. For many of them such posts served as no more than a temporary haven on the way to the United States but a few remained in Britain either permanently or for extended periods of time. Included amongst the latter was Arturo Duperier of the University of Madrid who together with his family had left Spain at the time of the Civil War and had settled for the time being in London. In 1939, as a result of Blackett's influence, he became interested in the relationship between the cosmic ray intensity and terrestrial magnetism and in the effect of atmospheric temperature and pressure on the intensity of the secondary radiation at ground level. During the war he set up a cosmic ray intensity recorder at Imperial College in the form of a small battery of Geiger counters arranged in threefold coincidence. Using the data from this recorder he developed Blackett's idea that the dependence of the ground level cosmic ray intensity on atmospheric temperature, which had been known for several years but not understood, was the direct result of the instability of the muon (Blackett, 1938). In particular Duperier was able to show by partial correlational analysis that in accounting for the temperature effect the best atmospheric parameter to use was the height of the 100 mb pressure level which is close to the average height of muon production (Duperier, 1945, 1948). An incidental but extremely interesting bonus to this systematic study of atmospheric effects was provided by the famous solar and interplanetary cosmic ray events of February and March 1942 which were registered by Duperier's counters in London (Duperier, 1942) by Ehmert in Germany (Ehmert, 1948) and by the world wide network of Carnegie Institution ion chambers. (Lange and Forbush, 1942)

Following these events a twofold coincidence counter recorder similar to Duperier's instrument was built by McCaig in Manchester and this latter recorder remained in operation until 1946. The large Forbush Decreases of February 5th and 7th 1946 were registered by both the London and Manchester recorders which showed a total decrease in intensity of just over 10% (Duperier and McCaig, 1946). This event was associated with a sunspot group having an area 4900 millionths of the solar hemisphere, the largest which had ever been recorded at Greenwich up to the time.

3. The Cosmic Ray Intensity Recorders in Manchester

Meanwhile Blackett had returned to Manchester in 1945 and had arranged for the production of a properly engineered and much enlarged version of the rather rudimentary counter arrays of Duperier and McCaig. These recorders were built by Cinema Television of London with Mr. L.C. Bentley as the engineer in charge. In this connection it is of interest to note that in June—July 1946 a Commonwealth Scientific Conference was held in London under the auspices of the Royal Society. The subject matter of this conference covered virtually the whole of science but two recommendations in respect of cosmic rays which were adopted by the Conference read as follows:

1. Further measurements of the variation with time of the cosmic ray intensity at selected stations at sea level and on mountains. Measurements in the Southern hemisphere are of particular importance.

2. Further measurements of the variation of cosmic ray intensity with latitude and longitude by experiments in aircraft over a wide range of height.

These recommendations were drawn up and adopted mainly as a result of Blackett's initiative and it was his intention that the new Cintel recorders should be used in implementing this programme.

The first of these recorders was, I believe, delivered to the Manchester Laboratory sometime in May 1946 and D.W.N. Dolbear who had just returned to Manchester as a postgraduate student, following his demobilization from the Royal Air Force, was given the job of running it. He had by way of introduction already taken over McCaig's intensity recorder following the latter's departure to a post with the Permanent Magnet Research Association. Dolbear had had the new recorder in operation only a matter of weeks when the sun, evidently fully aware of its obligations in the matter, produced the great solar flare of July 25 which in turn generated a 15–20% increase in the flux of ionizing cosmic ray secondaries in Manchester. This ground level event, as we would now call it, was followed by a Forbush Decrease of some 6–7% on July 27. The event of July 25 was only the third such solar flare effect to be observed on a world-wide scale and the first in peacetime. Not surprisingly it generated a great deal of interest.

It was at about this time that I returned to Manchester to start research on cosmic rays as a postgraduate student after five years in the Royal Air Force. Most of the cosmic ray work in Manchester was concerned with the study of the secondary radiation but my own interests lay in the geophysical and astrophysical aspects of the primary radiation rather than in high energy nuclear physics. In the summer of 1946 the cosmic rays still appeared to be a unique and isolated phenomenon quite unrelated to the general body of astronomy and it was this very anomalous character of the radiation which held a particular fascination for me. Consequently I had no difficulty in making up my mind to work on intensity variations. It so happened that Dolbear and I were old friends having been at school together and Blackett was very willing that we should now work together under his supervision. He probably felt some sympathy for a pair of relatively elderly students returning to pick up the threads after a long absence from physics. He had been in a rather similar position on leaving the Royal Navy after World War I to take up physics as an undergraduate at Cambridge. In having the benefit of his guidance Dolbear and I could and did count ourselves extremely fortunate.

4. The Ground Level Solar Flare Event of July 25 1946

However, to return to the solar event of July 25. Having heard that similar variations in intensity to those registered in Manchester had been observed in the Carnegie Institute ion chambers Dolbear wrote to Scott Forbush offering to publish the Manchester alongside his. As far as I can remember Scott never replied and Dolbear and I, after waiting several months, eventually published our data independently in Nature (Dolbear and Elliot, 1947).

Meanwhile in the late summer of 1946 Hannes Alfvén visited the laboratory and described to us a model which, he claimed, could explain the variations in intensity associated with the July 25 event. According to his picture the increase and subsequent decrease in intensity was brought about by the approach and passage of a solar plasma stream which was electrically polarized by its motion through the solar magnetic dipole field. Alfvén subsequently published this interpretation (Alfvén, 1946) but Dolbear and I were both convinced that the model didn't properly fit the facts and we proposed to say so in our own letter to Nature on the July 25 event. We were however dissuaded from doing so by older and perhaps wiser (although not in this case) hands who were reluctant to accept the scientific judgement of two very new research students when it ran counter to that of such a world renowned figure as Professor Alfvén.

Although Alfvén's electric field mechanism was in our view quite incapable of explaining the events of July 25 onwards it nevertheless seemed possible that it might account for the much smaller quasi-regular solar daily variation which has an amplitude of a few tenths of a percent. In the event this also turned out not to be so because, apart from anything else, the interplanetary magnetic field geometry is quite different from that of a dipole.

The increase in intensity of July 25 was correctly interpreted by Forbush (1946) as being due to the emission of energetic particles from the flare and it may now seem rather strange that this explanation was not immediately apparent to all concerned. In order to understand why it was not apparent it has to be remembered that at the time it was widely believed that the sun had a general dipole field with a pole strength of some 50 gauss. This belief was based on Hale's measurement of the Zeeman broadening of photospheric spectral lines and the solar wind not having yet been discovered it was assumed that the dipole field must extend outwards into space following the familiar inverse cube law. Such a field would be sufficient to contain charged particles in the energy range $10^9 - 10^{10}$ eV and generated on the solar surface to the close vicinity of the sun everywhere except at the poles. Consequently, in order to make the solar particle hypothesis plausible it was necessary to invent an escape mechanism. This involved the additional and somewhat ad hoc idea that sunspot magnetic fields could provide a "tunnel" enabling the charged particles to escape through the large scale solar field (Forbush et al., 1949). The correct explanation for the escape of the particles was, of course, quite different and only became apparent much later when the true geometry of the interplanatary magnetic field was established.

5. The Daily Variation Measured by Means of Inclined Telescopes

Although the events described above generated a great deal of interest and excitement at the time they were incidental to the main purpose for which the new intensity recorders were to be used which was the investigation of the solar daily variation. The intention was to try to distinguish between the daily variation of the secondary cosmic rays produced by atmospheric temperature and pressure changes and that which could be interpreted as a variation in intensity of the primaries at the top of the atmosphere. This was to be accomplished by inclining the two independent counter arrays at $45°$ to the vertical in the North-South plane. Similar experiments had been done by Alfvén and Malmfors in Stockholm in 1943 (Alfvén and Malmfors, 1943) and by Kolhörster in

Berlin in 1941 (Kolhörster, 1941) but this was not known to us when the Manchester work was begun. It has to be remembered in this connection that the war had seriously disrupted the flow of scientific information and literature and the return to normal occupied a period of several years.

The new monitors operating satisfactorily in this way had certainly got off to a flying start with the excitement of the July event but our delight with the new equipment was short-lived when in a matter of a month or two we were experiencing severe difficulty. The source of trouble was the GM counters which were of the internally quenched variety using argon as the filling gas with ethyl alcohol as the quenching agent. There were sixty such counters in our recorders compared with nine in Duperier's array which had served as the prototype. This represented a considerable and as it turned out a rather rash extrapolation in scale. Apart from the increase in number the counters in the new recorders had a much greater length to diameter ratio than those used by McCaig and Duperier and we soon discovered that the useful lifetime of any one of our counters was no more than two or three months. The reason for this was not far to seek. Each discharge in the counter initiated by a charged particle spread the full length of the counter and consequently the fraction of the available alcohol quench gas used per discharge is greater for a long thin counter than for a short fat one of the same area.

After several months of unequal battle in which we tried to refill counters with argon and alcohol faster than the cosmic rays could empty them we came to the conclusion that we had to find some other way. The solution eventually adopted was to use an electronic switch which reduced the voltage across the counter so rapidly as to prevent the discharge from spreading more than a fraction of the full length of the tube. The voltage was then held at the low level for a sufficiently long time for the positive ion sheath to reach the cathode so that there were no associated secondary discharges. This procedure extended the lifetime of a typical counter to three or four years (Elliot, 1949) and by the end of 1947 we had a directional intensity recording system which was sufficiently stable to measure the annual variation in cosmic ray intensity associated with the annual atmospheric temperature wave. It was therefore fully adequate for good measurements of the solar daily variation and we were able to confirm the earlier results of Alfvén and Malmfors and show that a significant part of both the 12hr and 24hr solar waves originated outside the atmosphere (Elliot and Dolbear, 1950). All subsequent versions on the intensity recorders which were built by Cintel for underground and airborne measurements made use of this external electronic quenching technique.

6. The Solar Magnetic Field

The questions relating to the solar magnetic field and its extension into interplanetary space have already been referred to in connection with the escape of the solar flare particles but there were other aspects too. Jánossy in 1937 had suggested that the so-called latitude cut-off, that is to say the absence of any increase in intensity of cosmic rays between latitude $50°$ and the poles, could be an effect of the solar dipole field (Jánossy, 1937). A solar dipole with a polar field strength of 50 gauss would suffice to deny access to the earth to cosmic rays with energy less than 2 GeV or so.

This cut-off in energy is, in the usual Störmer theory, a function of arrival direction with the consequence that for a mid-latitude station on the earth there will be a diurnal variation in intensity as the earth rotates relative to the Störmer forbidden cone which is itself fixed relative to the sun.

For a solar field as large as 50 gauss this daily variation would be quite large were it not for the earth's magnetic field which scatters cosmic rays into the forbidden directions and so reduces the magnitude of the effect (Alfvén, 1947). Because of this an equilibrium intensity is established in the forbidden directions where the scattering into these directions is just balanced by absorption on the Earth, Moon, Mars, Venus and the Sun. (Kane *et al.*, 1949). The net result is that at an altitude of 30000 ft, for instance, there should be a sharp drop in the intensity of the secondary radiation of 5 or 6% during the early morning hours compared with the normal daytime value. The size of this decrease depends on the magnitude of the solar dipole moment and is a function of the latitude and altitude at which it is measured.

Dolbear and I attempted to detect this decrease in intensity, which was calculated to occur between 0100 and 0400h, by means of an automatic intensity recorder flown in a Mosquito aircraft of the RAF. Tragically, after five flights this aircraft together with its crew of two and the on-board cosmic ray recorder were lost somewhere in the North Atlantic between Iceland and the coast of Scotland. In the course of these flights no statistically significant decrease in intensity was measured and it was concluded that the latitude cut-off at 50° could not be attributed to a solar dipole field (Dolbear and Elliot, 1950). A similar conclusion was reached by Vallarta and Pomerantz (1949) and by Singer (1952).

A further search for a day/night difference in intensity, this time at the height corresponding to the Pfotzer maximum, was carried out by Dawton and myself in

Fig. 1. Loading the airborne cosmic ray recorder into the bomb-bay of a Mosquito aircraft (1949).

Fig. 2. Following the progress of a balloon launched from the roof of the Physical Laboratories, Manchester. (l to r: Mrs Blackett, Miss Blackett, Professor P.M.S. Blackett, Mr. D.I. Dawton)

Fig. 3. Balloon telemetry reception equipment in operation at the Manchester University Athletic Ground (1952).

1952 using balloon sondes (Dawton and Elliot, 1953). The intensity difference produced by a solar dipole at this altitude is greater than at aeroplane altitudes but even so we were unable to establish in the course of these flights any systematic difference which could be ascribed to a general solar magnetic field.

Whilst this kind of experiment did not provide any positive evidence for the existence of a solar dipole the limited accuracy attained meant that a dipole moment as high as 6×10^{33} gauss cm^3 could not be excluded at that time. Although new and improved Zeeman measurements of the polar field of the sun were beginning to cast serious doubt on the existence of such a high field (see for example Thiessen, 1949) the results of this type of measurement could not be regarded as conclusive either

(Alfvén, 1951) and the reality or otherwise of a solar dipole field in space remained an open question until it was finally excluded by the in situ measurement of the properties of the solar wind in 1960/61.

7. The Solar Flare of November 19 1949 and the Manchester Neutron Monitor

Meanwhile there was no repetition of the spectacular solar particle event of July 1946 until Saturday November 19 1949 when the Carnegie ion chambers at Cheltenham and Godhavn recorded increases of some 40% above the cosmic ray background whilst at the Climax mountain station the measured increase was 180% (Forbush, Stinchcomb and Schein,1950). At Manchester, which on this occasion was in a less favoured impact zone, the increase at its maximum was a mere 12% and lasted for under an hour.

It will be remembered that the Manchester telescopes were being used at this time to study the solar daily variation in the north and south directions and in order to eliminate any systematic instrumental effects the two directional recorders were rotated through 180° at midday on alternate Saturdays. In carrying out this manoeuvre it was necessary to disconnect the recorders from the mains power supply and because of this the data for the bi-hourly interval 1100–1300h were normally discarded on these occasions. The solar flare increase on November 19 fell rather precisely within the time interval 1130–1230 — a remarkable manifestation of the well-known "marmalade-side down" law. (For the benefit of the uninitiated and the theoreticians this is the law of Nature which ensures that any piece of breakfast toast which is

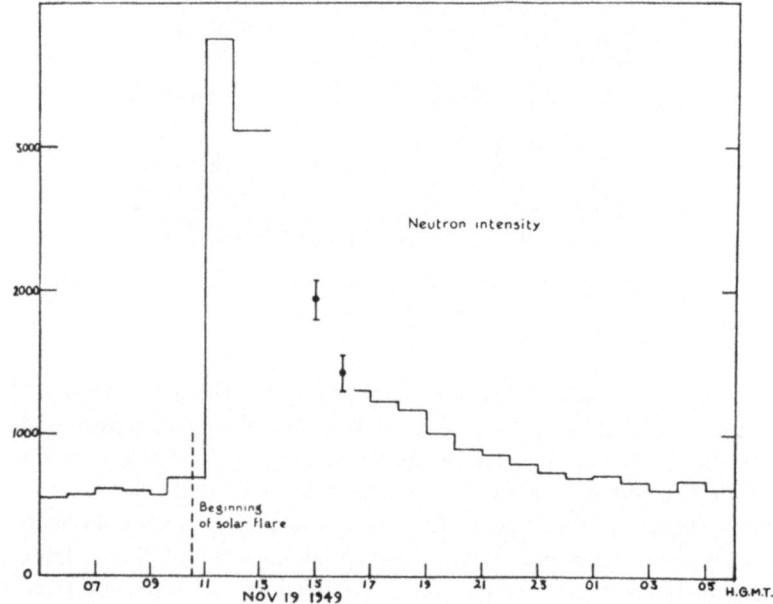

Fig. 4. The first solar flare event to be detected with a neutron monitor by Adams in 1949. The two points with error bars were short observations timed with a watch whilst he was trying to locate the "fault" in his equipment.

accidentally dropped invariably reaches the carpet sticky side down quite contrary to generally accepted notions about probability.)

On this same Saturday in another laboratory in the same building one of our fellow postgraduate students, N. Adams, was having trouble with his apparatus which consisted of a pair of proportional counters filled with boron trifluoride and located inside a 'pile' of graphite blocks. This experiment, which had nothing to do with the intensity variation work, was designed to measure the energy spectrum of the cosmic ray neutron component and the counters responded to incident neutrons of a few MeV in energy. Just before midday Adams was somewhat disconcerted to see the counting rate of his proportional counters rapidly increase to something like 3.5 times their normal rate. He, not unnaturally, suspecting that something had gone wrong with the apparatus switched it off and set about making a systematic check. Apart from this high counting rate, which persisted, he could find nothing amiss and he eventually switched the equipment on again and went home.

Although he didn't know it at the time he had just witnessed the first solar flare particle event to be registered by a neutron monitor.

When he returned to the laboratory on Monday morning everything including the counting rate was normal and it didn't occur to him to mention to anybody what appeared to have been no more than an obscure and temporary fault in his equipment. It was only some weeks later in the course of conversation with Dolbear that the real nature of the "malfunction" was recognized and the results of this unique observation published (Adams, 1950) and (Adams and Braddick, 1950).

8. The Study of Intensity Variations Resumed at Imperial College

In 1950 Dolbear left Manchester to take up a post with British Petroleum and the embryonic "Time Variations Group" now consisted of myself together with D.I. Dawton, J.K. Crawshaw and T. Thambyahpillai who had joined as new PhD students. Dawton worked on the high altitude balloon measurements already mentioned and on the relationship between the intensity of the electronic component of cosmic rays and atmospheric temperature and pressure. Thambyahpillai continued the work on the solar daily variation with counter telescopes and an early neutron monitor whilst Crawshaw undertook the search for a sidereal daily variation in the frequency of cosmic ray air showers of energy $10^{16}-10^{17}$ eV. As they obtained their Doctorates these three also left to take up posts elsewhere and were in turn replaced by M.N. Posener and I.J. Van Heerden who later accompanied me to Imperial College, London, following Blackett's appointment to the Chair of Physics there in 1953.

By this time Duperier had already returned to Spain and the study of intensity variations which he had initiated at Imperial College some fifteen years earlier was now resumed there. Thus in a way the wheel had come full circle.

9. In Retrospect

The immediate post-war period in Blackett's laboratory in Manchester had been one of quite extraordinary interest and activity. It began with Blackett's return to Manchester from his wartime activities and the re-assembly of the pre-war group of cosmic ray workers which included such luminaries as Braddick, Janossy, Lovell, Rochester, Sitte and Wilson. To this highly competent and experienced nucleus was added a new generation of younger researchers all eager to come to grips with their particular selection from the fascinating range of problems posed by the cosmic rays at that time. It was this period that saw the discovery of the unstable V-particles by Rochester and Butler and the initiation of work on radio astronomy at Jodrell Bank by Lovell. The latter activity began as a search for radar echoes from cosmic ray air showers using a surplus army radar set. It was a search that was based on a miscalculation as to the magnitude of the signal to be expected but which led to the establishment and development of one of the premier radio astronomy laboratories in the world. Blackett himself quickly became deeply involved in the problem of planetary and stellar magnetism having revived an earlier suggestion by Schuster that there was a fundamental relationship between magnetism and angular momentum. Although this idea was not supported by later experimental work it led him together with Runcorn and Clegg to the study and rapid development of the subject of paleomagnetism. This in turn led ultimately to the general acceptance of Wegener's hypothesis of continental drift and the present day concepts of plate tectonics.

It was a source of great satisfaction and pleasure to everyone in the laboratory when Blackett was awarded that Nobel Prize for Physics in 1948 and in retrospect it is hard to believe that there could have been any better place for a postgraduate student to start his research career than Manchester at that time.

REFERENCES

Adams, N. and Braddick, H.J.J.: 1950, *Phil. Mag.* **41**, 505.
Adams, N.: 1950, *Phil. Mag.* **41**, 503.
Alfvén, H.: 1946, *Nature* **158**, 618.
Alfvén, H. and Malmfors, K.G.: 1943, *Ark. Mat. Astr. O Fys.* **29A**, No. 24.
Alfvén, H.: 1947, *Phys. Rev.* **72**, 88.
Alfvén, H.: 1951, *Nature,* **168**,1036.
Blackett, P.M.S.: 1938, *Phys. Rev.* **54**, 973.
Dawton, D.I. and Elliot, H.: 1953, *J. Atmos. Terr. Phys.* **3**, 217.
Dolbear, D.W.N. and Elliot, H.: 1947, *Nature,* **159**, 58.
Dolbear, D.W.N. and Elliot, H.: 1950, *Nature,* **165**, 353.
Duperier, A.: 1942, *Nature,* **149**, 579.
Duperier, A.: 1945, *Proc. Phys. Soc.* **57**, 464.
Duperier, A.: 1948, *Proc. Phys. Soc.* **61**, 34.
Duperier, A. and McCaig, M.: 1946, *Nature,* **157**, 477.
Ehmert, A.: 1948, *Z. Naturforsch.* **3a**, 264.
Elliot, H.: 1949, *Proc. Phys. Soc. A.,* **62**, 369.
Elliot, H. and Dolbear, D.W.N.: 1950, *Proc. Phys. Soc. A.,* **63**, 137.
Forbush, S.E.: 1946, *Phys. Rev.* **70**, 771.
Forbush, S.E., Gill, P.S. and Vallarta, M.S.: 1949, *Rev. Mod. Phys.* **21**, 44.
Forbush, S.E., Stinchcomb, T.B. and Schein, M.: 1950, *Phys. Rev.* **79**, 501.
Janossy, L.: 1937, *Zeits. Phys.* **104**, 430.
Kane, E.O., Shanley, T.J.B. and Wheeler, J.A.: 1949, *Rev. Mod. Phys.* **21**, 51.
Kolhörster, W.: 1941, *Phys. Zs.,* **42**, 55.
Lange, I. and Forbush, S.E.: 1942, *Terr. Mag.,* **47**, 185, 331.
Singer, S.F.: 1952, *Nature* **170**, 63.
Thiessen, G.: 1949, *Observatory,* **69**, 228.
Vallarta, M.S. and Pomerantz, M.A.: 1949, *Phys. Rev.,* **76**, 1889.

COSMIC RAY ASTROPHYSICS AT CHICAGO (1947–1960)
(Some Personal Reminiscences)

John A. SIMPSON*

1

Cosmic ray research had a distinguished history at the University of Chicago in the 1930's through the work of A.H. Compton, M. Schein and their associates. Figure 1 is an old photograph of the participants who attended the International Cosmic Ray Conference at the University in the summer of 1939–a picture which reveals–through the now-famous names of those present–the broad, intellectual interest in both the elementary particle and the astrophysical/geophysical aspects of cosmic rays in the 1930's (Symposium, 1939). Soon thereafter these fields of research were disrupted by World War II, and it was not until well into 1946 that cosmic ray research again became established throughout the world.

It is to the history of my group in cosmic ray astrophysics at the University of Chicago, from its beginning in this post-war period through the beginning of our cosmic ray investigations in space and to the formation of the Laboratory for Astrophysics and Space Research that I have devoted this personal account.

My decision to undertake cosmic ray investigations and form a group at the University involved several accidental factors amid the events of the early 1940's. My formal education as an undergraduate at Reed College (1936–40), followed by graduate school at New York University (1940–43), overlapped with the beginning of the war. In 1942 while I was studying and teaching in New York, I was approached by Volney Wilson, an associate of Arthur H. Compton (who was by that time Director of the Metallurgical Laboratory, the code name for the part of the Manhattan Project located at the University of Chicago), to undertake the development of high speed radiation detection systems for the Chicago project. As a result of this work I was invited to Chicago in 1943 to participate in the war effort there as a scientific group leader in the Ryerson Laboratories. Upon arriving I was amazed to encounter Enrico Fermi, Leo Szilard, Eugene Wigner and many other distinguished physicists and chemists for the first time, and to find that a sustained nuclear chain reaction had already been achieved on campus the previous December (1942). My laboratories occupied the third floor of Ryerson where my groups devised and developed the new kinds of instrumentation needed, for example, to analyze Plutonium and associated radioactive products from the nuclear reactors being constructed at Hanford, Washington; or to solve some critical problems encountered at one of the laboratories at Chicago, Oak Ridge or Hanford.

During the summer months of 1945 it became clear, especially to those of us associated with the project, that the war must soon come to an end. The University of Chicago was still host to the Metallurgical Laboratory. Some of the senior pre-war faculty at the University, especially William Zachariasen, S.K. Allison and Walter Bartky, were turning their thoughts to the post-war rebuilding of physics at the University. Meetings

* Enrico Fermi Institute and Department of Physics, University of Chicago, Chicago, U.S.A.

Y. Sekido and H. Elliot (eds.), Early History of Cosmic Ray Studies, 385–409.
© *1985 by D. Reidel Publishing Company.*

Fig. 1. Participants at the Cosmic Ray Conference (Symposium on Cosmic Rays, 1939) convened at the University of Chicago in the summer of 1939. The identification of participants is given by numbers in the over lay of this photograph as follows:

1. H. Bethe	18. W. Bothe	35. W. Bostick[+]
2. D. Froman	19. W. Heisenberg	36. C. Eckart
3. R. Brode	20. P. Auger	37. A. Code[+]
4. A.H. Compton	21. R. Serber	38. J. Stearns (Denver?)
5. E. Teller	22. T. Johnson	39. J. Hopfield
6. A. Baños, Jr.	23. J. Clay (Holland)	40. E.O. Wollan*
7. G. Groetzinger	24. W.F.G. Swann	41. D. Hughes[+]
8. S. Goudsmit	25. J.C. Street (Harvard)	42. W. Jesse*
9. M.S. Vallarta	26. J. Wheeler	43. B. Hoag
10. L. Nordheim	27. S. Neddermeyer	44. N. Hillberry[+]
11. J.R. Oppenheimer	28. E. Herzog (?)	45. F. Shonka[+]
12. C.D. Anderson	29. M. Pomerantz	46. P.S. Gill[+]
13. S. Forbush	30. W. Harkins (U. of C.)	47. A.H. Snell
14. Nielsen (of Duke U.)	31. H. Beutler	48. J. Schremp
15. V. Hess	32. M.M. Shapiro[+]	49. A. Haas? (Vienna)
16. V.C. Wilson	33. M. Schein*	50. E. Dershem*
17. B. Rossi	34. C. Montgomery (Yale)	51. H. Jones[+]

*Then research associate of Compton.
+Then graduate student of Compton.
(The author is indebted to M.M. Shapiro for this photograph and for identifying the participants.)

were held with members of the Department in May 1945 to approve a slate of new faculty appointments which became known when the war ended. Later, Chancellor Hutchins arranged the financial support for establishing a new institute within which physicists and chemists could freely pursue fundamental research in areas associated with the nuclear sciences in an interdisciplinary environment while enjoying joint faculty appointments in the teaching departments. Thus, the Institute for Nuclear Studies was created August 1945.*

It was an exciting and an irresistable attraction for me as a young physicist to find my name on the initial list of proposed faculty along with Fermi, Teller, Urey, Libby, Maria and Joseph Mayer and many other senior scientists whom I admired. It was to be my first opportunity to establish my own research. However, many of us in the laboratory had become deeply concerned during 1943—45 with the use of the newly created bomb, and the domestic and international implications of the nuclear age for mankind. We had organized the Atomic Scientists of Chicago which became known publicly after the release of the bomb in Japan. As chairman of this group I found myself torn between the alternatives of my great desire to pursue free research and teaching in my newly appointed position as instructor at the University and the agonizing realization that this was a unique moment in history when we could strive for civilian control of atomic energy and press for international control of the bomb. (For an account of this period see Simpson (1981) and Smith (1965). I chose the latter alternative and it was not until late 1946 that I could return from a leave of absence to begin my research work at the University as an instructor.

The fledgling Institute already was rapidly growing, lacking only equipment and adequate facilities for the large numbers of faculty and staff who had been arriving over the previous months. My first laboratory and office were combined in a small room under the staircase of Eckhart Hall.

Although I found in my notebooks a wide range of questions covering many areas in physics in which I had become interested, I had narrowed my choices down to neutrino physics and cosmic rays. It was not long before I decided to investigate some outstanding problems in cosmic ray physics. My attraction to this subject was stimulated initially by a sophomore course in modern physics at Reed College, but it had been dormant in my mind until 1946 when I was free to choose my new research directions.

It was interesting, but unfortunate for me, that I really had no opportunity for discussions of cosmic ray physics with two of the early pre-war leaders in the field that I had known. While in graduate school at New York University Serge A. Korff joined the faculty. I benefited very much from Serge's extensive experimental knowledge, especially of classical astrophysics and particle detectors. Nevertheless, with our thoughts at that time mainly on new developments in nuclear physics, we never had substantive discussions on cosmic rays.

Again, even though the Comptons generously gave me a room on the top floor of their home near campus when I arrived in Chicago in 1943, where I lived until they went to St. Louis, there was no occasion under the stress of the times to discuss cosmic rays. It was only after the war that I became aware of the early and leading role which Compton and his colleagues had played in establishing cosmic ray phy-

* Later called the Enrico Fermi Institute.

sics at the University of Chicago, and only recently did I learn from Bruno Rossi that Compton decided to undertake cosmic ray research after hearing Rossi's talk on cosmic rays in Rome, 1931. Later Marcel Schein, after coming to Chicago at Compton's invitation of the late 1930's, continued with W. Jesse, E.O. Wollan and students through and after the war until his death in 1961, to work mainly on the high energy interactions and production of particles in cosmic ray physics.* Bruno Rossi also was at Chicago in this prewar period.

Consequently I came to the subject in 1946 with little background but with an abundance of enthusiasm. Although I decided to establish my laboratory and work independently of Marcel Schein's cosmic ray group, I was received warmly by Marcel and his colleagues, participating in their seminars and meeting the cosmic ray physicists who were visiting them from time to time. For example, it was through Marcel Schein that I came to know Sandoval Vallarta in those early years. It was to become clear over the next few years that there would be no real conflict of scientific objectives between Schein's group and my evolving laboratory.

Thus with few constraints on my research decisions in 1946–47, except need for funds and technical support, I decided to try to study the energy spectrum of the primary proton component through measurements of the secondary nucleonic component as a means of investigating the origin of the cosmic radiation. I started by investigating the properties of the nuclear interactions linking the primary cosmic ray protons and the

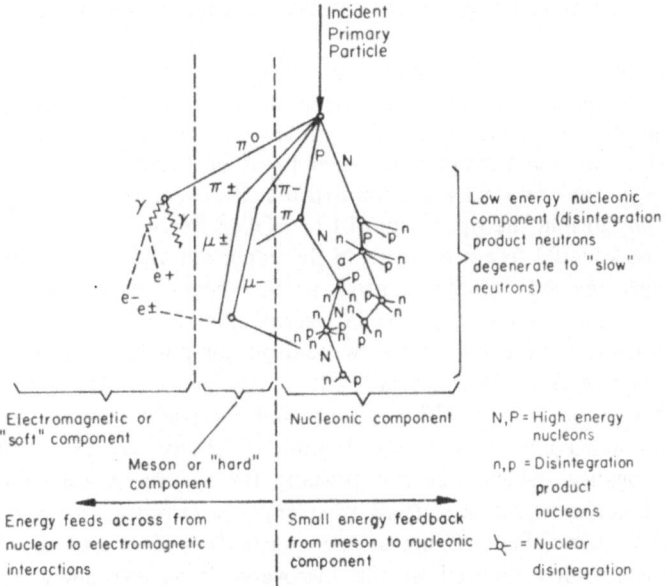

Fig. 2. Schematic diagram of the secondary components generated in the atmosphere by an incident high energy cosmic ray proton.

---

* Schein and his collaborators, Jesse and Wollan (Schein *et al.*, 1941), had just before the war reported the results of an especially significant experiment carried on a balloon flight from campus which demonstrated that the bulk of cosmic rays were protons – a result which is often overlooked in the literature due to the disturbed conditions prevailing in and after 1941.

observed secondary nucleons in the atmosphere (Fig.2) which accompanied the meson component. Already reports of "star events" in nuclear emulsions had appeared (Occhialini and Powell, 1947), and it was clear that the secondary neutron component could be a readily measurable indicator of the production of nuclear interactions in the atmosphere. My first measurements were to be the altitude dependence, then the latitude dependence, of the nucleonic component as reflected in the fast and slow neutron fluxes. This approach enabled me to utilize several instrumental developments evolving from my earlier work on the Manhattan Project. I knew, for example, that Boron enriched to at least 96% Boron-10 was available to make BF_3 proportional counters and I had applied them to some problems involving the detection of delayed fission neutrons from the nuclear reactors. The pre-war paper by Bethe *et al.* (1940) was valuable for determining the baseline of neutron flux levels, and, while I was undertaking the construction of the apparatus, the group of Agnew, Bright and Froman at Los Alamos (Agnew *et al.*, 1947) reported the first measurements of the altitude dependence of cosmic ray neutrons in "free space".

Through the Office of Naval Research I learned that a group of B29 bombers had been equipped with special engines and propellers for high altitude research flights* and were available with Air Force flight crews at the Naval Research Station at China Lake (Inyokern), California, a vast beautiful desert area. Already Carl Anderson and Victor Neher at the California Institute of Technology had started to utilize these aircraft, and I soon found myself planning for their use in my work. An annual budget of $1500 for my research was made available through an existing ONR grant for cyclotron research at the University. With my technical assistant, A. Hoteko, to help me build the apparatus—which included provision for measuring independently both the fast neutron and slow neutron fluxes and the intensity of the meson component through a multiple coincidence telescope with absorber—we were ready by late 1947. I felt uneasy with the casual manner in which I was outfitted by the Air Force with a parachute, oxygen tank and mask and warned that if it were necessary to jump from the aircraft I was *not* to pull the ripcord until I had fallen freely to below 15–20 thousand feet — otherwise I would freeze to death in the upper atmosphere! But somehow at the time these dangers seemed remote in contrast to the expectations for new scientific results.

In flights from the desert naval base in December 1947 we measured not only the altitude dependence for the fast and slow neutron component, but also obtained our first evidence that there was a significant latitude dependence extending poleward to at least $52°$ geomagnetic latitude. This meant that we could observe, by way of the secondary nucleonic component, a contribution of cosmic ray primaries down to at least ~ 1 GV rigidity, which was not possible for the meson-electron component. Returning to Chicago in January 1948 we planned a campaign to cross the geomagnetic equator *via* Lima, Peru, flying at the constant pressure altitude of 30000 feet (312 grams/cm^2) which resulted in the discovery of an extremely large ($\gtrsim$350%) neutron latitude dependence (Simpson, 1948, 1951) — much larger and extending to much higher geomagnetic latitude than the well-known meson component latitude effect (Fig. 3). Rossi and his collaborators reported in 1948 (Bridge *et al.*, 1948) the local star production in ionization chambers as a function of latitude, and, later in 1949,

* Normally the B29 aricraft was limited to 30000 feet altitude, but with special engines—as we later found—could reach 41–42000 feet for up to 15–20 minutes before becoming unstable or destroying the engines from overheating.

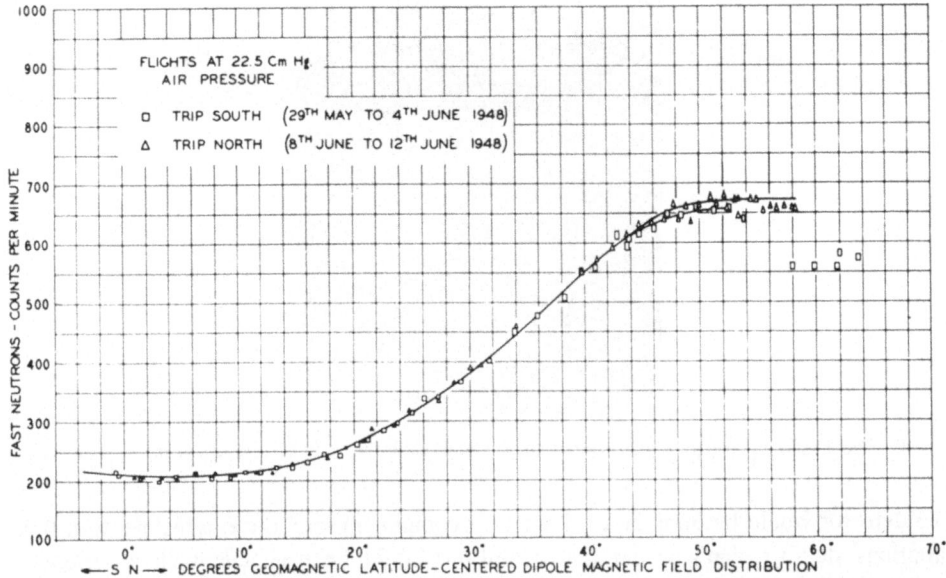

Fig. 3. The neutron intensity as a function of geomagnetic latitude (Simpson, 1951). Note the appearance of time-dependent variations at high geomagnetic latitude.

L.C.L. Yuan carried out neutron latitude measurements which were in substantial agreement with these studies (Yuan, 1949).

The flux levels at geomagnetic latitudes above the "knee" derived from flights over the desert base in December 1947 and January 1948 at constant pressure-altitude displayed changes of intensity that could only be accounted for by intensity variations of the primary radiation (e.g., Fig. 3). Thus, it became clear to me that the two factors of:

1) the large geomagnetic latitude dependence and sensitivity to primaries extending to below 1 GV magnetic rigidity, and

2) the observed large intensity variations of the neutron component with time;
could lead to developing a unique means for monitoring the variations of the low energy primary radiation with time in the magnetic rigidity range essentially inaccessible by means of the penetrating meson component. Furthermore, the long lifetime of the cosmic ray neutrons eliminated the problem of atmospheric temperature variations which introduced variations in meson component intensity.

But what about local neutron production effects and possible changing conditions with time for the slowing down of the neutrons in the local environment? Could these variations perturb the observations? All of these factors were in my thoughts when one day at Inyokern early in 1948 I conceived of a neutron monitoring system which would be analogous to a sub-critical nuclear reactor in which lead would substitute for uranium as the producer of neutrons, and paraffin wax would substitute for a carbon or heavy water moderator and tamper. Thus the fragmentation of a lead nucleus by the incident high energy secondary nucleons of cosmic ray origin would yield a multiplicity of fast neutrons which would then become thermalized in the surrounding paraffin wax and be detected using BF_3 proportional counters embedded in the "pile."

Fig. 4. The first neutron pile constructed in the rear cabin of a B-29 aircraft (January–February 1948).

The detector would be immune to the meson component, and the external neutron flux variations due to changing external conditions could be negligible with a sufficient enclosure of wax. Finding that I could purchase at the local Navy post exchange an adequate supply of paraffin wax packaged for sale for canning food, I decided to construct a pile in the B29 utilizing the lead from my meson telescope and the BF_3 counters from my existing experiment in the aircraft. We constructed the pile shown in Fig. 4 in the B29 rear cabin, holding the makeshift assembly together for flight with cargo straps secured to the floor boards. It worked and, for me, this was the beginning of investigations that led into our research in cosmic ray astrophysics.

Returning to the University early in 1948, I came to our regular Thursday Institute Seminar to report on what I had learned. This is a post-doctoral/faculty seminar where informal reports and discussions of research in progress or new work are freely reported and questioned. When I arrived Willard Libby already was speaking and describing the progress he had made in perfecting methods for C^{14} dating. He discussed his unsuccessful efforts to obtain the correct global average of C^{14} based on the available neutron fluxes which had been reported in the literature − all for high geomagnetic latitudes. I was able to tell the seminar of our discovery of the large neutron latitude effect which accounted for the approximately two-fold discrepancy in Libby's analysis (Libby, 1955). That it so promptly would find an application to another field of science was personally satisfying.

Until my 1948 observations of large intensity changes with time in the neutron component, I had not taken a serious interest in the origin of the cosmic ray time-dependent variations of secondary intensity reported in the literature. Of special significance was the work of Scott Forbush who, by beautiful analyses of the Compton ionization chamber data (Forbush, 1954) had discovered all of the intensity variations in the secondary radiation and obtained important correlations of the meson intensity with geomagnetic and solar parameters. H.V. Neher and workers in Germany also had been measuring intensity changes in the upper atmosphere using ionization chambers and counters on balloons, which revealed many of the same correlations. Clearly, there was a challenge to investigate the physics underlying these cosmic ray phenomena.

At that time it was widely accepted that the changes with time of the meson component intensity, in addition to the variations arising from temperature-induced changes in the height of their production by cosmic rays in the upper atmosphere, were the result of changes in the magnetic rigidity cutoff arising from a time dependent equatorial ring current. Furthermore, the prevailing view of the interplanetary medium was largely one of a static, vacuum region within which discrete, fast streams of ions and electrons issued from local regions above the solar photosphere, either as 27-day recurring streams resulting in geomagnetic effects at Earth, or as bursts of radiation from solar flares which led to geomagnetic storms.

But were the observed intensity changes to be accounted for only by geomagnetic effects, or was it possible that the cosmic radiation intensity changes were occurring in the interplanetary medium, possibly under the control of solar phenomena? In some respects, my views in 1948–49 of the alternatives were an oversimplification, but they served as a starting point for our experimental investigations of the physical phenomena underlying cosmic ray intensity changes in the solar system. Taking into account the newly found properties of the low energy nucleonic component, especially its sensitivity or response to changes in the low energy primary intensity, I decided in 1948 to develop a neutron intensity monitor (Fig. 5) suitable for continuous operation at several locations over a wide range of geomagnetic latitudes (Fig. 6), and undertake a series of fast and slow neutron intensity measurements at different times in aircraft over a latitude range that included the so-called knee of the latitude curve (Fig. 6). Figures 7 and 8 show a cross-section and photograph, respectively, of the monitor which we evolved in 1949-50 and, later, which became the standard neutron monitor for world-wide investigations during the International Geophysical Year (Simpson, 1958). It soon became evident from these measurements that the principal intensity variations were not due to the geomagnetic cutoff variations but, instead, were consistent with primary cosmic ray intensity changes in the interplanetary medium. At about that time Alfvén's ideas that magnetic fields on an astrophysical scale could be "frozen in" and transported in moving plasma were beginning to emerge — ideas which much later became basic to interplanetary dynamics. Indeed, following Alfvén's visit to Chicago to discuss his views on cosmic ray origins in the solar system, Fermi undertook in 1949 the development of his interstellar acceleration mechanism for the origin of cosmic rays (Fermi, 1949).

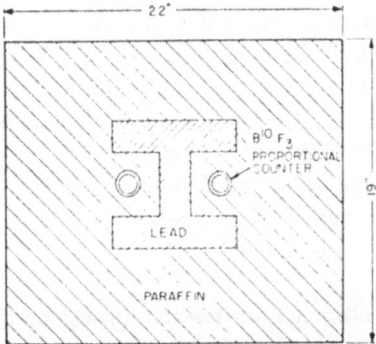

Fig. 5. Cross-section of neutron intensity monitor — basic design used in aircraft measurements and for early (1949) observations at Chicago and Climax.

Fig. 6. The distribution of neutron intensity monitors arranged to form a "spectrograph" for determining the magnetic rigidity dependence of intensity variations. The routes
of the high altitude aircraft are also shown (circa 1951).

These ideas also became pervasive in our considerations of various models proposed to explain the primary cosmic ray intensity variations on a solar system scale.

Another event which further stimulated our work at this time was the flare of 19 November 1949. Having worked out a new design of monitor installed on an aircraft in 1949 for more or less continuous measurements at sea level and at high altitude, we were shocked one morning in October to find that both the aircraft and the instruments were destroyed by fire and later saddened that we had missed the famous November flare. Nevertheless, the news that Adams and Braddick at Manchester (Adams and Braddick, 1950), as they were building a pile to study local neutron production, had observed a dramatic flux increase again demonstrated that large yields of secondary nucleons could be obtained from low energy primary nuclei.

In 1950 my students S.B. Treiman, J. Firor, W. Fonger and I had worked out our designs, built monitors, and had chosen high altitude sites at Climax, Colorado, Sacramento Peak, New Mexico and Huancayo, Peru, for installation of the new equipment in addition to a Chicago monitor. Later Sandoval Vallarta invited me to establish a laboratory in Mexico City. The neutron multiplicity investigations of the Cocconis were helpful to us (G. Cocconi and V.C. Tongiorgi, 1949). By 1951 an entire network forming a magnetic spectrometer was in place. Both the Huancayo and Climax monitors continue today to provide continuous records from the early years (Fig. 9). Prior to 1953 we had been making frequent changes in the detectors and operating procedures, such as snow removal, and, therefore, we did not publish records prior to that date.

In the period 1949–51, although our research provided convincing evidence that the principal changes of low energy cosmic ray intensity were due to changes in the primary spectrum in the interplanetary medium at 1 AU, I was–in some of my publications–still unable to decide between an interplanetary modulation effect (i.e., decreases

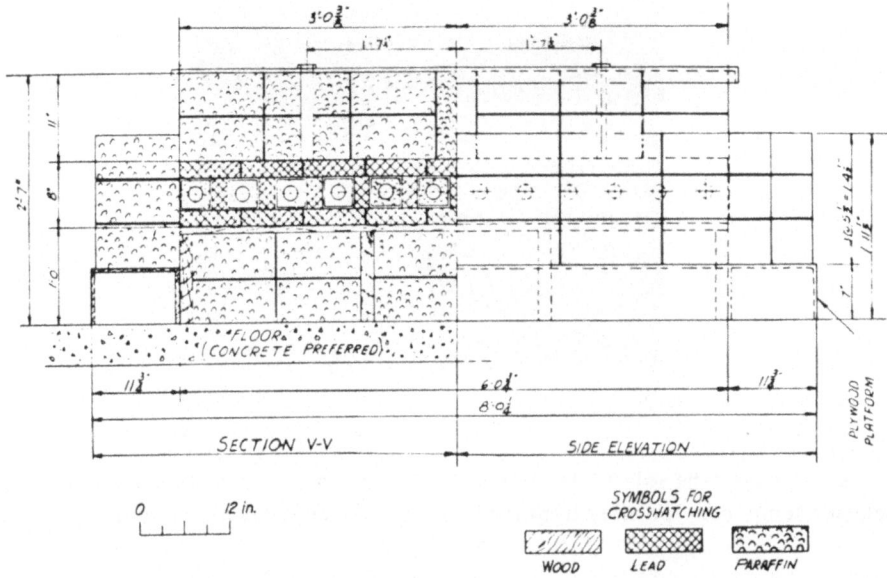

Cosmic-radiation neutron pile, side elevation.

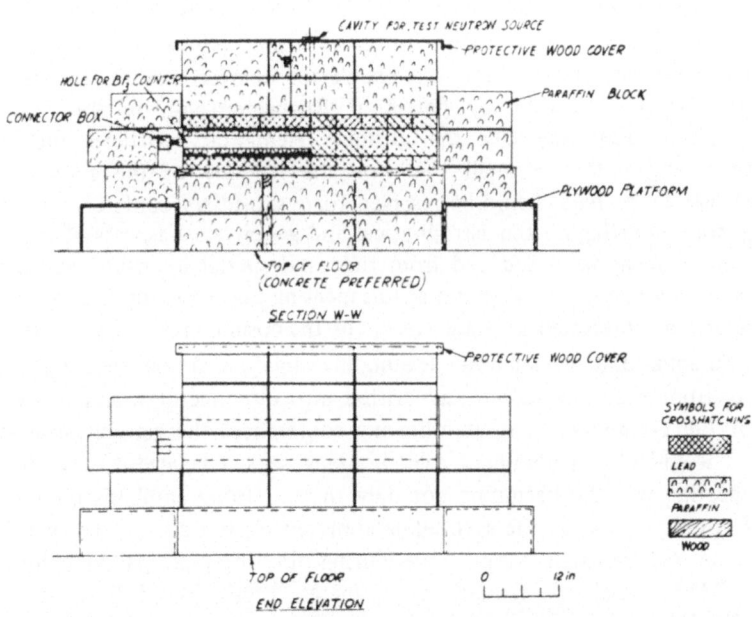

Cosmic-radiation neutron pile, end elevation.

Fig. 7. The design used since 1950, and adopted for investigations during the International Geophysical Year (Simpson, 1958) at approximately 60 locations.

Fig. 8.   J.A. Simpson with a typical monitor, circa 1951.

of intensity below the galactic level) and a time-dependent acceleration of a low energy nucleonic component, possibly from the Sun. Thus the experiments continued.

2

Gregor Wentzel* had been following with interest the work of our little group and took the initiative in arranging for me to attend the Third International Conference on Cosmic Rays to be held in the delightful Pyrenees town of Bagnères De Bigorre during the summer of 1953. Since the main thrusts of cosmic ray research at that time were in the areas of meson production, the search for heavier mesons, the heavy nuclei component discovered earlier by Peters at Rochester and by the University of Minnesota group, and especially the unstable elementary particles—the Organizing Committee arranged that a small room on the balcony of the casino where the conference was to be held would be reserved for a subsidiary conference for the 15 to 20 investigators reporting on the astrophysical and geophysical aspects of cosmic rays — especially those phenomena deduced from time and spatial intensity variations of the various cosmic ray components. It was at this meeting that I had my first opportunity to meet those already interested in these aspects of the cosmic rays, especially H. Elliot, W. Bothe, V. Sarabhai and several other leading investigators. As our meetings progressed, participants from the main casino hall drifted into our meetings, and it was clear by the end of the conference that at following conferences the astrophysical aspects of cosmic rays would play a dominant role. Looking back, I believe it is fair to conclude that this meeting was the beginning not only of the strong emphasis on astrophysical aspects of cosmic rays at all the succeeding conferences, but also of the significant role which cosmic ray research would have during the International Geophysical Year 1957–59. P.M.S. Blackett's remarks at Bagnères foreshadowed these developments (Blackett, 1953).

   "... The subsidiary conference on "Time Recording and Geophysical Aspects of Cosmic Rays" have sent in a report to us to request that certain steps be taken

* Gregor Wentzel joined the Institute and Department of Physics in 1948.

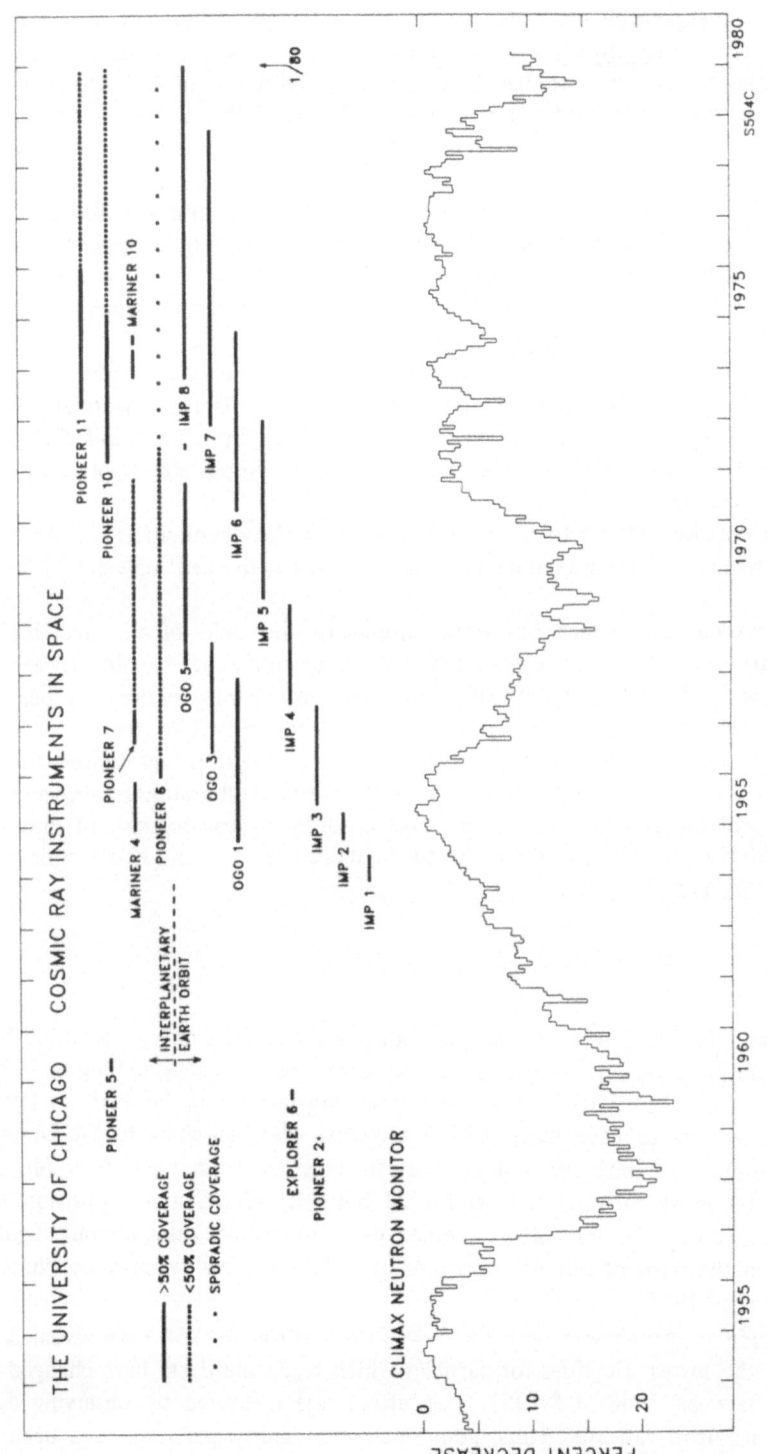

Fig. 9. Continuous record of neutron intensity 1953 through 1980. Peak intensities are times of minima in successive solar activity cycles. Data from 1949–52 were not published due to changing designs, and methods of correcting for snow cover, normalizations, etc. Bars show the times when University of Chicago instruments were in space (e.g., see Fig. 19).

towards the standardization of cosmic ray recording equipment in general and in particular in relation to the International Geophysical Year which is due to take place in a few years time. It has been proposed (this would be subject to confirmation by our parent body, the International Union) that a sub-commission be formed to study, watch and control these matters. It would draw up plans to standardize recording equipment and would organize the collection and interpretation of these results. If such a sub-commission is set up, Professor Vallarta has kindly agreed to act as chairman and Dr. Simpson as secretary. The membership of this committee will be considered in detail later.

So we hope that this conference will give rise to an organization which should prove exceedingly valuable in this very important aspect of cosmic rays.

The decision to keep the subject matter of the main conference narrow has, I think, been fully justified. I agree with the remarks of our General Secretary in holding that it has resulted in an extraordinarily fine, disciplined and fertile discussion which would have been quite impossible if the whole subject of cosmic rays had been covered.

The second decision related to the subject matter of the next conference. As it is usual to attempt to have a conference every two years, the next one should be in 1955.

Since the present conference has been limited to the field of the unstable elementary particles, it has been suggested that the next conference should concern itself mainly with the other aspects of cosmic rays, which may perhaps be best designated under the title of "Geophysical and Astronomical Aspects." This would include the time variations of cosmic rays and their various causes, the primary spectrum, the propagation of the rays through the atmosphere and underground, and theories of origin of the rays. In order to discuss many of these subjects profitably, it will be desirable to invite some of our Astronomical colleagues to take part."

3

At Chicago it was already evident to me that our small but eager group was taxed by the magnitude of the program we had set for ourselves. So I began searching in 1952 for an associate who could work with me over the long term. One day early in 1953 Maria Mayer called me to have lunch with a physicist who had come to the United States from Germany to look for a research post. This luncheon with Peter Meyer turned out to be most fruitful, and within 24 hours I offered him a position to collaborate in our group. He immediately joined us in our studies using aircraft flights and balloons. On the basis of our investigations over the next few years we concluded (Meyer and Simpson, 1955):

"High altitude measurements over the years 1948 through 1954 have revealed the fact that the lowest rigidities for particles which reach the Earth have changed significantly between 1948 and 1951. This effect was measured by observing a shift of $3°$ in cutoff latitude. Thus, additional, low-rigidity particles have been admitted to the top of the atmosphere since 1948. The measurements were obtained by observing nucleonic component intensity and were independently confirmed by vertical counter telescope measurements.

"In addition, we found that the power-law spectrum of the form

$$j = C \ (p/z)^{-n}$$

assumed to be represented in the low-rigidity spectrum in 1948 by $n \sim 2$ must have a new exponent $n \sim 2.7$ in 1951, with an interval of particle rigidities $\sim$ 1–4 BV. The change in cutoff appears to be associated with a change in the shape of the primary spectrum. Finally, this change in spectrum and cutoff is accompanied by an increase in total cosmic-ray intensity amounting to more than 13% between 1948 and 1951. Also, there may have been additional intensity decreases during times of no spectrum change (probably after 1951) which could further increase the total intensity . . ."

"From these experimental facts it is deduced that the low-rigidity cutoff in the primary spectrum is not of terrestrial origin."

"We are led, therefore, to consider the cutoff as a mechanism operative within the solar system . . . If it should subsequently develop, as may well happen, that the long-time changes of total cosmic-ray intensity shown by Forbush (1954) to be related to the general level of solar activity, are also related to the change of cutoff which we report here, this would be further evidence that the mechanism is a property of the solar system."

"At the present time, therefore, we have the possibility that the cutoff mechanism, and its change with time, is a property of a volume in space having roughly the dimensions or scale length of our solar system. It is quite possible that a distribution of magnetic fields may be found in the interplanetary space of the solar system which will prevent low-energy particles present within the galaxy from entering the solar system near the position of the Earth's orbit. For example, if the level of general solar activity controls the configuration of outlying magnetic fields—such as outgoing clouds of ionized matter containing magnetic fields— the eleven year changes of cosmic-ray intensity and the changes in the low-rigidity cutoff reported here, could have a common origin."

At the same time my search for the region on the Sun responsible for the 27-day recurring intensity variations (after only partitially successful attempts at correlations with the intense coronal green line emission regions) focussed on the work by H.W. and H.D. Babcock who had developed a method for making extensive solar magnetic field maps, based on the Zeeman effect, which resulted in a series of composite, "snap shots" of the polarity distribution of solar magnetic fields above the photosphere. This led to our joint paper showing the relationship of unipolar magnetic field regions (UM) on the Sun* with cosmic ray intensity modulation (Simpson *et al.*, 1955):

"The remarkable associations with unipolar magnetic regions indicated in Figs. 3 and 4 (here shown as Fig. 10) lead to the tentative suggestion that these UM regions are the source or starting point for the development of a large-scale phenomenon within the solar system; this phenomenon being capable of producing both the observed changes in cosmic-ray intensity and the recurring geomagnetic storms. From this point of view, the UM regions would correspond to the

* Called Unipolar because the field is of one sign and no magnetic flux has been found returning to the Sun.

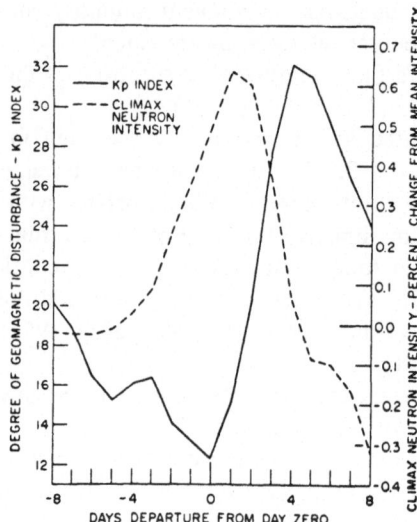

Fig. 10. For seven consecutive solar rotations the unipolar solar magnetic field passed central meridian, designated as day zero. The decrease of primary cosmic ray intensity occurred
~3–4 days later due to interplanetary modulation
(Simpson, Babcock, and Babcock, 1955).

hypothetical and long-sought-for "M" regions invoked by Bartels and others to explain the recurring geomagnetic storms."

"... then it is unlikely that the influence of the modulating mechanism could be restricted to the small volume of space located in the vicinity of a special solar region, such as a UM region, since the solid angle subtended—even by the entire Sun—is too small to account for the magnitude of the 27-day variations as they were measured in earlier years of high solar activity. This leads to the further assumption that there exists an extensive, long-lived volume in interplanetary space extending far beyond the Earth's orbit with its origin in the UM region. A stream of neutral ions, coronal streamers or outgoing turbulent magnetic fields might fill such a volume in space so as to provide the modulating mechanism for changing the primary particle intensity observed at the Earth and subsequently induce severe disturbances in the Earth's external magnetic field. Since the material or fields leaving the Sun probably require 1–2 days to reach the Earth, the relationships of the three phenomena (shown here in Fig. 10) are consistent with these general views."

"In the past, special forms of neutral ion beams have been proposed by Chapman and Ferraro, Alfvén and others to account for the magnetic storms. A more recent proposal by Morrison (1956) assumes turbulent magnetic fields in the form of an outgoing "beam" which acts as a scatterer for cosmic-ray particles and, therefore, temporarily reduces the primary cosmic-ray intensity within the beam. However, detailed attempts to work out any of these mechanisms have so far not been satisfactory."

Fig. 11. Faculty of the Enrico Fermi Institute and the Department of Physics on the occasion of S.K. Allison's first retirement in 1957 from the directorship of the Fermi Institute. Left to right, first row: Harold C. Urey, Cyril Smith, Allison, Maria Mayer, Herbert Anderson, Warren Johnson, Marcel Schein; second row, Joseph E. Mayer, T.R. Hogness, R.H. Dalitz, John Simpson, Alex Langsdorf; third row, Riccardo Levi-Setti, Y. Nambu, Valentine Telegdi, Clyde Hutchison, S. Chandrasekhar; fourth row, Roger Hildebrand, William H. Zachariasen, Edward Anders; top row, Albert Crewe, Gregor Wentzel, Nicholas Metropolis, Peter Meyer, Courtenay Wright, Nathan Sugarman, Ulrich Kruse, Carl York, Eugene Parker.

It now appears that this was a precursor of the recurring fast solar wind stream correlations with the corotating interaction regions and with cosmic-ray intensity modulation which were found in the 1960's and 1970's.

Again, it was clear that our investigations required extensive theoretical work which involved nonlinear hydromagnetic theory, and I began searching for another participant in our studies. In 1955 I was attracted to Eugene N. Parker, then at the University of Utah, who joined our group and eagerly went to work on various models and ideas to explain our experimental conclusions on the modulation of the cosmic rays. It was not many years later that Meyer and Parker were invited to join our faculty. Collaborations among us have been most fruitful from time-to-time throughout these many years. Indeed, the extraordinary interest in astrophysics and cosmo-chemistry among the Institute faculty (Fig. 11) was a splendid interdisciplinary environment in which to carry out cosmic ray research at Chicago in the 1950's.

4

From the early 1950's we were also hoping to "capture" a solar flare event with both our neutron monitor network (Fig. 6) and the instrumentation being carried on balloons which we were launching from the University campus (Fig. 12) in order to obtain the

Fig. 12. The author with assistants Hoteko and Lapinski preparing a balloon launching in Stagg Field,
circa 1952. (The stands in the background housed the first
sustained nuclear reaction in 1942).

solar flare proton energy spectrum and use the protons as probes of the interplanetary
medium. However, it was only after the beginning of the new solar activity cycle that
a flare occurred whose particle fluxes were of sufficient magnitude for this purpose.
This was the flare of 23 February 1956. We had included special circuits in the
Chicago and Climax neutron monitors which could detect a rapid and large intensity
increase and trigger alarms. Although I was entertaining friends at dinner, and Peter

Fig. 13. Peter Meyer preparing for the balloon-born neutron measurements for the flare of 23 February 1956.

Meyer was away dancing with his wife, we soon converged on Stagg Field to prepare and launch a balloon flight to obtain the fast neutron production in the atmosphere from the incoming flare protons as a function of altitude (Figs. 13 and 14). At that time the event also was recorded by the University of Chicago network (Figs. 6 and 15) and onboard the ice-breaker U.S.S. Arneb which was in the Wellington, New Zealand harbor returning from an Antarctic sea-level latitude survey (Simpson, 1956) with one of our neutron monitors under the direction of Rochus Vogt from our laboratory. Meyer, Parker and I were elated over what we were learning for the first time from the experimental observations and reported, in part (Meyer *et al.*, 1956):

"The experiments lead to a model for the inner solar system which requires a field-free cavity of radius greater than the Sun-Earth distance enclosed by a continuous barrier region of irregular magnetic fields (B (rms) $\approx 10^{-5}$ Gauss) through which the cosmic-ray particles must diffuse to reach interstellar space. This barrier is also invoked to scatter flare particles back into the field-free cavity

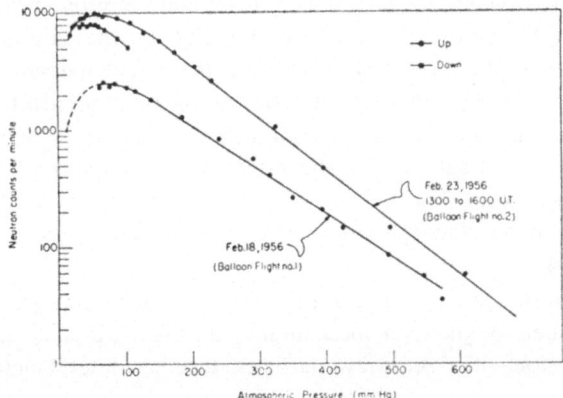

Fig. 14. The altitude dependence of fast neutron intensity during the solar flare of 23 February 1956 (Meyer, Parker, and Simpson, 1956).

and to determine the rate of declining intensity observed at Earth. The diffusion mechanism is strongly supported by the fact that the time dependence $t^{-3/2}$ represents a special solution of the diffusion equation under initial and boundary conditions required by experimental evidence. The coefficient of diffusion, the magnitude of the magnetic field regions, the dimensions of the barrier and cavity, and the total kinetic energy of the high-energy solar injected particles have been estimated for this model. Recent studies of interplanetary space indicate that the conditions suggested by the experiments may be established from time-to-time in the solar system. The extension of the model to the explanation of earlier cosmic-ray flare observations appears to be satisfactory."

"The solar flare event was superposed by chance upon a large but typical intensity decrease of nonsolar cosmic rays which began several days prior to February 23. Hence, the flare particles have been used as probes to explore the intensity modulation mechanism responsible for this decrease of background cosmic-ray intensity."

Although it subsequently turned out that the "field-free" cavity aspect of our model was wrong (solar flare particles were able to propagate rapidly along interplanetary field lines) and the magnetic field diffusion (barrier) region was not limited to the few astronomical units beyond Earth as we believed at that time (our current experiments on Pioneer 10, at 25 AU, at the time of this writing, reveal that we are still deep in the diffusion region), it was the first time that not only diffusive propagation was clearly demonstrated,* but also the first time that the solar flare proton spectrum and the total flare energy appearing in accelerated charged particles had been obtained. Soon thereafter R. Lüst, who was spending the year in our laboratory, and I re-evaluated the propagation in the magnetic fields between the Sun and Earth (Lüst and Simpson, 1957) during the initial phase of the event which indicated substantial interplanetary magnetic fields could be present. By this time we had obtained much of what we needed to know in order to begin to understand the interplanetary dynamical processes involving charged particle modulation.

The onset of the new solar cycle after solar minimum in 1954 also gave Meyer and myself the opportunity to make further high altitude aircraft measurements of the neutron intensity from which we concluded (Meyer and Simpson, 1957):

"The low-rigidity cutoff for primary particles in the cosmic-ray spectrum appeared in 1956. From these new results and the earlier measurements in 1948, 1951, and 1954, it is clear that the shift of the low-rigidity cutoff to a very small value is restricted in time to an interval within which solar activity reached a minimum in the 11-year solar cycle. This effect was accompanied by other changes in the primary spectrum; namely (a) the total cosmic-ray intensity and (b) the exponent for the power law spectrum, both passed through maxima near the solar minimum in 1954."

"The 1956 results further support the view that these changes in the primary spectrum have their origin in a mechanism controlled by solar activity — most likely the diffusion of cosmic-ray particles through interplanetary disordered

* Morrison (1956) and Parker (1956) at that time had been studying the question of how high energy particles behave in propagating through magnetic field irregularities.

magnetic fields transported by plasma clouds of solar origin. If this is so, then only for a brief period near solar minimum is there the possibility of access to the true galactic spectrum for particles below approximately 30 BV."

Clearly, by this time we had convinced ourselves that we at least had a qualitative explanation of the physical conditions resulting in the inverse correlation of ion chamber intensity with the 11-year solar cycle which had been reported by Forbush (1954).

Parker (Fig. 16) rapidly developed the theory for the expansion of the solar corona leading to the formulation of the solar wind concept and transport of solar magnetic fields throughout the interplanetary medium with the average direction of the magnetic field being aligned along an Archemedian spiral. His classic work and a review of the experimental basis for the theory which included modulation of the cosmic rays by the 11-year solar activity cycle and shocks from solar flares, is to be found in his 1963 monograph (Parker, 1963).

5

The years 1957–59 were also the years of the International Geophysical Year during which I was involved as the Reporter for cosmic rays on the international committee organizing the IGY (CSAGI). Both the U.S.S.R. and the U.S. had stated to us their intent to launch Earth satellites and, in Barcelona 1956, the Soviet representatives privately had briefed both Herbert Friedman and myself on their planned telemetry frequency, the satellite orbit and other details of a satellite launch for the next year. I reported their plans to our U.S. officials in 1956 upon my return to the states. In spite of this, the U.S. appeared to be taken by surprise in 1957 when Sputnik first entered space. It had been a continual source of frustration for many of us in the United States to realize that the U.S. had the rocket power and capability for launching satellites, but that their use for scientific purposes had been withheld from

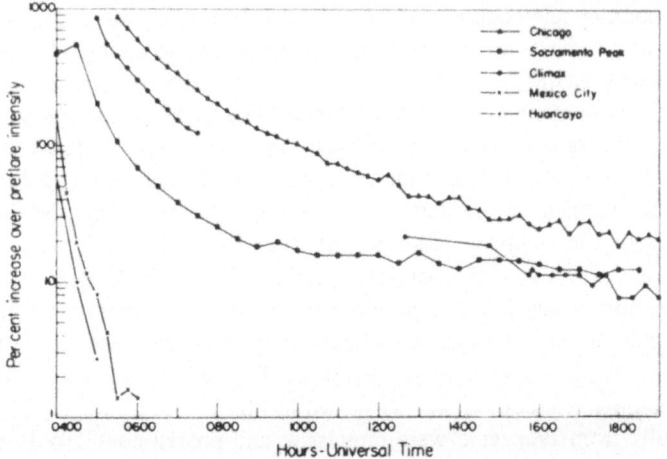

Fig. 15. The time dependence of neutron intensity obtained by the University of Chicago network of station 23 February 1956 (Meyer, Parker, and Simpson, 1956).

Fig. 16. E.N. Parker discussing theoretical ideas with I. Lerche and student P. Isenberg in a recent photo.

us. Soon thereafter (December 1957), however, I was able to obtain a private grant from Lawrence Kimpton, then Chancellor of the University of Chicago, to work towards establishing an engineering group drawn from the Laboratories for Applied Science which had been managed by the University during the emergency years of the Korean War. As a result, Meyer and I worked with some of their staff to design and build satellite experiments. Our initial goal was to prove that the modulation of galactic cosmic rays was not geocentric but was holiocentric by determining the radial gradient of cosmic-ray intensity both outward from Earth and away from Earth orbit under conditions when the cosmic rays were heavily modulated. The depressed cosmic-ray intensity at that phase of the solar activity cycle was ideal for our purposes (Fig. 9). We designed our experiment between December 1957 and March 1958, and by May 1958 had been awarded half the payload capability on Pioneer 2. It was at that time that I invited C.Y. Fan to join Meyer and myself to carry forward these space experiments. A cross-section of the Pioneer detection system, composed of a triple coincidence proportional counter telescope encased in a lead shield, is shown in Fig. 17.

Being novices on the launching of major rocket-borne experiments, we had to learn rapidly many new techniques during the summer of 1958. Without adequate vibration and shock test equipment we devised a scheme to drop our apparatus out a third floor window of our laboratory where it would land in a toy sandbox! The test worked, and our experiment was launched on Pioneer 2 on 8 November 1958. Unfortunately, the U.S. Air Force experienced another rocket failure, but not before we discovered that the inner edge of the trapped radiation belt—that Van Allen had discovered earlier in 1958—contained protons with energies exceeding 75 MeV. At least we had started along the direction of developing a program for conducting experiments in space.

The satellite Explorer 6 brought us success in 1959 as we prepared for the first deep space mission, to be launched in 1960 as Pioneer 5, carrying the same triple coincidence counter telescope design as shown in Fig. 17 and photograph in Fig. 18. The scientific results from Pioneer 5 were impressive and proved conclusively that not only was the 11-year solar modulation cycle heliocentric (Fan *et al.*, 1960, p.272), but that it was solar flare shocks in interplanetary space which produced the large sudden intensity decreases of cosmic radiation (Fan *et al.*, 1960, p.269) which had been observed as inten-

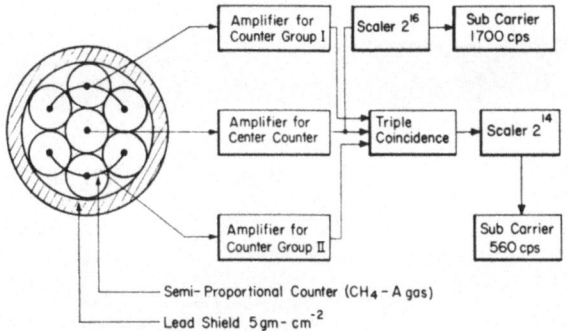

Fig. 17. Cross-section of the triple coincidence proportional counter telescope used for the first deep space mission – Pioneer 5 in 1960 (Fan *et al.*, 1960, P.269, 272)

sity decreases in ionization chambers by Forbush. We noted (Fan *et al.*, 1960, P.272):

" ... Any electromagnetic modulating mechanism required to account for the 11-year intensity variations is located principally outside the orbit of Earth. Such results from Pioneer 5 place strong constraints upon acceptable heliocentric models for the 11-year cosmic ray intensity variation."

We nearly lost our measurements of the solar flare shock effects at Pioneer 5. When the flaring region was reported I called late at night to alert the U.S. tracking team to record the data. Not receiving any response I decided to call A.C.B. Lovell directly and urged him to cover the events which were about to occur with the Jodrell Bank radio telescope which had been modified not only to assist in the launching of Pioneer, but also to record telemetry data at various times. It also was during 1959–60 that Lovell's radio telescope came under severe criticism for the large debts it had created during construction. His reaction to my request was immediate – we obtained the crucial data and Jodrell Bank, when it became known that Great Britain had contributed to the success of the U.S. space mission, was soon out of debt (Lovell, 1968)!

These experiments in space in 1958–60 were bringing to a close at Chicago a charming and exciting era where we could directly and personally carry out all aspects of an experiment with our students, and devise simple, theoretical models to account for our experimental results. We were about to enter an era which offered greatly expanded opportunities for experiments in satellites and space probes and direct entry into astrophysical environments but required a change in our style for conducting research by demanding the efforts of an entire group having diverse and specialized talents. The era of quick turn-around from one experiment to the next was giving way, at least partially, to long-term commitments for space experiments. Nevertheless, we worked out a style for carrying out our research at Chicago such that our many intelligent and creative students, who always play such an important role in our work, could continue to make essential contributions to these investigations. To accomplish this, we established in 1962 the Laboratory for Astrophysics and Space Research within the Enrico Fermi Institute, which opened in 1964 as an interdisciplinary laboratory including both theoretical and experimental work. In 1959 I also initiated the development of semi-conductor sensors for charged particle telescopes which has been crucial for our research in space. Consequently, it was possible to carry out over 30 experiments in space and

Fig. 18. C.Y. Fan and the author discussing the Pioneer-5 experiment at a Conference, 1960.

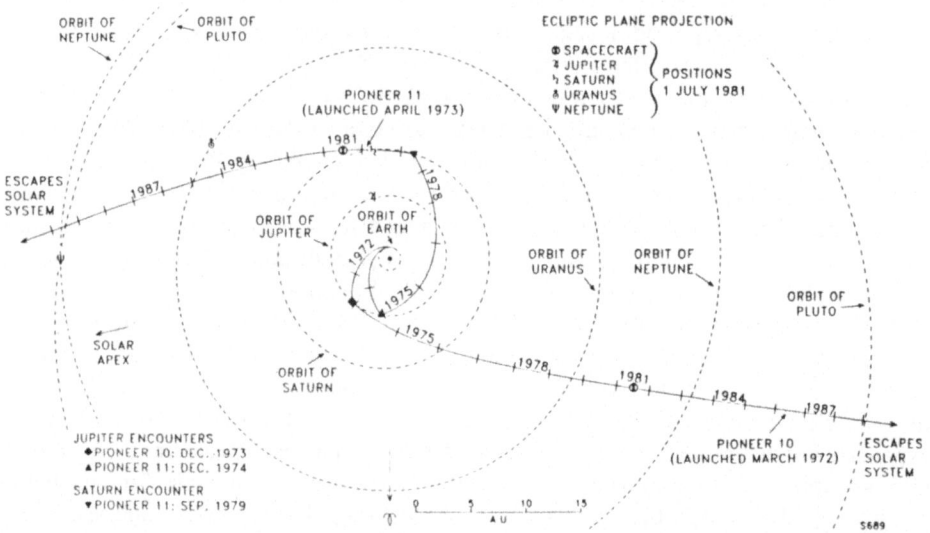

Fig. 19. The Pioneer-10 and Pioneer-11 trajectories and locations at the time of writing this account.

about 100 balloon flights over the last 22 years (Fig. 9) and to participate in missions in search of the magnetospheres of the neighboring planets Mercury, Venus, Mars, Jupiter and Saturn. Pioneer 10, at the time of my writing, is now at 25 AU and transmitting data continuously from our experiments on its way out of the solar system (Fig. 19). It was my hope throughout the earlier years that the evolution of ideas arising from direct experimental studies of charged particles, plasmas and magnetic field interactions on a solar system scale, as well as the discoveries of the hierarchy of accelerated charged particles on all astrophysical scales (from planetary magnetospheres, solar flares, the heliosphere and the Galaxy), would guide our thinking about cosmic ray phenomena occurring on the large scale of the galaxy. Our satellite and balloon experiments, coupled with the theoretical developments in space plasma physics, give us confidence that this may be so.

REFERENCES

Agnew, H.M., Bright, W.C. and Froman, D.: 1947, *Phys. Rev.* **72**, 203.

Adams, N. and Braddick, H.J.J.: 1980, *Phil. Mag.* **41**, 503.

Bethe, H.A., Korff, S.A. and Placzek, G.: 1940, *Phys. Rev.* **57**, 573.

Blackett, P. M. S.: 1953, Closing Remarks Congress International Sur Le Rayonnement Cosmique, Bagneres De Bigorre, France.

Fan, C.Y., Meyer, P. and Simpson, J.A.: 1960, *Phys. Rev. Lett.* **5**, 272; 1960, *Phys. Rev. Lett.* **5**, 269.

Fermi, E.: 1949, *Phys. Rev.* **75**, 1169.

Forbush, S.E.: 1954, *J. Geophys. Res.* **59**, 525, and references therein.

Libby, W.F.: 1955, Radiocarbon Dating, University of Chicago Press, Chicago.

Lovell, A.C.B.: 1968, The Story of Jodrell Bank, Harper and Row, New York.

Lüst, R. and Simpson, J.A.: 1957, *Phys. Rev.* **108**, 1575.

Meyer, P. and Simpson, J.A.: 1955, *Phys. Rev.* **99**, 1517.

Meyer, P. and Simpson, J.A.: 1957, *Phys. Rev.* **106**, 568.

Meyer, P., Parker, E.N. and Simpson, J.A.: 1956, *Phys. Rev.* **104**, 768.

Morrison, P.: 1956, *Phys. Rev.* **101**, 1397.

Occhialini, G.P.S. and Powell, C.F.: 1947, *Nature* **159**, 93.

Parker, E.N.: 1956, *Phys. Rev.* **103**, 1518.

Parker, E.N.: 1963, Interplanetary Dynamical Processes, Interscience Publishers, New York.

Schein, M., Jesse, W.P. and Wollan, E.O.: 1941, *Phys. Rev.* **59**, 615.

Simpson, J.A.: 1948, *Phys. Rev.* **73**, 1389; 1951, *Phys. Rev.* **83**, 1175

Simpson, J.A., Babcock, H.W. and Babcock, H.D.: 1955, *Phys. Rev.* **98**, 1402

Simpson, J.A.: 1956, Antarctica in the International Geophysical Year; Geophysical Monograph, 1 (National Academy of Sciences, U.S.A.).

Simpson, J.A.: 1958, Annals of the International Geophysical Year, *IV*, Part VII, 351, Pergamon Press, London.

Simpson, J.A.: 1981, *Bull. Atomic Sci.* **37**, 26.

Smith, A.K.: 1965, A Pearl and Hope – The Scientists Movement in America 1945–47, University of Chicago Press, Chicago.

Symposium on Cosmic Rays: 1939, *Rev. Mod. Phys.* **11**, 122.

Tongiorgi, G. and Tongiorgi, V.C.: 1949, *Phys. Rev.* **76**, 318.

Yuan, L.C.L.: 1949, *Phys. Rev.* **76**, 1267.

ON THE BIRTH AND DEVELOPMENT OF
COSMIC RAY ASTROPHYSICS

Vitaly L. GINZBURG*

At the present time and indeed ever since the fifties, the astronomical aspect (including investigations of different variations) has played a dominating role in the study of cosmic rays. This is explained, firstly, by the creation of accelerators with which cosmic rays cannot compete for the purposes of high-energy physics. (I mean, of course, at those energies reached by accelerators). Secondly, the development both of astronomy itself and of equipment for the investigation of primary cosmic rays has increased the possibilities of new discoveries in cosmic ray astrophysics and has stimulated greatly a correspondingly increased interest in the subject. There is hardly any doubt that the situation will remain the same in the foreseeable future, and this will be touched upon again at the end of the paper.

This book should presumably reflect the history of the birth and development of cosmic ray astrophysics or, as it is now often called, high-energy astrophysics although the term "cosmic ray origin" is even more often used by tradition. To argue about terminology is hardly fruitful and not very important either but it is still necessary to come to an agreement in this respect. Below I shall use the term cosmic ray astrophysics when speaking of the field of investigations devoted to determining the characteristics, the role etc. of relativistic charged particles (cosmic rays) in space, in astronomy. High-energy astrophysics is wider, it includes uncharged high-energy particles, i.e. gamma rays and neutrinos (with an energy, say, higher that 10-20 MeV). Sometimes X-rays are included here too. As to the origin of cosmic rays, this term, as I think, is reasonable to apply only to a narrower problem, i.e. to clarifying the origin of cosmic rays observed near the Earth.

In spite of the fact that, as has been said, the terminology is rather conditional, it seems more correct to give essentially different answers to the questions, when cosmic ray astrophysics was born and when the problem of the cosmic ray origin appeared. Indeed, the very discovery of cosmic rays about 70 years ago already raised the question of their origin. But how can we speak of astrophysics, if for almost 15 years after their discovery of V.Hess**, that is to say until about 1927, there were still doubts as to the extraterrestrial origin of cosmic rays. Over the same period cosmic rays (irrespective of their origin) had been believed to be gamma rays. It was only later that the discovery of geomagnetic effect and the use of equipment on balloons made it possible to establish the nature of primary cosmic rays: in about 1939–41 it became

---

* P.N. Lebedev Physical Institute, Acad. Sci., Moscow, USSR.
<Ed.> According to the author's request, the original manuscript written in Russian is kept by one of the editors (Y.S.) at Cosmic Ray Research Laboratoy, Nagoya University.

**The question of the discovery of cosmic rays will evidently be touched upon in other papers of the present collection. Proceeding from the data known to me (Hillas, 1972; Dorman, 1981), I think it is well established (and, probably, commonly accepted) that cosmic rays were discovered by V. Hess in 1912.

Y. Sekido and H. Elliot (eds.), Early History of Cosmic Ray Studies, 411–426.
© 1985 by D. Reidel Publishing Company.

clear that these are mainly protons, and in 1948 the nuclei of a number of elements were discovered in the composition of primary cosmic rays (Hillas, 1972; Dorman, 1981).

Nevertheless, cosmic rays remained something of secondary importance for astronomy, and this is quite natural: comsic rays were observed only near the Earth and due to the high degree of their isotropy no evidence existed as to the character and location of their sources. This situation is analogous to the one which would exist in optical astronomy if one did not observe separate stars and nebulae but analyzed only the spectrum of all the sources taken together. True, before 1950 there did appear some papers anticipating future developments. Among them one should mention the supposition by Baade and Zwicky (1934) of the connection of supernova flares with the formation of neutron stars and with cosmic ray generation; we should also mention the paper by Fermi (1949) devoted to the acceleration of cosmic rays when they propagate in interestellar magnetic fields.

However, the situation changed quickly and radically in $1950-53$, when the connection was established between cosmic rays (or, more precisely, their electron component) and cosmic radioemission of a synchrotron origin. As a result, there now appeared the ability to acquire information on cosmic rays far from the Earth both inside our Galaxy and outside its limits. Thus, cosmic rays turned out to be a universal phenomenon and a source of important astronomical information since they radiate radiowaves and also electromagnetic waves of other wavelengths (true, in the fifties only optical synchrotron radiation was observed apart from radioemission). Another, not less important circumstance also became clear: there are many cosmic rays in the Universe in the sense that their energy and pressure in some regions are very essential in the general balance and for the dynamics of interstellar medium, supernova remnants, radioemitting "clouds" in radio galaxies etc. In other words, it turned out that cosmic rays play an important role in astronomy, that "elements" of the Universe are not only stars, planets, electromagnetic radiation, nonrelativistic interstellar and intergalactic plasma but cosmic rays also are an essential ingredient. This was, naturally, just how cosmic ray astrophysics, one of the branches of astronomy, was born.

The aim of this book is "to take the evidence" of those who have participated in the process of cosmic ray studies. As for me, this means that I should first of all dwell on the birth of cosmic ray astrophysics in $1950-53$. As is well known, participants of the events often do not notice many things or see them in the wrong light. Therefore, it is as a rule only a historian of science basing his studies on all the available materials (and first of all on publications) who can reconstitute an objective picture of events. The birth and the development of cosmic ray astrophysics is no exception in this respect but, on the contrary, stands in particular need of the judgement of a detached on-looker. The point is that, firstly, in this case there were many participants in the events. Secondly, as far as I can judge, some of those participants are of different opinions as to their own and other people's role. Thirdly, quite erroneous statements on this subject can be met in the literature. In such a situation I wish to make only a few remarks and to give the relevant references. I hope all this will help historians of science.

Cosmic radioemission was discovered in 1931 by K. Jansky, whose publications appeared in 1932 – 35. The second radioastronomer was G. Reber, his first results were published in 1940 (for some interesting historical details and references see Kraus, 1967). Solar radioemission was registered in 1942 – 43 in several places, in particular, by Reber. Finally, immediately after the second world war radioastronomy began rapidly developing in different countries. Note that the first (after the Sun) identified source of cosmic radioemission was the Crab Nebula (J. Bolton, 1948; see Kraus, 1967).

There were two reasons for the progress in radioastronomy. The first was the perfection of radio-equipment, whose sensitivity had reached a quite fantastic level. The second one was the existence of a much more intensive cosmic radioemission than had been supposed. For example, even in the last century there was no doubt as to the presence of some solar radioemission corresponding, say, to the radiation of a black body with a temperature of the photosphere $T \approx 6000°$. Actually, however, in the meter waveband the intensity of radioemission from the quiet Sun is two orders greater than this and the intensity of sporadic solar emission a few more orders of magnitude higher still compared to that expected from an equilibrium photosphere. The situation is the same, as is now known, with X-ray astronomy.

Here is evidently just the moment for an autobiographical remark. I graduated from the physical faculty of the Moscow University in 1938 as a specialist in optics but immediately changed to theoretical physics-quantum electrodynamics, the theory of higher spin particles etc. But when in 1941 the war broke out, I like many my colleagues, began looking for an application of my efforts to a more practical field and so rather occasionally, I devoted myself to radio wave propagation in ionosphere. These studies, together with other topics, occupied me for a number of years and the results are summarized in the book (Ginzburg, 1970). It is just the work on radio wave propagation in ionosphere that brought me to radioastronomy.

As early as in the thirties the well known Soviet radiophysicists L.I. Mandelstam and N.D. Papaleksi had already thought of the radiolocation of the Moon, and then, probably, of the radiolocation of other celestial bodies. In any case, at the end of 1945 or at the beginning of 1946 N.D. Papaleksi asked me to clarify the conditions of radio wave reflection from the Sun. While solving this problem I saw at once that meter waves, to say nothing of longer ones, must be reflected above the photosphere – in the corona. Simultaneously it became clear that radio waves are noticeably absorbed in the corona and, consequently, must be radiated by it. The effective temperature of such an equilibrium radiation may reach the temperature of coronal plasma which is estimated at $T \sim 10^6$ (Ginzburg, 1946). A nonequilibrium (sporadic) radioemission may, of course, be still stronger just as is observed.

In March and November of 1947 I published the reviews (Ginzburg, 1947) devoted to solar and galactic radioemission. In particular I indicated in that review that galactic radioemission at 4.7 m and in a shorter waveband can be associated with thermal radiation by the interstellar gas at a temperature $T \approx 10000°$. Jansky's data (the effective temperature $T_{eff} \approx 1.5 \cdot 10^5$ degrees at a wavelength $\lambda = 14.6$ m) were cited in (Ginzburg, 1947) but no great confidence could be attached to them. This was connected with the need to confirm these data and with the impossibility of reconciling them with thermal radiation of interstellar gas. But Jansky was right and by

1948 or thereabouts is became quite clear that nonsolar cosmic radioemission includes a nonthermal component. At that time it was already established that an effective temperature of sporadic radioemission of the Sun may reach colossal values of the order of 10^{13} degrees. Therefore, it was easy to suppose the existence of stars still more active in the radio-frequency band. So, quite naturally, there arose and developed a "radio-star hypothesis" which connected nonthermal cosmic radiation with the presence in the Galaxy of a rather large number of "radio stars" (Unsöld, 1949, 1951, 1955; Ryle, 1949, 1950; Shklovskii, 1952).

Gradually it became clear, however, that to explain the observations one needed to postulate a huge number of "radio stars" which possess quite unusual properties and spatial distribution. The consequences associated with the hypothesis of radio stars were not confirmed and, what is even more important, there was appeared the alternative hypothesis which was late to prevail, and which associated nonthermal radioemission with the synchrotron mechanism.

Electromagnetic radiation due to the relativistic particle motion in a magnetic field, i.e. synchrotron radiation was analyzed in detail by Schott (1912) as early as 1912. The practical importance of this radiation became apparent, however, only in the forties, in connection with the creation of cyclic electron accelerators (particularly the synchrotron), and a large number of articles appeared which repeated the results of Schott and developed them. The papers (Artsimovich and Pomeranchuk, 1946; Schiff, 1946; Vladimirskii, 1948; Schwinger, 1949) are cited as examples.

It seems today that in such a situation the idea of applying the synchrotron mechanism to cosmic rays should have appeared in 1946 – 47 the more so as Pomeranchuk considered the radiation of ultra-relativistic electrons as they moved in the Earth's magnetic field as early as 1939 (Pomeranchuk, 1939). Post factum, however, many ideas and hypotheses often seem obvious. In reality, however, to apply the synchrotron mechanism in astronomy one should have known the theory of synchrotron radiation and realized possible conditions for its emergence in certain concrete situations far from the Earth. In the event it was not until 1950 that Alfvén and Herlofson used the synchrotron mechanism to explain radiation from radio stars (Alfvén and Herlofson, 1950) (this article is reproduced in (Hillas, 1972)). Specifically, in (Alfvén and Herlofson, 1950) a more or less ordinary star is considered which is surrounded by some magnetosphere (or a "trapping field") in which relativistic electrons move.

Such an assumption corresponds to the models of the solar origin of cosmic rays and, as we now know, has nothing in common either with the solar system or with discrete sources of cosmic radioemission-supernova remnants etc. So the value of (Alfvén and Herlofson, 1950) was not in the choice of the model but in the fact that attention was paid (as far as I know, for the first time) to a possible connection between cosmic radioemission and cosmic rays. This line was continued in the paper by Kiepenheuer (1950), where there is a reference to (Alfvén and Herlofson, 1950) and the intensity of synchrotron radiation is estimated which should appear in interstellar space in the field $H \sim 10^{-6}$ Gauss for the relativistic electron concentration $N_e \sim 3 \cdot 10^{-11}$ cm^{-3}.

Both the papers (Alfvén and Herlofson, 1950; Kiepenheuer, 1950) appeared in "Physical Review" in the form of rather short letters and did not attract much attention. Apart from brevity, a still more important role here was apparently played by other factors: most of the astronomers were unacquainted with the synchrotron mechanism, the notes were published not in an astronomical journal, the radio-star hypothesis, as has already been mentioned, was attractive (or, in any case, popular). I on the contrary noticed the papers (Alfvén and Herlofson, 1950; Kiepenheuer, 1950), was acquainted with the synchrotron mechanism and did not see particular grounds for supposing the existence of a huge number of radio stars. In short, using formulae of the paper (Vladimirdkii, 1948), I immediately repeated in more detail all the estimations of (Alfvén and Herlofson, 1950; Kiepenheuer, 1950). As far as I know, my article (Ginzburg, 1951) which was submitted for publication in October, 1950 was the first and for a rather long time the only response to the papers (Alfvén and Herlofson, 1950; Kiepenheuer, 1950). As has been mentioned, (Ginzburg, 1951) suggested no new ideas (I do not touch upon the discussion contained in (Ginzburg, 1951) referring to cosmic ray radioemission in the Earth's magnetic field — that is another question), but, of course, promoted the "introduction" of the synchrotron mechanism into astronomy. And this process turned out to be very long and not simple even in the USSR, where I made reports and generally popularized the synchrotron theory in every possible way.

In such cases, however, talks do not suffice — one should have demonstrated the defects of the alternative approach (the radio star hypothesis) and found some bright arguments in favour of the synchrotron mechanism. I may permit myself to express a supposition that the absence of success in this respect on my part is not so much for the lack of imagination as for almost a complete lack of knowledge of astronomy — it so happened that I was then unacquainted not only with the university but even with the school course of astronomy. The only astronomical object I had been acquainted with before was the Sun. Therefore, it was only to the Sun that G. G. Getmanzev, my post graduate of those days, and I applied the synchrotron theory (Getmanzev and Ginzburg, 1952). But in the case of the Sun there were other possibilities as well, and it was not easy to separate the synchrotron component from the total radiation. I would like to mention also the work by Getmanzev (1952)* performed at my suggestion. Therein he obtained an important result concerning synchrotron radiation for electrons with a power-law spectrum. The papers by Getmanzev (1951, 1952) and also the analysis of the possibility of using radio-wave diffraction by the Moon to determine the dimensions of the sources (Ginzburg, 1947; Getmanzev and Ginzburg, 1950) promoted the development of radioastronomy but did not bring any change even into the minds of astronomers acquainted with our work. So as late as 1952 I.S. Shklovskii not only continued advocating the radiostar hypothesis but also considered the synchrotron hypothesis inadmissible (Shklovskii, 1952).** As an argument confirming such a conclusion Shklovskii presents the following: a magnetic field

---

* Dr. G.G. Getmantsev died on April 30, 1980 in his 55th year.

**In (Shklovskii, 1952), p.445 we find: " Apart from the notion that the sources of galactic radioemission are the interstellar ionized gas and "radio stars" (in the above-mentioned sense), there also exists another concept developed most completely by V. L. Ginzburg (this is Ginzburg, 1951 from the literature cited here, V. L. G.). According to (Ginzburg, 1951), the source of galactic radioemission may be a bremsstrahlung radiation of relativistic electrons in interstellar magnetic fields. We are not in a position to make here a detailed analysis of this hypothesis, which seems to us for some reasons to be inadmissible."

$H \sim 10^{-5} - 10^{-6}$ Gauss exists only in ionized gas clouds, whereas outside the clouds "the magnetic field strength will be in any case by a few orders of magnitude lower" (Shklovskii, 1952). At that time, before the appearance of the work by Pikel'ner (1953) mentioned below, such an erroneous idea had probably been spread. I dwell here upon the paper by Shklovskii (1952) in more detail only for the reason that it is just the paper to have been repeatedly mentioned in the West for the main one to express and develop the hypothesis of the synchrotron nature of cosmic radioemission. To some extent such a misunderstanding is connected, of course, with poor knowledge of Russian.

The next very important step in this direction was made in 1952 by I.M. Gordon (due to some circumstances his paper (Gordon, 1954) was published only in 1954, but he made reports in 1952, and the main content of (Gordon, 1954) has been known in the USSR at least since the end of 1952). Gordon applied the synchrotron mechanism to optical radiation in solar flares and even discussed the possibility of synchrotron X-ray radiation. The main thing is that he paid attention to the importance and the possibility of polarization measurements — the presence of polarization is characteristic of synchrotron radiation and consequently the detection of polarization can provide decisive evidence one way or the other.

For the particularly important results of this period one should refer the work of S.B. Pikel'ner (1953) who, as a matter of fact, introduced the concept of the gas galactic halo and "cosmic ray halo." He stressed also that the magnetic field with $H > 3 \cdot 10^{-6}$ Gauss exists in the entire Galaxy. Note by the way that, in my opinion, S.B. Pikel'ner, who died in 1975 on the 55th year of his life, was the most prominent Soviet astronomer-theoretician, he left a lasting work in astrophysics (see the obituary (Pikel'ner, 1976). The contribution made by Pikel'ner is not limited to his publications, for he helped disinterestedly everybody who needed his advice, and there was no end to those who came to this noble and kind man.

From the work of Pikel'ner (1953) it became particularly clear that the above-mentioned objection by Shklovskii to the application of the synchrotron mechanism to the Galaxy was quite groundless. And just with the reference to the paper (Pikel'ner, 1953) Shklovskii changed his mind drastically and in the article (Shklovskii, 1953a) now stated that the attempt to explain nonthermal galactic radioemission "by a summary effect of galactic stars turns out quite groundless." This is how the synchrotron hypothesis won in the USSR.

But in the West the radio star hypothesis was used for a few more years. Moreover, my paper submitted to the Manchester symposium on Radio Astronomy (1955) was not even included in the Proceedings of the symposium (Radio Astronomy, 1957). Meanwhile, the paper by Unsöld defending the radio star hypothesis was published there (Radio Astronomy, 1957) (see also Unsöld, 1955). And only in the next, Paris symposium on Radio Astronomy (1958) was the synchrotron mechanism, so to say, officially accepted and my report published (Paris Symposium on Radio Astronomy, 1959). By the way I had before published two papers in English on the synchrotron mechanism and on the origin of cosmic rays (Ginzburg, 1956, 1958).

The study of the Crab Nebula turned out to be very important for the development of high-energy astrophysics. This object was the first identified discrete source of cosmic radioemission (Kraus, 1967), and since the angular dimension of the Crab makes up several minutes of the arc, its very powerful radioemission definitely could not be regarded as thermal (see, for example, (Getmanzev, 1951)). On the contrary, it was quite natural to associate radioemission from the Crab with the presence in it of relativistic electrons. However, even Gordon, who had already turned to considering synchrotron radiation over the entire spectrum, including radiation at optical wavelengths, did not apply this approach to Crab Nebula and this (the supposition of the synchrotron origin of optical radiation) was done by Shklovskii (1953b). The paper (Shklovskii, 1953b) was undoubtedly of considerable importance but it should be mentioned that the only source of information about the synchrotron mechanism cited therein is the article by Shklovskii himself (Shklovskii, 1953a). Neither is there (in Shklovskii, 1953b) any suggestion to measure the radiation polarization, although in the literature this suggestion is often ascribed to Shklovskii and even called a "brilliant prediction". This suggestion was, in fact, made by Gordon (1954). Moreover, Shklovskii objected to the possibility to measure polarization. The discussion of this topic took place during the conference on the cosmic ray origin held in Moscow in May, 1953 (Proc. 3rd Conf. Questions Cosmogony, 1954). I shall quote here some remarks made at this conference (Proc. 3rd Conf. Questions Cosmogony, 1954).

I.M. Gordon. I would like to put such a question: in December in Leningrad I suggested an analogous scheme with respect to the continium emission in flares, where for a possible nature of the continium and verification of the hypothesis, according to which such a spectrum is due to relativistic electron radiation, I planned a rather simple experiment — the measurement of light polarization. There is no doubt that if the continium of Crab Nebula is indeed caused by radiation of relativistic electrons in a magnetic field, then the radiation will be partially polarized due to anisotropy of the magnetic field. We could not unfortunately perform such an experiment with flares, since we are short of adequate equipment, whereas Crab Nebula is a very suitable object for this purpose. Maybe such an attempt has been made? (Proc. 3rd Conf. Questions Cosmogony, 1954, p. 253).

I.S. Shklovskii. It is much more difficult to undertake such a verification for Crab Nebula than for flares. Taking into account the fact that the magnetic field in the region of supernova remnants is extremely inhomogeneous and some scales of homogeneity in the field have the dimension of the order of 10^{16} cm, the appearance of such an effect can hardly be expected at all, since polarization will be statistically averaged (Proc. 3rd Conf. Questions Cosmogony, 1954, p. 254).

V.L. Ginzburg. And, finally, the last point-polarization of radiation. I do not agree with what I.S. Shklovskii answered to I.M. Gordon, and I believe that it is interesting to investigate the polarization of optical radiation from supernova remnants. Subject to the magnetic field orientation, polarization will be different, the radiation intensity will also vary. If the dimensions of the quasi-homogeneous field regions are very small, the effect will be small too, but, of course, it is worth measuring. I think that such attempts may also be undertaken in the radio band, although there are much less possibilities here. Maybe one would succeed in using the fact that the source in the Taurus in covered by the Moon, and then by the diffraction on the edge of the Moon one can enlarge the angular resolution of radio astronomical apparatus by about two orders of magnitude (Proc. 3rd Conf. Questions Cosmogony, 1954, p. 260).

I.M. Gordon. The latter remark refers to radiation polarization. Polarization of optical light (here I cannot agree with I.S. Shklovskii) must undoubtedly exist because the spherical symmetry, which most probably took place during the matter ejection was unlikely to vanish completely due to turbulence (Proc. 3rd Conf. Questions Cosmogony, 1954, P. 268).

I.S. Shklovskii. First of all about polarization of the visible radiation of Crab Nebula. I cannot agree that the polarization can be found. It is practically very difficult to register polarization of several per cent even for such a powerful source as the Sun. Here we are dealing with an object of the 9th magnitude. Knowing that the dimension of the region with a homogeneous magnetic field hardly exceeds 0.01 of the whole system, I think it is difficult to expect a somewhat noticeable polarization (Proc. 3rd Conf. Questions Cosmogony, 1954, p. 276)

I hope that the inclusion here of this information is justified in view of some erroneous statements encountered in the literature and, so to say, "adapted by repetition only". The decision to present the facts* given here was also stimulated by the circumstance that in the paper (Shklovskii, 1962) Shklovskii himself stated (p. 36, 37): "It is of particular importance that on the basis of this theory (the synchrotron theory. V.L.G.) we succeeded in predicting several essentially new and important phenomena, the existence of which was then unknown, and which, soon after the predictions, were discovered in specially undertaken observations (polarization of optical and radioemission of Crab Nebula, ...)."

Not to be misunderstood I should add that in spite of what has been said, I consider the contribution made by I.S. Shklovskii to high-energy astrophysics, to be substantial. Besides, as for myself, I have never claimed much in this respect and always referred to the earlier work I knew, in particular to the papers Alfvén and Herlofson (1950) and Kiepenheuer (1950). Thus, the above-mentioned remarks are not at all aimed at supporting my own priority.

To make the picture complete we should add that the measurements of polarization of the optical radiation from the Crab (Vashakidze, 1954; Dombrovskii, 1954; Oort and Walraven, 1956) were quite successful and also very important from the point of view of the formation of "astronomical public opinion". But as I have always thought, an extraordinary amount of attention has been paid to the Crab not so much for the essence of the matter but due to the bright history of the discovery and investigation of the envelope of the supernova which flared in 1054 of our era. In other words, if the Crab Nebula did not exist at all, the development of high-energy astrophysics would in practice change very little, since the Crab, although "the first among equal" is not a unique object — but just another example of a supernova remnant, a pulsar etc.

We have already mentioned at the beginning of this paper that the turning point in the understanding of the astrophysical role of cosmic rays was reached when their connection with cosmic radioemission was established. To fully describe the modern

---

* These facts have been in brief and partially implicity presented in the introduction to (Ginzburg and Syrovatskii, 1965). Note that the expediency of polarization measurements is also mentioned in my paper (Ginzburg 1953a) submitted for publication before the conference (Proc. 3rd Conf. Questions Cosmogony).

state of high-energy astrophysics and to follow its development over 30 years especially would require a whole encyclopaedia. In 1963, when there was much less material, S.I. Syrovatskii and I tried to solve this problem in the book (Ginzburg and Syrovatskii, 1964). We have probably managed to do it to some extent since this book is rather widely known, and somewhat surprisingly for me I come across references to it even in current literature. The preparation of the second edition, which we repeatedly thought of, was hampered by several circumstances. One of them has been the abundance of new materi-al*. Another is the impossibility to answer several "damned questions" which arose from the outset (I mean round about say 1953). The latter remark should be explained.

Once we had obtained information about cosmic rays far from the Earth, the situation with regard to the whole problem of the origin of cosmic rays observed near the Earth changed radically. It became clear, for example, that one cannot use models of the solar origin of the main part of cosmic rays (Selected paper on cosmic ray origin theories, 1969). On the other hand, there appeared grounds to choose the halo galactic model in which cosmic rays originate in the Galaxy, occupying a rather extensive halo. Supernova stars were put forward as the main sources of cosmic rays with still more grounds than before (Baade and Zwicky, 1934; Selected paper on cosmic ray origin theories, 1969; Haar, 1950)** Such a "galactic halo model" was developed in detail (Proc. 3rd Conf. Questions Cosmogony, 1954; Ginzburg and Syrovatskii, 1964; Selected paper on cosmic ray origin theories, 1969; Shklovskii, 1953c, 1960; Ginzburg, 1953a, b, 1954; Hayakawa, 1956) and the paper (Ginzburg, 1953b, 1954) written in 1953 ended like this: " ... the main thing is done here, and the picture outlined above will not undergo drastic changes, like it had to suffer up to recently, until radio astronomical methods were used for the clarification of this range of questions. With the development of radio astronomy, and also cosmic electro-dynamics, the question of cosmic ray origin became a truly astrophysical problem and overstepped the limits of mainly hypothetical constructions inaccessible for control by observation. For this reason and also bearing in mind the progress made in cosmic ray physics one could be sure that further development of the theory of cosmic ray origin will be a rapid one."

Such an optimistic prognosis, which I repeated many times later on, was however confirmed only in general. In some particular branches uncertainty remained for decades! So, objections were raised to the existence of the radio halo (Radio Astronomy and the Galactic System, 1967), metagalactic models were still discussed (*Phil. Trans. Roy. Soc.* A277, 317, 1975), the hypothesis of the role of supernovae as the dominating sources of cosmic rays was argued. The latter question is to a certain extent quantitative (a certain role in cosmic ray generation may be played by various stars and types of processes) and is still being discussed up to the present (16th Intern.

---

* The book (Ginzburg and Syrovatskii, 1964) contains more than 500 references. If one tried today to cover the material with the same degree of completeness, the number of references would increase by many times.

**A new argument here was radio data testifying to the presence in supernova remnants of large numbers of relativistic electrons (Shklovskii, 1953c; Ginzburg, 1953a; Hayakawa, 1956). I should stress that the appearance of my paper (Ginzburg, 1953a) was stimulated by the paper by Shklovskii (1953). I cannot say whether or not Shklovskii knew about the papers (Baade and Zwicky, 1934; Haar, 1950); in (Ginzburg, 1953a) the review (Haar, 1950) is cited.

Cosmic Ray Conference, 1979). I personally consider as before that supernovae are dominating galactic sources, but if the role of other stars turned out to be substantial, this would only enrich the picture. The questions of the halo and metagalactic models (in the latter ones all the cosmic rays observed near the Earth come from the Metagalaxy) are on the contrary qualitative, they determine the very type of the model. For the physical reasons (we speak of energetic estimates and dynamical considerations) presented in detail, for example, in (Ginzburg and Syrovatskii, 1964; Radio Astronomy and the Galactic System, 1967) (see also *Phil. Trans. Roy. Soc.* A277, 1975, p. 463) I did not for a single moment doubt either the existence of the halo or the validity of the galactic model. But the standpoint of the opponents of these models was also justifiable, they required observations not talks. However, it turned out to be extremely difficult to determine the energy density of cosmic rays outside the Galaxy (in the intergalactic space) and to find the "trapping region" of cosmic rays in the Galaxy (and thus to establish the existence and the dimensions of the cosmic ray halo or at least radio halo — the halo of relativistic electrons). Post factum this is more or less clear: radio data indicate only the average number of radiating electrons along the line of sight, and one can only relate this number to the spatial dimension of the radiating region surrounding the solar system by way of recalculations which were insufficiently convincing bearing in mind the accuracy and thoroughness of the sky survey is of that time. As to the main, proton-nuclear component of cosmic rays, there were not direct ways to estimate its intensity far from the Earth. Tehrefore, my optimism was probably based to a certain extent on underestimating the difficulties of observations, which is typical of the theoreticians.

Of course, success came from time to time. In 1961 relativistic electrons were finally discovered directly in the composition of primary cosmic rays observed near the Earth (only the upper limit — about 1% of protons, for their intensity was known earlier). In 1965 a thermal relict radioemission with a temperature of about 3 K was discovered, and soon a conclusion was drawn that the electron component of cosmic rays must originate in the Galaxy (electrons with an energy $E \gtrsim 10^{10}$ eV cannot reach the Earth even from the nearest galaxies due to inverse Compton losses).

But all this was not enough for impatient theoreticians (including, myself). I remember, in particular, that in 1967 (it I am not mistaken) I gave a talk on the origin of cosmic rays in the California Institute of Thechnology. Since the audience was wide, I spoke of supernovae, radio halo etc. Then somewhat impatiently R.Feynman asked me something like this: well, but we know all this, is there anything new? I felt ashamed and after that I did not love for about 10 years to speak about cosmic rays. Since then S.I. Syrovatskii and I have worked much less in this field. I have always had many other interesting things to do in various fields*; whilst S.I. Syrovatskii, apart from his interests in cosmic ray astrophysics, has worked in the field of magnetic hydro-dynamics and afterwards took a great interest in solar physics. In his last years, when he was already seriously ill, S.I. Syrovatskii went on developing very successfully and actively the theory of current sheets and a number of other problems. The life of this talanted physicist-theoretician ceased in September, 1979 in his 55th year (Syrovatskii, 1980).

---

* The list of practically all my publications up to 1977 can be found in (Ginzburg, 1978a).

At the symposium on radio astronomy in 1966 the following exchange of remarks took place (see Radio Astronomy and the Galactic System, 1967, p. 436).

H. Alfvén. Ginzburg has said that it is absolutely clear that cosmic radiation plays a decisive role in the Galaxy. I am not at all sure about this, because what we observe and what we conclude from observation are so different ... . It may very well be that 99% of the cosmic radiation is a local phenomenon confined to our environment in the same way as the Van Allen radiation belts are confined to the Earth's magnetic field.

Ginzburg answers: The arguments against the solar or local origin of cosmic rays are numerous ... The radio-astronomical evidence is quite strong. According to this, relativistic electrons are present in a gigantic region outside the solar system; although the halo is open to discussion, there is no question about the presence of cosmic-ray electrons in the disk, in densities comparable to those near the Earth ... . However, I agree that is extraordinarily difficult to disprove anything.

Alfvén: To disprove anything is very difficult, but also to prove it.

Ginzburg: Fortunately it is possible to do something. I have worked in the field for some years, and I can say in the course of time the argument slowly improves. So I hope during my lifetime I shall see the full victory of these things.

Alfvén: I hope you will live very long.

Now that I am writing this article in March, 1980, I am in my 64th year and I do not by any means think that it is "too long". But the proofs of the validity of the galactic halo model already exist, and I have lived to see this. The limits of this paper do not make it possible to discuss this point in more detail. I shall restrict myself to a brief explanation and some references to the literature.

The most important success for the essence of the matter in high-energy astrophysics for the last decade is the appearance of observational gamma-astronomy (we do not here touch upon X-ray astronomy, it is "not typical" of high-energy astrophysics and may be related to it only conditionally). One of the most important sources of γ-ray in space is the decay of π^0-mesons produced in collisions of cosmic rays (protons and nuclei) with interstellar gas particles. It is clear that the intensity of such γ-rays is proportional to the gas concentration n and to the intensity of cosmic rays I_{cr}, and after some re-calculation to their energy density W_{cr}. Thus, gamma-astronomy offers the possibility to determine the density W_{cr} far from the Earth (the gas concentration n being known), as in the case of synchrotron radiation where one can find the energy density of the electron component of cosmic rays $W_{cr,e}$ (the strength of the magnetic field H being known).

In metagalactic models of the cosmic ray origin the density everywhere in the Galaxy and around it is approximately the same. Therefore, in the framework of such models one can predict a flux of γ-rays from the decay of π^0-mesons coming from the Magellanic Clouds (Ginzburg, 1972) or, say, in the direction of the galactic anticentre (Dodds *et al.*, 1975; Wolfendale, 1979). Unfortunately, the Magellanic Clouds have not yet been observed in γ-rays. But measurements in the direction of the anticentre have been carried out and evidence rather convincingly a decrease of the density W_{cr} in the regions nearer to the galactic boundaries than the solar system (Dodds *et al.*, 1975; Wolfendale, 1979). More accurate measurements must, of course, specify the quantitative results, but as far as I know, no arguments exist that would cast doubt on the above-mentioned decrease of the density W_{cr}. That is why I consider metagalactic models to be disproved.

The proof of the existence of the radio halo and, thus, of a not smaller (and rather of a more strongly pronounced) halo of cosmic rays is based on the observation of on edge galaxies NGC 4631 and NGC 891 (Ekers and Sancisi, 1977; Allen *et al.,* 1978). It is worth mentioning that there were some earlier attempts to observe the galaxy NGC 4631 (Pooley, 1969) but the radio halo was not then observed (I remember, I was somewhat discouraged by this, the more so as I promoted such observations; the cause of the failure in the discovery of the halo in (Pooley, 1969) is not quite clear to me). Treatment of radio data for our Galaxy is less obvious but also now leads to the conclusion that a radio halo exists (Bulanov *et al.,* 1976; Webster, 1978). Looking backward I think that the delay in solving the problem of the existence of radio halo was in the first instance the result of misunderstandings (including also the confusion in the definition of radio disk and radio halo; see (Ginzburg, 1978b, 1979; Ginzburg and Ptuskin, 1976)) and, second, due to the difficulty of observing the radio halo at centimeter and even decimeter waves; at meter waves observations with a high angular resolution are difficult and in practice have not been carried out for edge-on galaxies (for some additional remarks in this respect see Radio Astronomy and the Galactic System, 1967, p. 365; Ginzburg, 1978b, 1979; Ginzburg and Ptuskin, 1976; 16th Intern. Cosmic Ray Conference, 1979, vol. 2, p. 148).

Thus, by the end of the 70s the largest blank spots in the general picture of the cosmic ray origin had vanished at last, and an important stage in the development of cosmic ray astrophysics was over. In most cases such general conclusions are of course somewhat conditional and besides, we shall be quite sure of their validity only within a decade. However, I have never advocated caution just for the sole fear of making a mistake, and this also refers to predictions.

The present paper could of course be entirely limited to the past, but I would not like to do that. History of science like general history is of course interesting by itself. At the same time, one of the strongest stimuli for our attempts to fall back on the history is still the desire by extrapolation to look into the future. There is a grain of truth in the remark imbued with certain bitterness: "History teaches us only that history teaches us nothing". But this refers rather to social life and politics than to science. In any case, while writing this article, I was thinking about the present and the future not less than about the past. I would like to hear the talks that will be made on August 7, 2012, when, I hope, the centennial of the discovery of cosmic rays will be celebrated (on August 7, 1912, V. Hess made his most successful flight, and this date is best of all suited for the "birthday" of cosmic rays). But, alas ... people of my generation can hardly hope even for a smaller thing — to come to know the state of cosmic ray studies at the beginning of the next centry — on January 1, 2001. At the same time, as we know from our past experience, it is quite possible to imagine the general perspective of the development of high-energy astrophysics for 20—30 years ahead. Unexpected events will happen, of course, and this is one of the most attractive features of science. However, under conditions, when the creation of some grandiose installations (e.g. satellite observatories of HEAO type, the deepunderwater neutrino station DUMAND etc.) takes not less than a decade, a prognosis for 20—30 years does not seem Utopian. So, what can one expect in the foreseeable future? I shall mention very briefly several key problems and tendencies*.

a) The new installations which are being created for satellites and high-altitude

* See next page.

balloons and partially also the ones which now exist will already allow us in the next decade and, in any case, before 2000 to gain much new information on the chemical and isotope composition of cosmic rays (including radio-active ^{10}Be and other nuclei) and on the spectra of different components (including electrons, positrons and antiprotons).

b) The use of all these data for establishing the composition of cosmic rays in the sources and the character of their propagation in interestellar space etc. requires the development of theory and some calculations beyond the scope of a widely used and, I am sure, quite unrealistic leaky box model (Ginzburg, 1978b, 1979; Ginzburg and Ptuskin, 1976; Ginzburg et al., 1980; Owens and Jokippi, 1977; Cavalho and Haar, 1979; Wallace, 1980).

c) The theory of particle acceleration in space including various plasma effects is being vigorously developed at the present time although it must be admitted that this has been in progress for a long time (Fermi, 1949; Ginzburg and Syrovatskii, 1964; Selected paper on cosmic ray origin theory, 1969). Here one should clarify the role of acceleration in interstellar medium by shock waves generated in supernova flares, to analyse some effects in young supernova remnants during explosions and near pulsars. This is connected with the entire problem of cosmic ray sources in the Galaxy (the contribution from supernovae and other stars, the role of acceleration in the region near the galactic centre). Also related is the problem of cosmic ray acceleration and propagation in radio galaxies and quasars.

d) In spite of remarkable achievements of radio astronomy as a whole, the progress in some branches has been slow. This refers particularly to the meter wave band. Meanwhile, remote regions of the halo of galaxies and galactic clusters must be most clearly discernible at relatively long wavelengths. An analysis of the frequency dependence of brightness distribution in the radio halos over a wide wave band from centimeters to 3–10 meters will apparently allow valuable information to be obtained on the halos and on the character of cosmic ray propagation in the halo (diffusion, galactic wind and convection etc. (16th Intern. Cosmic Ray Conference, 1979, Vol. 2, p. 148).

e) The spectrum of cosmic rays is extended at least up to energies of about 10^{20} eV. A superhigh-energy range ($E > 10^{16} - 10^{20}$ eV) is accessible to investigation only by way of observing air showers. The chemical composition of cosmic rays in this region is unknown, the origin of particles is unclear. The most probable now seems the model in which particles with $E < 10^{19}$ eV originate mostly in the Galaxy and those with an energy $E > 10^{19}$ eV – in the Local Supercluster. But this is unclear, and the question is, generally speaking, open. To solve this problem one has to make further

---

* Immediately before the present paper I have written another one intended for the collection of papers (Liber Amicorum for Jan H. Oort, 1980) in honour of the 80th jubilee of Jan Oort, one of the most prominent astronomers of our century. Simultaneously, the paper "The origin of cosmic rays (introductory remarks)" is being written, which is to be submitted for the IAU-IUPAP symposium on the origin of cosmic rays (Italy, June, 1980). The Proceedings of this symbosium (Origin of Cosmic Rays, 1981), as may be expected will rather completely reflect the modern state of the problem (see also 16th Intern. Cosmic Ray Conference, 1979). By virtue of what has been said, some overlap of the present paper and the previous ones (Liber Amicorum for Jan H. Oort, 1980; Origin of Cosmic Rays, 1981) turned out to be inevitable.

cumbersome measurements of anisotropy (this is however, of interest also at lower energies), to study the structure of showers etc.

Apart from their role in astrophysics, superhigh-energy cosmic rays are (and will probably be for many decades) also of interest for physics. As is known, from 1927–29 and to the early fifties the cosmic rays played an exclusive role in high-energy physics, they helped to discover position, $\mu^{\pm}$-leptons, $\pi^{\pm}$-mesons, $K^{0,\pm}$-mesons and a few hyperons (Hillas, 1972; Dorman, 1981). Later on physical investigations passed over to accelerators, but for energies inaccessible to accelerators, cosmic rays remain, naturally, the only source of particles. In the 80s one can, as far as we know, reckon on the use of collidding beams of protons with a maximum energy $E = 10^{12}$ eV in each beam. In a recalculation from the centre of mass system to a laboratory frame of reference, this is equivalent to the use of protons with an energy $E = \dfrac{2E_c^{\,2}}{Mc^2} \approx 2 \cdot 10^{15}$ eV. Consequently at energies $E > 2 \cdot 10^{15}$ eV cosmic rays will for a long time remain the only source of particles. True, there are very few particles with such energies (for example, the intensity of particles with energy $E \geqslant 10^{16}$ eV makes no more than $10^2 \dfrac{\text{particles}}{\text{km}^2 \text{ ster.hr.}}$ but the difficulties are compensated to a certain extent by an unprecedented increase of technical possibilities.

f) The role of gamma-astronomy has already been mentioned above. The current decade will lead, as we expect, to progress in this field analogous to that in X-ray astronomy in the 70s. A new generation of gamma-telescopes will make it possible not only to confirm the results obtained on the satellites SAS-2, COS-B etc. but also to investigate a large number of discrete sources, including Magellanic Clouds. The already established fact that γ–luminocity of quasar 3C 273 and of some galactic sources is very high (for 3C 237 the luminosity $L_\gamma(50 \leqslant E_\gamma \leqslant 500 \text{ MeV}) \approx 2 \cdot 10^{46}$ erg/sec, for the pulsar PSR 0532 in the Crab $L_\gamma(E_\gamma \geqslant 100 \text{ MeV}) \approx 3.5 \cdot 10^{34}$ erg/sec) seems rather significant.

One can expect very interesting results also from the study of γ-lines, γ-rays with an energy $E_\gamma \gtrsim 10^{11}-10^{12}$ eV detected by the flashes of Cherenkov radiation in the atmosphere etc.

g) To the key problems of high-energy astrophysics and also astrophysics as a whole one should refer the study of high-energy neutrinos. This field of study, if we mean experiment, is just emerging. However, an underground detection of neutrinos from supernova flares in the Galaxy is already realistic. The creation of deep-underwater optical and (or) acoustic systems (the DUMAND project etc.) is expected to allow us to determine the direction of arrival of neutrinos of $E_\nu \gtrsim 10^{12}$ eV from remote extragalactic sources with a rather high angular resolution of nearly $1°$. Since such neutrinos can be generated only by cosmic rays (protons and nuclei) with a sufficiently high energy the potential importance of these and related γ-ray measurements is obvious. It should be mentioned here that it is just combined measurements of a high-energy neutrino flux and a γ-ray flux are a perspective method to investigate the cores of quasars and active galactic nuclei (Berezinsky and Ginzburg, 1981). In general, I am sure that neutrino astronomy (both at high energies and at energies $E_\nu \lesssim 10-20$ MeV), along with gravitational wave astronomy, is the fundamental and not yet used reserves of astronomy.

Apart from the above-mentioned problems a)–g), one could point out some other ones but we believe that what has been said testifies already to a richness of the current problems and available possibilities. So, there is every reason to suppose that within the next 20–30 years the relative weight of high-energy astrophysics in astronomy will only increase, and in any case, not be diminished.

And what will happen afterwards? Of course, many unclear details will remain and, what is even more important, quite new problems and questions are likely to appear of which we have no notion today. At the same time I think it quite possible and even probable that there will be practically no other sources of astronomical information except those already known (electromagnetic waves, cosmic rays, neutrinos and gravitational waves). In this case some saturation and a qualitative change in the character of the development of astronomy is inevitable. How interesting it would be to go back to this question, say, in 30 years!

In conclusion I would like to appologize to the reader for the frequent use in this paper of personal pronouns (I, me, etc). In English, it is true, it does not seem to sound quite so sharply egocentric as in Russian. In any case, if it is possible to avoid the use of personal pronouns in scientific papers, I was unable to do it here.

The author is obliged and grateful to Prof. H. Elliot for his help with the English translation of this article.

REFERENCES

Alfvén, H. and Herlofson, N.: 1950, *Phys. Rev.* 78, 616.

Allen, R.J., Baldwin, J.E. and Sancisi, R.: 1978, *Astron. Astrophys.* 62, 397.

Artsimovich, L.A. and Pomeranchuk, I.Ya.: 1946, *Zh. Eksperim. Teor. Fiz.* 16, 379.

Baade, W. and Zwicky, F.: 1934, *Nat. Acad. Sci. USA* 20, 259; 1934, *Phys. Rev.* 46, 76.

Berezinsky, V.S. and Ginzburg, V.L.: 1981, *Mon. Not. RAS.*, 194, 3.

Bulanov, S.V., Dogel, V.A. and Syrovatskii, S.I.: 1976, *Astrophys. Space Sci.* 44, 267.

Cavalho, J.C. and Haar, D. ter: 1979, *Astrophys. Space Sci.* 61, 3.

Dombrovskii, V.A.: 1954, *Doklady Acad. Sci. USSR* 94, 1021.

Dorman, I.V.: 1981, Cosmic Rays (historical account), Nauka, Moscow. (in Russian)

Dodds, D., Strong, A.W. and Wolfendale, A.W.: 1975, *Mon. Not. RAS* 171, 569.

Ekers, R.D. and Sancisi, R.: 1977, *Astron. Astrophys.* 54, 973.

Fermi, E.: 1949, *Phys. Rev.* 75, 1169.

Getmanzev, G.G.: 1951, *Uspekhi Fiz. Nauk* 44, 527.

Getmanzev, G.G.: 1952, *Doklady Acad. Sci. USSR* 83, 557.

Getmanzev, G.G. and Ginzburg, V.L.: 1950, *Zh. Experim. Teor. Fiz.* 20, 347.

Getmanzev, G.G. and Ginzburg, V.L.: 1952, *Doklady Acad. Sci. USSR* 87, 187.

Ginzburg, V.L.: 1946, *C.R. (Doklady) Acad. Sci. USSR* 52, 487.

Ginzburg, V.L.: 1947, *Uspekhi Fiz. Nauk* 32, 26; 1947, 34, 13.

Ginzburg, V.L.: 1951, *Doklady Acad. Sci. USSR* 76, 377.

Ginzburg, V.L.: 1953a, *Doklady Acad. Sci. USSR* 92, 1133.

Ginzburg, V.L.: 1953b, *Uspekhi Fiz. Nauk* 51, 343.

Ginzburg, V.L.: 1954, *Forschritte der Phys.* 1, 659.

Ginzburg, V.L.: 1956, *Nuovo Cim. Suppl.* 3, 38, (see also Hillas, 1972).

Ginzburg, V.L.: 1958, *Progress in Elem. Particle and Cosmic Ray Physics. Amsterdam* 4. 339.

Ginzburg, V.L.: 1970, *The Propagation of Electromagnetic Waves in Plasmas*, Pergamon Press, Oxford

Ginzburg, V.L.: 1972, *Nature (Phys. Sci.)* 239, 8.

Ginzburg, V.L.: 1978a, *Bibliography of Acad. Sci. USSR scientists*, Phys. ser. N.21, Science Publ, Moscow.

Ginzburg, V.L.: 1978b, *Sov. Phys. – Uspekhi.* **21**, 155.
Ginzburg, V.L.: 1979, *The Large-Scale Characteristics of the Galaxy,* p.485. Reidel Publ., Dordrecht, Holland.
Ginzburg, V.L. and Ptuskin, V.S.: 1976, *Rev. Mod. Phys.* **48**, 161.
Ginzburg, V.L. and Syrovatskii, S.I.: 1965, *Ann. Rev. Astron. Astrophys.* **3**, 297.
Ginzburg, V.L. and Syrovatskii, S.I.: 1964, *The Origin of Cosmic Rays,* Pergamon Press, Oxford.
Ginzburg, V.L., Khazan, Ya. M. and Ptuskin, V.S.: 1980, *Astrophys. Space Sci.* **68**, 295.
Gordon, I.M.: 1954, *Doklady Acad. Sci. USSR* **94**, 813.
Haar, D. ter: 1950, *Rev. Mod. Phys.* **22**, 119.
Hayakawa, S.: 1956, *Progress Theor. Phys.* **15**, 111.
Hillas, A.M.: 1972, *Cosmic Rays,* Pergamon Press, Oxford.
Kiepenheuer, K.O.: 1950, *Phys. Rev.* **79**, 738.
Kraus, J.D.: 1967, *Radio Astronomy,* McGraw-Hill Book Company, New York.
Liber Amicorum for Oort, Jan H.: 1980, *Oort and the Universe,* p. 129, Reidel Publ., Dordrecht, Holland.
Oort, J.H. and Walraven, Th.: 1956, *Bull. Astron. Inst. Netherlands* **12**, 285.
Origin of Cosmic Rays: 1980, IAU-IUPAP Symposium, Bologna, Italy, Reidel Publ., Dordrecht, Holland.
Owens, A.J. and Jokipii, J.R.: 1977, *Astrophys. J.* **215**, 677.
Paris Symposium on Radio Astronomy, 1959, Stanford Univ. Press., Stanford, Calif.
Pikel'ner, S.B.: 1953, *Doklady Acad. Sci. USSR* **88**, 229.
Pikel'ner, S.B. (Obituary): 1976, *Uspekhi Fiz. Nauk* **119**, 377.
Pomeranchuk, I.Ya.: 1939, *Zh. Eksperim. Teor. Fiz.* **9**, 915.
Pooley, G.G.: 1969, *Mon. Not. RAS* **144**, 143.
Proc. 3rd Conf. Questions Cosmogony (Origin of Cosmic Rays): 1954, Acad. Sci. USSR, Moscow.
Radio Astronomy, 1957, Manchester Symposium, Cambridge, England.
Radio Astronomy and the Galactic System: 1967, IAU Symposium, **31**, (ed. Woerden, H. van) Academic Press, New York.
Ryle, M.: 1949, *Proc. Phys. Soc. (London)* **62A**, 491; 1950, *Rep. Progr. Phys.* **13**, 184.
Selected paper on cosmic ray origin theories (ed. Rosen, S.): 1969, Dover Publication, New York.
Schiff, L.: 1946, *Rev. Sci. Instr.* **17**, 6.
Schott, G.A.: 1912, *Electromagnetic Radiation,* Cambridge, Univ. Press, Cambridge.
Schwinger, J.: 1949, *Phys. Rev.* **75**, 1912.
Shklovskii, I.S.: 1952, *Astron. Zh.* **29**, 418.
Shklovskii, I.S.: 1953a, *Astron. Zh.* **30**, 15.
Shklovskii, I.S.: 1953b, *Doklady Acad. Sci. USSR* **90**, 983.
Shklovskii, I.S.: 1953c, *Doklady Acad. Sci. USSR* **91**, 475.
Shklovskii, I.S.: 1960, *Cosmic Radio Waves,* Harvard Univ. Press, Cambridge, U.S.A.
Shklovskii, I.S.: 1962, *Uspekhi. Fiz. Nauk* **77**, 3.
16th Intern. Cosmic Ray Conference, 1979, Conference papers, v.1–14. Kyoto, Japan.
Syrovatskii, S.I. (Obituary): 1980, *Uspekhi Fiz. Nauk* **131**, 73.
Unsöld, A.: 1949, *Zs. f. Astrophys.* **26**, 176.
Unsöld, A.: 1951, *Phys. Rev.* **82**, 857.
Unsöld, A.: 1955, *Zs. f. Phys.* **141**, 70.
Vashakidze, M.A.: 1954, *Astron. Circular. N* **147**, 11.
Vladimirskii, V.V.: 1948, *Zh. Eksperim. Teor. Fiz.* **18**, 392.
Wallace, J.M.: 1980, *Astrophys. Space Sci.* **68**, 27.
Webster, A.: 1978, *Mon. Not. RAS.* **185**, 507.
Wolfendale, A.W.: 1979, *Pramana.* **12**, 631.

RECOLLECTION OF EARLY COSMIC RAY RESEARCH

Hannes ALFVÉN*

When in the summer 1979 I was invited to tell my recollections from the early days of the study of Cosmic Radiation I was glad to accept to do it. The deadline was more than a year away and that seems always to be close to infinity. However, this inspired me to think not only of the history of astrophysics but also of its state today. Isn't it natural that I considered the latter to be more interesting? I began to summarize my views — which in part are somewhat controversial — of the present state of the science of "Cosmic Plasma" in a small monograph with this name and I believed that this should take me three months. This turned out to be essentially correct — however this did not include the polishing, references, etc. and in spite of the fact that I had very good collaborators, this took me three more months. And then there was little time for a historical investigation of what I thought about Cosmic Radiation half a century ago. So I have to confine myself to telling two anecdotes to which I shall add a general comparison of astrophysics at that time and the present astrophysics.

Half a century ago I started — like many other young physicists — in nuclear physics. As I was more interested in experiments than theory, my first work in this field was concentrated on electrical methods to detect nuclear radiations. That lead me to work on Geiger-Müller tubes and I constructed giant tubes which were suited for measuring Cosmic Rays. I found this field of investigation to be a fascinating field because Cosmic Rays had energies far larger than the radiations from radio-active substances, indeed up to 10^{10} or perhaps even 10^{11} eV! I began to speculate about their origin and made an ingeneous theory which in reality was so silly that I do not even mention it here. For unknown reasons "Nature" agreed to publish a letter about it, but as soon as it was published I found that the idea was completely unreasonable.

This must have been around 1933, because I remember very well that it was before the first international conference I ever attended which was in London 1934. At that conference I met A.H. Compton, who immediately mentioned that he had read my letter and thought it interesting. I said that I was ashamed of having published something so completely silly to which he answered: "Don't give it up too easily". As he was one of the great authorities on Cosmic Radiation this was an enormous encouragement to me. In retrospect this may have been the trigger which brought me into astrophysics. It happened at a time when almost everybody was running towards nuclear physics. It saved me from the guilt associated with atomic bombs and nuclear energy which every nuclear physicist of today must feel at the bottom of his heart.

Next episode is dated 1948. In the meantime, there had been a war, the atomic bombs had been exploded and hence, science had changed its character for ever. My scientific work in astrophysics had continued. I had presented a theory of cosmic ray acceleration by electromagnetic effects around double stars (*Zeitschrift für Physik*, **105**,

*Royal Institute of Technology, Stockholm, Sweden.

Y. Sekido and H. Elliot (eds.), Early History of Cosmic Ray Studies, 427–431.
© *1985 by D. Reidel Publishing Company.*

319, 1937 and **107**, 579, 1937). With the excpetion of Swann's famous but unrealistic "cygnotron" (as Vallarta called it) I think this was the first attempt to explain the high energies as due to electromagnetic affects. However, processes of this kind could not supply enough intensity if Cosmic Radiation filled the whole universe, as was the generally accepted view. I then pointed out that this dilemma could be solved if Cosmic Radiation was confined to our galaxy. This required a galactic magnetic field of at least 10^{-10} gauss. (It should be remembered that the highest Cosmic Radiation energies known at that time were only 10^{11} eV.)

I tried to publish this paper in − I think − "Nature" and other journals which at that time were generally read, but it was not accepted. "Is that guy completely crazy? Doesn't he know that space is empty which means that there can be no particles carrying currents. And we all know that the Earth's magnetic field derives from a permanent magnet in its centre. Hasn't he understood how rapidly it decreases with distance? Does he believe that there are a number of magnets floating around in the galaxy?" The only journal which accepted it was Arkiv for Fysik, to which it was communicated by my teacher Professor Manne Siegbahn. Certainly this was a very reputable journal but not very much read.

So with this as a background, I had the following experience in 1948. I was starting my first transatlantic journey and on my way I attended a conference in Birmingham. I came a little late to the meeting and entered a lecture hall where Edward Teller spoke about the origin of Cosmic Rays. I had shown beyond any doubt − I thought − that Cosmic Radiation was a galactic phenomenon generated by some electromagnetic effects and trapped inside the galaxy where its intensity of course was uniform. Teller claimed that it was a local phenomenon, generated and trapped inside what we today call the heliosphere. The whole audience roared with laughter, and so did I. It was the first time I heard Teller and did not know that this dynamic personality always makes everybody laugh − independent of whether he speaks about his dear atomic bombs or astrophysics.

When I came to the USA, quite a few people asked me about Teller's new idea. I said that everybody laughed at his lecture and that this was nothing to be taken seriously. I made the Grand Tour of the USA − of course quite an experience to anybody who sees it for the first time. I spent some time at Cal Tech and then Teller came there. To my great surprise he repeated the same silly lecture. Now I objected to it and after his lecture we had a long discussion with the result that he invited me to continue the discussion at his seminar in Chicago on my way back.

I must admit that he defended his views in a clever way, and I must admit that it was not at all so obvious that Cosmic Radiation was a galactic phenomenon as I had thought. After a few days of thinking I actually got convinced that he was essentially correct. I sent him a picture postcard where I said that I agreed in essential respects, and even had found some new arguments in favour of the heliospheric confinement of Cosmic Rays. I think that these were connected with the magnetohydrodynamic waves which I had published my first paper about in 1942.

When a few weeks later I came to Chicago Teller introduced me at this seminar with the words: "I need only tell you that this is the guy who wrote the picture postcard." Teller and I later wrote a joint paper about the local origin of Cosmic Rays (*Phys. Rev.*, **75**, 892, 1949). After some years Teller changed his views − and has since then believed in the standard views of Cosmic Radiation as a galactic phenomenon. This is now so generally accepted that the radiation is referred to as the "Galactic Radiation". It is a little paradoxical that I now should be one of the rather few supporters of Teller's theory, whereas the Galactic theory which essentially derives from my views in the late 1930's is generally accepted. In fact it has become so sacrosanct that all my attempts for decades to start a serious discussion have led to nothing.

My brief visit to Chicago in 1948 had also another consequence. One of the members of the seminar was Enrico Fermi, who got interested in the origin of Cosmic Rays. After the seminar he asked me to explain what the magnetohydrodynamic waves were. Since I published my first paper in 1942 very few people — with Lyman Spitzer and Martin Schwarzschild as the most prominent exceptions — had believed in them. I got letters from colleagues who asked me whether I had not understood that this was nonsense. If they existed, Maxwell would have described them and it is quite clear that he has not. Hence it is impossible that they could exist! I completely agree that they are nothing remarkable. (I considered them mainly as a spin-off of my theory of sunspots, which I thought to be much more important.) To discover them would have been an appropriate pastime for Maxwell on a tedious Sunday afternoon, but as a matter of fact he seems to have preferred other pastimes.

Fermi listened to what I said about them for five or ten minutes, and then he said: "Of course such waves could exist". Fermi had such an authority that if he said "of course" today, every physicist said "of course" tomorrow. Actually he published a paper in which he explained them in such a clear way that no one could doubt their possible existence. What I had not succeeded to do in six years was done by Fermi in what was only the introduction to a presentation of the famous Fermi mechanism of Cosmic Rays acceleration (the importance of which I have not yet been able to understand).

From my personal point of view my early work on Cosmic Rays was very important. Contrary to almost all astrophysicists my education had taken place in a laboratory where my hands got dirty when I studied electrical discharges. Hence it was natural to me to describe plasma phenomena in terms of particles moving in magnetic fields in the same way they do in the laboratory. The cosmic ray phenomena in space led me to approach the whole astrophysics in this way. Since then the astrophysics of low density plasmas had been to me a science in which all phenomena ultimately should be described from the point of view of the individual particles. Instead of treating hydromagnetic equations I prefer to sit and ride on each electron and ion and try to imagine what the world is like from its point of view and what forces push them to the left or to the right.

This has been a great advantage because it gives me a possibility to approach the phenomena from another point than most astrophysicists do, and it is always fruitful to look at *any* phenomenon under two different points of view. On the other hand it has given me a serious disadvantage. When I describe the phenomena according to this formalism most referees do not understand what I say and turn down my papers. With the referee system which rules the US science today, this means that my papers rarely

TABLE I. Change in views during half a century.

	Early views		Present views	
	E	D	E	D
Measured intensity of C R is typical for				
Universe	X			
Galaxy		X	X	
Heliosphere				X
In a highly conductive plasma the magnetic field lines are frozen-in		X	X	
It has no sense to speak of frozen-in field lines	X			X
Sunspots are produced by hydromagnetic waves originating in the core (M.N., **105**, 3; 382, 1945)		X		X
Sunspots are produced by phenomena in the upper layer of the sun	X		X	
Essential features of the structure of the Solar System cannot be understood without introducing plasma processes		X		X
Plasma processes were not very important when the solar system formed	X		X	
The Universe				
is matter-antimatter symmetric		X		X
		↑		
contains almost no antimatter	X	X	X	
The Universe is				
infinite	X	X		X
	↓			
finite	X		X	
The Universe was created at an				
instantaneous creation *ex nihilo*	X		X	
	↑			
is "ungenerated and indestructable"	X	X		X

"Early views" refer to 25 – 50 years ago.

E   means views held by the establishment.

D   means views held by a small group of dissidents (to which the author belongs).

C R   intensity refers to $W < 10^{13}$ eV. Higher energies can of course not be confined to the heliosphere.

The arguments for the views P D are given in detail in the monograph "Cosmic Plasma" Reidel 1981.

are accepted by the leading US journals. Europe, including the Soviet Union, and Japan are more tolerant of dissidents.

    The hydromagnetic approach to cosmic plasma is probably rather good for high-density plasmas like plasmas in the stellar interiors, but if applied to low density plasmas in cosmos it is often terribly misleading. From the thermonuclear crisis 20 years ago and the similar crisis in magnetospheric physics the last five years, we should learn that it is unrealistic to approach a low density plasma by considering it as a fluid. Indeed, the plasma has taught me that it is so complicated that it should be regarded more as a living being than a dead mechanical system. (Langmuir, who baptized it understood this and borrowed its name from the blood plasma.) This creature does not understand differential equations and vectors and tensors and does not care for such nonsense. It always finds new ways to cheat the mathematical physicist. This means that unless a scientist purges his brain of all such stuff he has little chance of understanding this naughty and whimsical child who loves to revolt against what the theoreticians have prescribed that it should do.

The mentioned conditions and quite a few other factors have lead to a disagreement between a very strong establishment (E) and a small group of dissidents (D) to which the present author belongs. This is nothing remarkable. What is more remarkable and regrettable is that it seems to be almost impossible to start a serious discussion between E and D. As a dissident is in a very unpleasant situation, I am sure that D would be very glad to change their views as soon as E gives convincing arguments. But the argument "all knowledgeable people agree that. . ." (with the tacit addition that by not agreeing you demonstrate that you are a crank) is not a valid argument in science. If scientific issues always were decided by Gallup polls and not by scientific arguments science will very soon be petrified for ever.

An attempt to summarize the state is given in Table I.

ONE NIGHT IN 1961

Epilogue

The period of "Early History" gradually passed away in 1950's. At Kyoto Conference in 1961 papers remarkably increased by many participants of the second generation while a sentiment of reminding the past gleamed in those of the first generation who were pioneers once challenged each lonely horizon.

One night was devoted to the enjoyment of Japanese traditional noh play. The program was a fantasia "Matsukaze" where the soul of a love remembered with longing her lover many years after they died. After the noh theater was closed, C.F. Powell stayed silently in the quiet air of autumn night and then told me by heart a verse of Shakespear's sonnet.

After compiling the reminiscences of this book in 1981, I learned the sad series of the sonnets and feel that its deep loneliness might have been impressive for the pure mind of Cecil.

> *Thou art thy mother's glass, and she in thee*
> *Calls back the lovely April of her prime;*
> *So thou through windows of thine age shalt see,*
> *Despite of wrinkles, this thy golden time.*

Shakespeare's Sonnet 3

Y.S., one of Editors

Y. Sekido and H. Elliot (eds.), Early History of Cosmic Ray Studies, 433.

NAME INDEX

Adams, N., 382, 383, 394
Adams, R. V., 128
Ageno, 240
Agnew, H. M., 390
Aitken, 11
Akinov, V. V., 371
Alexander, G., 330
Alfvén, H., 201–203, 378–380, 382, 393, 400, 414, 415, 418, 421, 427
Alichanov, 226, 228
Alichanyan, 226, 228
Allen, R. J., 422
Allison, S. K., 385, 401
Alvarez, L., 68
Alvial, G., 175
Amaldi, E., 222, 312, 326
Amati, D., 349, 350
Ammar, R. G., 329
Anders, E., 401
Anderson, C. D., 14, 23, 49, 50, 94, 96, 108, 117, 123, 125, 128, 129, 131, 132, 135, 140, 145, 146, 153–157, 161, 163, 226, 250, 265, 288, 309, 310, 312, 319, 387, 390
Anderson, G., 268
Anderson, H., 401
Anderson, K., 268, 271, 274
Anderson, O., 172
Anderson, S. D., 341
Andronikashvilli, 228, 231
Anna Akhmatova, 352
Annis, A., 317
Antropoff, 79
Aoki, 220
Appa Rao, M. V. K., 315
Appleton, E. V., 104
Arakawa, H., 195
Araki, G., 292, 293
Aristotle, 9
Armenteros, R., 251, 311, 313, 315, 316, 319, 323
Arnoldy, R., 268
Artsimovich, L. A., 414
Asano, Y., 189, 193
Aseikin, 235
Astbury, 313
Astier, A., 251

Astin, A. V., 256
Aston, 65
Auger, P., 48, 49, 51, 68, 192, 213–216, 228, 249, 265, 267, 295–297, 299, 343, 387
Azimov, S. A., 226, 234, 334

Baade, W., 412, 419
Babayan, K. P., 234, 369, 370
Babcock, H. D., 399
Babcock, H. W., 399
Bagge, E. R., 161, 162
Bailey, D. K., 256
Baldauf, 27
Baldo, M., 315
Bannik, B. P., 334
Baños, 387
Baradzei, 232
Barbasso, 62
Barkas, W. H., 326
Barker, K. H., 305, 309
Barkla, C. G., 99, 104, 108
Banóthy, J., 181–183, 193, 194, 296, 297
Barry, 365
Bartels, S. J., 168, 400
Bartky, W., 385
Bassi, P., 240, 241
Batusov, Ya. A., 334
Beck, 65
Becker, H., 122
Behounek, 23
Belenky, S. Z., 227, 348
Belowa, B., 39
Benedetti, S., 67
Benndorf, H., 8, 28, 29, 205
Bennet, R. D., 168
Bennett, W. E., 100
Bentley, L. C., 376
Bercha, S., 330
Berezinsky, V. S., 424
Bergwitz, 18, 23, 75, 78
Bernardini, G., 53–57, 297
Bethe, R. A., 51, 94, 124, 125, 152, 153, 163, 209, 260, 293, 308, 343, 387
Beutler, H., 387
Bhabha, H. J., 10, 51, 102, 103, 152, 156–158, 209, 210, 216, 288, 299, 343, 347
Birger, 228

435